Nanotechnology in the Life Sciences

Series Editor

Ram Prasad
Department of Botany
Mahatma Gandhi Central University
Motihari, Bihar, India

Nano and biotechnology are two of the 21st century's most promising technologies. Nanotechnology is demarcated as the design, development, and application of materials and devices whose least functional make up is on a nanometer scale (1 to 100 nm). Meanwhile, biotechnology deals with metabolic and other physiological developments of biological subjects including microorganisms. These microbial processes have opened up new opportunities to explore novel applications, for example, the biosynthesis of metal nanomaterials, with the implication that these two technologies (i.e., thus nanobiotechnology) can play a vital role in developing and executing many valuable tools in the study of life. Nanotechnology is very diverse, ranging from extensions of conventional device physics to completely new approaches based upon molecular self-assembly, from developing new materials with dimensions on the nanoscale, to investigating whether we can directly control matters on/in the atomic scale level. This idea entails its application to diverse fields of science such as plant biology, organic chemistry, agriculture, the food industry, and more.

Nanobiotechnology offers a wide range of uses in medicine, agriculture, and the environment. Many diseases that do not have cures today may be cured by nanotechnology in the future. Use of nanotechnology in medical therapeutics needs adequate evaluation of its risk and safety factors. Scientists who are against the use of nanotechnology also agree that advancement in nanotechnology should continue because this field promises great benefits, but testing should be carried out to ensure its safety in people. It is possible that nanomedicine in the future will play a crucial role in the treatment of human and plant diseases, and also in the enhancement of normal human physiology and plant systems, respectively. If everything proceeds as expected, nanobiotechnology will, one day, become an inevitable part of our everyday life and will help save many lives.

More information about this series at http://www.springer.com/series/15921

Basma A. Omran

Nanobiotechnology: A Multidisciplinary Field of Science

Springer

Basma A. Omran
Researcher in Microbiology
Petroleum Biotechnology Laboratory
Processes Design and Development Department
Egyptian Petroleum Research Institute (EPRI)
Nasr City, Cairo, Egypt

ISSN 2523-8027 ISSN 2523-8035 (electronic)
Nanotechnology in the Life Sciences
ISBN 978-3-030-46070-9 ISBN 978-3-030-46071-6 (eBook)
https://doi.org/10.1007/978-3-030-46071-6

This Springer imprint is published by the registered company Springer Nature Switzerland AG
The registered company address is: Gewerbestrasse 11, 6330 Cham, Switzerland

Some things are worth waiting for… I keep asking myself what it takes to be superior. I realized that desire, dedication, determination, focusing on my goal shall be the keys.

I dedicate this book to the person who meant and continues to mean the world to me, my Mother. This is for you, Mom. Thanks for always being there for me in each step of the way. I lovingly dedicate this book to my husband, Mohamed Omar, my little boy, Mazen, and to my newborn baby, Yassin. Thanks for your love, patience, and endless support when I needed you. You kept pushing me to accomplish this work, I love you dears. This book is dedicated to my sister, Lamis, and brother, Mohamed. I love both of you to the moon and back. Finally, this work is dedicated to the soul of my beloved father who is always in my heart. May Allah rest your soul in peace… To Allah we belong and to Allah we do return.

I love all of you and I am doing my best to deserve your love.

—Dr. Basma A. Omran

Preface

This book provides students, scientists, and chemical engineers with a recent and updated overview of the field of nanotechnology with precise and detailed data of nanobiotechnology. Nanobiotechnology is the combination of biotechnology and nanoscience. Nanobiotechnology is one of the most exciting and dynamic disciplines that has emerged over the last few decades. Academia and researchers have invested huge efforts regarding the achievement of fundamental research related to nanobiotechnology and its versatile applications. New insights and grand challenges have emerged. However, nanobiotechnology is widely expected to have a substantial impact on commercial applications in the very near future.

Chapter 1 shows the historical origin of nanotechnology, and the concept of nanotechnology. It also glances through the different types of nanomaterials and their classification. Additionally, it highlights the distinctive, superior, and outstanding features of nanomaterials. These characteristic properties involve optical, magnetic, electronic, mechanical, and catalytic properties. It discusses in detail the physical and chemical synthetic methodologies which are employed during the synthesis processes. Moreover, the different nanomaterial characterization techniques are discussed in detail, and the chapter ends with the history of nanobiotechnology.

Chapter 2 illustrates the world of prokaryotes, particularly bacteria and actinomycetes. It focuses on the biomimetic synthesis of nanomaterials via bacterial cells and actinomycetes (bio-nano-factories). It shows the advantages of these two biological entities as effective producers of nanoparticles. This chapter demonstrates common extracellular and intracellular synthesis approaches. It reviews and discusses the latest updated reported investigations regarding the biosynthesis of nanoparticles via bacterial and actinomycetes cells with elucidation of the involved synthesis mechanism.

Chapter 3 discusses "myconanotechnology," one of the most interesting branches of nanobiotechnology, which is concerned with nanoparticle genesis using different fungal and yeast strains. This chapter deals with the biogenic synthesis of nanoparticles via eukaryotic machineries including fungi and yeast. It describes the superior advantages of using fungi as nanoparticle biological producers over the rest of the other microbial strains. It discusses the latest published research articles regarding this point.

Chapter 4 demonstrates the use of algae (micro- and macro-) as efficient nanoparticle biosynthesizers. It illustrates the high diversity of algal strains. This branch of science is referred to as "phyconanotechnology." This chapter discusses the green synthesis of different cyanobacterial strain (blue-green algae), diatoms, and micro- and macro-algae. It also highlights the mechanisms involved in the biosynthesis process with different applications in which algal derived nanoparticles can have beneficial roles in carrying them out.

Chapter 5 discusses the science branch of "phytonanotechnology," which is mainly related to the biological fabrication of nanoparticles via extracts of different plant parts as well as agricultural wastes. Phytonanotechnology possesses numerous advantages for nanoparticles synthesis over microbial synthesis of nanoparticles. These advantages are mentioned in detail. This chapter also shows the varieties of agro-industrial wastes and how they can be exploited to synthesize valuable products like nanoparticles. It also emphasizes the different bioactive molecules which are involved in the process of metal salt bioreduction, capping, and stabilization procedures. Moreover, possible mechanisms which are behind the phytosynthesis of nanoparticles are mentioned as well. Eventually, the drawbacks and some recommendations concerning phytonanotechnology are highlighted.

Chapter 6 discusses the numerous applications in which biologically derived nanoparticles can be exploited. These applications include the biomedical fields starting from their anticancer, antimicrobial, antiparasitic, antidiabetic, anti-inflammatory potentials as well as in wound healing. Additionally, it involves management of environmental pollution and wastewater treatment from hazardous products like dye effluents, organic pollutants, phenolic compounds, and pathogenic microorganisms. Furthermore, biologically driven nanoparticles proved their remarkable effects as antibiofoulers and inhibitors of biofilm formation. Additionally, their possible applications in the agricultural sector as well as in food industries are recognized. Finally, risk assessment and challenges of the nanobiotechnology field are highlighted.

Nasr City, Cairo, Egypt Basma A. Omran

Acknowledgment

First, thanks to Allah. I always feel indebted to Allah, the most merciful and the most graceful. I would like to take this opportunity to express my gratitude towards people who have been a source of inspiration while writing this book.

I express my deep sense of gratitude and reverence to my PhD supervisors. I am indebted to many people who helped me to bring this book to light. I would like to acknowledge the generous assistance of Mr. Eric Stannard, Senior Editor, Botany, Springer for believing in my proposal and for encouraging me throughout this project. A debt of gratitude is owed to Mr. Arun Siva Shanmugam and Ms. Kalaiselvi Ramalingam for their patience during the production process of the book and in bringing this book to fruition.

Special thanks go to Egyptian Petroleum Research Institute (EPRI), which has been a source of inspiration. I really adore this place and I am proud to belong to such an institution.

Finally, I would like to express my sincere gratitude to my mother, sister, brother, my beloved husband, and my sons Mazen and my newborn baby Yassin for their encouragement and endless support which always inspired me.

Contents

Abbreviations

4-NP	4-Nitrophenol
AAS	Atomic Absorption Spectroscopy
AC	Activated Carbon
AFM	Atomic Force Microscope
AgCl NPs	Silver Chloride Nanoparticles
AgNPs	Silver Nanoparticles
AGS	Human Gastric Carcinoma
ALA	Antileishmanial Assay
AP	Aminophenol
AR	Acid Red
ARB	Accumulative Roll-Bonding
ATCC	American Type Culture Collection
ATP	Adenosine Triphosphate
ATR-FTIR	Attenuated Total Reflection Spectroscopy
AuNPs	Gold Nanoparticles
AW-AgNPs	*Acalypha wilkesiana* Silver Nanoparticles
BC	Before Christ
BET	Brunauer-Emmett-Teller
BI	Biocompatibility index
BSLA	Brine Shrimp Lethality Assay
BSNPs	Biogenic Silica Nanoparticles
CBB	Coomassie Brilliant Blue R-250
CCD	Central Composite Design
CCDF	Cyclic Closed-Die Forging
CEC	Cyclic Extrusion Compression
CFF	Cell-Free Culture Filtrate
CFU	Colony Forming Unit
CHNF	Chitosan Nanofibers
CHO	Chinese Hamster Ovary Cells
CLSM	Confocal Laser Scanning Microscope
CMSP	Carboxymethyl Sago Pulp

CNF	Cellulose Nanofibers
CNTs	Carbon Nanotubes
CoNPs	Cobalt Nanoparticles
CPK	Creatine Phosphokinase
CP-MB	Creatine Kinase-Myocardial Bound
CR	Congo Red
CSE-AgNPs	Coconut Shell Extract-AgNPs
CTDW	Cationic Textile Dye Waste
CTnT	Cardiac Troponin T
CV	Cyclic Voltammetry
CVD	Chemical Vapor Deposition
DAP	Diastolic Arterial Pressure
DCT	Direct Contact Test
DDMP	2, 3-Dihydro-3,5- dihydroxy-6-methyl- 4H-pyran-4-one
DL	Dalton lymphoma
DLS	Dynamic Light Scattering
DNA	Deoxyribonucleic Acid
DPPH	2, 2-diphenyl-1-picrylhydrazyl
DPV	Differential Pulse Voltammetry
DSC	Differential Scanning Calorimetry
DSMZ	Deutsche Sammlung von Mikroorganismen und Zellkulturen
DTA	Differential Thermal Analysis
EA	Elemental Analyzer
EAB	Electrochemically Active Biofilm
ECAP	Equal Channel Angular Pressing
ECG	Electrocardiogram
ED	Electron Diffraction
EDC	Endocrine Disrupting Compound
EDX	Energy Dispersive X-Ray Spectroscopy
EDXRF	Energy Dispersive X-Ray Fluorescence Spectroscopy
EE	Ethanol Extracts
EM	Electron Microscopy
ENM	Engineered Nanomaterial
EPS	Extracellular Polymeric Substances
ESBL	Extended Spectrum β-Lactamase
ESEM	Environmental Scanning Electron Microscope
FAgNPs	*F. chlamydosporum* AgNPs
fcc	Face Centered Cubic
FEG-SEM	Field Emission Gun Scanning Electron Microscopy
FESEM	Field Emission Scanning Electron Microscopy
FETEM	Field Emission Transmission Electron Microscopy
FSP	Flame Spray Pyrolysis Technique
FTIR	Fourier Transform Infrared Spectroscopy
FTIR-ATRR	Fourier Transform Infrared Spectroscopy-Attenuated Total Reflectance

GC/MS	Gas Chromatography/Mass Spectroscopy
GE	Graphite Electrode
GeNPs	Germanium Nanoparticles
GO	Graphene Oxide
GQDs	Graphene Quantum Dots
GRx	Glutathione Peroxidase
GSP	Gold Shaping Protein
HAADF	High-Angle Annular Dark-Field
HBV	Hepatitis B virus
HCV	Hepatitis C Virus
HEK293	Human Embryonic Kidney
HFB4	Human Normal Melanocytes
HIP	Hot Isostatic Pressing
HIV	Human Immunodeficiency Virus
HPLC	High Performance Liquid Chromatography
HPT	High Pressure Torsion
HRSEM	High Resolution Scanning Electron Microscope
HRTEM	High Resolution Transmission Electron Microscope
HUVEC	Human Umbilical Vein Endothelial Cells
HVPE	Hydride Vapor Phase Epitaxy
IC50	Half Maximal Inhibitory Concentration
ICP	Inductively Coupled Plasma
ICP-AES	Inductively Coupled Plasma Atomic Emission Spectroscopy
ICP-MS	Inductively Coupled Plasma-Mass Spectroscopy
IMTECH	Institute of Microbial Technology
IONPs	Iron Oxide Nanoparticles
IP	Intra Peritoneal
IR	Infrared Spectroscopy
ITS	Internal Transcribed Spacer
IUPAC	International Union of Pure and Applied Chemistry
kD	kilodalton
K-doped GO	Potassium-Doped Graphene Oxide
kHz	Kilohertz
LC50	Half Lethal Concentration
LCNF	Lignocellulose Nanofibers
LD50	Median Lethal Dose
LDH	Lactate Dehydrogenase
LIBS	Laser-Induced Breakdown Spectroscopy
LLE	Lavender Leaf Extract
LPE	Liquid Phase Epitaxy
LPS	Lipopolysaccharides
MA	Mechanical Alloying
MAP	Mean Arterial Pressure
MB	Methylene Blue
MBC	Minimal Bactericidal Concentration

MBE	Molecular Beam Epitaxy
MCFF	Mycelial Cell Free Filtrate
MCS	Mechanochemical Synthesis
MCT	Mycological Collection of Trichocomaceae
MDR	Multi-Drug Resistant
MFC	Minimal Fungicidal Concentration
MFCs	Microbial Fuel Cells
MG	Malachite Green
MIC	Minimal Inhibitory Concentration
MIONPs	Magnetic Iron Oxide Nanoparticles
mm	Millimeter
MO	Methyl Orange
MOFs	Metal Organic Framework
MOVPE	Metal Organic Vapor Phase Epitaxy
MPL	Multiphoton-Excited Luminescence
MR	Methyl Red
MRSA	Methicillin-Resistant *Staphylococcus aureus*
MS	Mass Spectrometry
MTCC	Microbial Type Culture Collection
MTT	(3- (4, 5-Dimethylthiazol-2-Yl) -2, 5-Diphenyltetrazolium)
NaA-ZNPs	NaA Zeolite Nanoparticles
NADH	Nicotinamide Adenine Dinucleotide Hydrogen
NADPH	Nicotinamide Adenine Dinucleotide Phosphate Hydrogen
NCBI	National Center for Biotechnological Information
NCCU	Neurosciences Critical Care Unit
NCI	National Cancer Institute
NCIM	National Collection of Industrial Microorganisms
NFMC	National Facility for Marine Cyanobacteria
NMR	Nuclear Magnetic Resonance
NNI	National Nanotechnology Initiative
NOX	(NADPH) Oxidase
NPs	Nanoparticles
NTA	Nanoparticle Tracking Analyzer
ODS	Oxide-Dispersion Strengthened
OFAT	One-Factor-at-a-Time Technique
PAgNPs	*P. chrysogenum* AgNPs
PBS	Phosphate Buffer Saline
PCM	Polish Collection of Microorganisms
PdNPs	Palladium Nanoparticles
PES	Photoelectron Spectroscopy
PhR	Phenol Red
PL	Photoluminescence
PL	Photoluminescence Spectroscopy
PLD	Pulsed Laser Deposition
PMF	Proton Motive Force

PMT	Photomultiplier Tube
POME	Palm Oil Mill Effluent Treatment
PTCC	Persian Type Culture Collection
PtNPs	Platinum Nanoparticles
PVA	Polyvinyl alcohol
PVD	Physical Vapor Deposition
PVP	Polyvinylpyrrolidone
PW	Post-Distillation Water
QDs	Quantum Dots
RCS	Repetitive Corrugation and Straightening
Rc-ZnO NPs	*Rubus coreanus* Zinc Oxide Nanoparticle
rDNA	Ribosomal Deoxy Ribonucleic Acid
ROS	Reactive Oxygen Species
RP-HPLC	Reverse-Phase High-Performance Liquid Chromatography
rpm	Round Per Minute
rRNA	Ribosomal Ribonucleic Acid
RS	Raman Spectroscopy
RT	Room Temperature
RUBP	Ribulose-1, 5-bisphosphate
SAED	Selected Area Electron Diffraction
SBFSEM	Serial Block-Face Scanning Electron Microscopy
SC	Subcutaneous
SDS-PAGE	Sodium Dodecyl Sulfate Polyacrylamide Gel Electrophoresis
SEM	Scanning Electron Microscope
SeNPs	Selenium Nanoparticles
SERS	Surface Enhanced Raman Spectroscopy
SeS NPs	Selenium Sulfide Nanoparticles
SIM	Simavastin
SNOM	Scanning Near-Field Optical Microscopy
SOD	Superoxide Dismutase
SPD	Severe Plastic Deformation
SPR	Surface Plasmon Resonance
SRB	Sulphate Reducing Bacteria
SSMR	Super Short Multi-Pass Rolling
STEM	Scanning Transmission Electron Microscope
STS	Severe Torsion Straining
TEM	Transmission Electron Microscope
TGA	Thermogravimetric Analysis
TGDTA	Thermo Gravimetric Differential Thermal Analysis
TiNPs	Titanium Nanoparticles
TiO_2	Titanium Oxide
TLC	Thin Layer Chromatography
TPH	Total Petroleum Hydrocarbons
TSC	Trisodium citrate
UFG	Ultrafine Grains

UHV	Ultrahigh Vacuum
UHV-CVD	Ultrahigh Vacuum Chemical Vapor Deposition
UPS	Ultraviolet Photoelectron spectroscopy
US	United States
USNPs	Ultra-Small Nanoparticles
UV/Vis	Ultraviolet Visible Spectrophotometry
VSM	Vibrating Sample Magnetometry
WB	Wheat Bran
WE	Water Extract
WHO	World Health Organization
WR	Watermelon Rind
XANES	X-ray Absorption Near-Edge Spectroscopy and Extended X-ray
XAS	X-ray Absorption Spectroscopy
XPS	X-Ray Photoelectron Spectroscopy
XRD	X-ray Diffraction
ZnO/rGO	ZnO/reduced graphene oxide
ZOI	Zone of Inhibition
ZrO_2 NPs	Zirconia Nanoparticles
ZrO_2	Zirconium dioxide

Chapter 1
Fundamentals of Nanotechnology and Nanobiotechnology

1.1 Introduction

Nanotechnology is a very interesting scientific discipline and has versatile practical applications. It depends majorly on manufacturing novel, and promising materials at the level of the nanoscale (Rastogi et al. 2018). The Japanese Professor Norio Taniguchi was the first to coin the term nanotechnology (Taniguchi et al. 1974). Particles with dimensions of 100 nm or less are referred to as nanoparticles and are considered as the building blocks of nanotechnology. For instance, the length of a carbon-carbon bond is in the range of 0.12–0.15 nm, and the diameter of the DNA double helix is approximately 2 nm. Conversely, the length of the mycoplasma which is the smallest cellular forms of bacteria is nearly 200 nm. For instance, the human hair diameter is around 80,000 nm (Guo 2013). These nanostructured materials are characterized by extraordinary properties such as the controlled shape, size, crystallinity, and composition. Nanotechnology covers fields starting from biology to materials science and chemistry to physics. The controlled size, shape, composition, crystallinity, and structure-dependent properties of nanoparticles govern the unique properties of nanotechnology. However, physical and chemical methodologies used for nanoparticle synthesis such as sol-gel technique, chemical vapor deposition, precipitation, hydrothermal synthesis, and microemulsion proved to be toxic, hazardous, not eco-friendly, expensive, and harsh conditions are needed such as high temperature and/or pressure to complete the synthesis process. Thus, significant efforts were devoted for finding alternative techniques to guarantee the safe and clean synthesis of nanoparticles. This was successfully achieved by the progress of the nanobiotechnology field.

B. A. Omran, *Nanobiotechnology: A Multidisciplinary Field of Science*,
Nanotechnology in the Life Sciences, https://doi.org/10.1007/978-3-030-46071-6_1

1.2 Fundamentals of Nanotechnology

1.2.1 Definition and Concept of Nanotechnology

The prefix "nano" is originated from the Greek word "*nanos*" which means "dwarf" or very small. Nanotechnology is defined as "the science that allows the manipulation of matter at atomic or molecular levels in the range of 1–100 nm" (Omran et al. 2018a, El-Gendy and Omran 2019). Thus, this would aid in producing innovative materials with extraordinary properties (Mohan Bhagyaraj and Oluwafemi 2018). Nanotechnology is a multidisciplinary science that combines several scientific fields including biotechnology, biology, chemistry, physics, medicine, pharmacy and engineering, etc. It is worth noting that the principles of quantum physics contributed in the revolutionary nature of nanotechnology (Basu 1997). The behavior of materials at the nanoscale is under the control of quantum laws rather than those of classical physics (Mohan Bhagyaraj and Oluwafemi 2018). These extraordinary features of nanoscaled materials offer new properties that make them superior and more favorable than those of bulk materials.

1.2.2 Historical Arise of Nanotechnology

Nanotechnology is the science which is concerned with the development and manipulation of materials at the nanoscale level. One nanometer is one billionth of a meter (10^{-9} m) which is extremely small (Ball et al. 2019). According to Balzani (2005), the U.S. National Nanotechnology Initiative (NNI) defined nanotechnology as the science that focuses on the research and the development of materials at the atomic or molecular levels to generate structures that can be employed in versatile applications. Paul and Chugh (2011) reported that the history of nanotechnology science began during the eras of the fourth and the fifth centuries BC in China and India. The traditional medical practitioners managed to synthesize gold colloids referred to as "Swarna Bhasma" which possessed potential therapeutic capabilities. Dykman and Khlebtsov (2012) demonstrated Paracelsus treated mental disorders and syphilis in Europe during the middle ages via colloidal gold. Ball et al. (2019) mentioned that in 1618, Doctor Francisco Antonii wrote a book on how colloidal gold can be prepared and the different applications in which it can be exploited. Furthermore, in 1857, Michael Faraday published the first scientific book on colloidal gold (Ball et al. 2019). In 1867, James Clerk Maxwell released his frontier observations about a technology that can manipulate individual molecules. In the early twentieth century, major advances were performed in ultramicroscopes. In 1914, Zsigmondy managed for the first time to observe and investigate nanostructures of a size of 10 nm (Zsigmondy 1914). Yet, the Nobel laureate Richard Feynman inspired the world in nanomaterials' science by his lecture "There's Plenty of Room at the Bottom" in 1959 at Caltech, USA (Feynman 1960). In his famous speech, he clarified that there might be an ability to manipulate and control materials in a very small scale. The biggest popularization in nanomaterial science began when the American engineer Eric Drexler published his book in 1986 which was entitled *Engines of Creation: The Coming Era of Nanotechnology*. Engineer Eric Drexler was known as the "founding father of nanotechnology" (Drexler 1981). In 1991, Drexler established the molecular nanotechnology field by his doctoral dissertation which was entitled "Nanosystems, Molecular Machinery, Manufacturing, and Computation." His work represented an inspiration for his scientific colleagues to further explore other manufacturing systems of nanotechnology which helped engineers and scientists in different disciplines to scale down the use of materials to the nanoscale levels. The term "nanotechnology" was first coined by the Japanese Professor Norio Taniguchi in Tokyo Science University in 1974. Two of his published works were related to nanotechnology, and they were *Nanotechnology: Integrated Processing Systems for Ultra-Precision and Ultra-Fine Products* and *On the Basic Concept of Nanotechnology* (Mohan Bhagyaraj and Oluwafemi 2018). Richard E. Smalley was another scientist whose work attributed majorly to the nanotechnology science. Smalley was a Nobel laureate, and he got the Nobel Prize in chemistry in 1996 for discovering Buckminsterfullerene. He worked in Rice University in Texas and was a professor of physics, chemistry, and astronomy and the founding director of a center for nanoscale science and technology since 1996 till 2002. His work focused mainly on different types of fullerenes for instance C60, C70, etc. He encouraged their use in various conducting applications based on his studies (Nelson and Strano 2006).

Certain developments aided in extensive progress in nanotechnology including the discovery of carbon nanotubes in 1991, the design of a nano-robotic system in 1997, and the design of the first nanochemical apparatus for DNA in 1998. The first industrial report on nanotech revolution and the first nanotech industrial conference were in 2001 and 2002, respectively (Mohan Bhagyaraj and Oluwafemi 2018). Currently, nanotechnology has become one of the most important technologies in all areas of science particularly modern materials science (Visweswara Rao and Hua Gan 2015).

1.3 History of Nanomaterials

Since thousands of years ago, humans started to use nanomaterials. Heiligtag and Niederberger (2013) reported that more than 4500 years ago, people reinforced ceramic matrixes by the inclusion of natural asbestos nanofibers. Also, Walter et al. (2006) demonstrated in his publication in 2006 that the Ancient Egyptians used nanomaterials for more than 4000 years ago. The Ancient Egyptians formulated a hair dye by a chemical process to synthesize PbS NPs with a size of approximately $\approx$5 nm diameter PbS NPs for hair dye. Likewise, the first synthetic pigment which was referred to as the "Egyptian blue" was fabricated and used by Egyptians near the third century BC (Johnson-McDaniel et al. 2013). Egyptian blue is a multifaceted mixture of SiO_2 and $CaCuSi_4O_{10}$. Different countries like Egypt, Greece, and Mesopotamia extensively used the Egyptian blue for decorative purposes which was then discovered during the archaeological explorations.

The chemically synthesized NPs occurred in centuries back to the thirteenth and fourteenth centuries BC. Schaming and Remita (2015) reported that both Egyptians and Mesopotamians made glass using metals which was considered as the starting era for the use of metallic nanoparticles. Afterwards, at the end of the Bronze Age (1200–1000 BC), red glass was explored in Frattesina di Rovigo (Italy) which was found to be colored by the incorporation of copper nanoparticles (CuNPs) (Artioli et al. 2008). Likewise, CuNPs in addition to cuprous oxide (cuprite Cu_2O) were discovered in the Celtic red enamels which belonged to 400–100 BC period (Brun et al. 1991). Furthermore, a very popular Romanian work piece was discovered and was an evidence in using metallic NPs. The Lycurgus Cups which belong way back to the fourth century are Roman glass cups, and they are built up from a dichroic glass which exhibits different colors, for instance, green and red when the light passes from the front and behind, respectively, as was conducted by Leonhardt (2007). Nonetheless, recent studies discovered that the Lycurgus Cups contained Ag-Au alloy NPs, with a ratio of 7:3 along with 10% of CuNPs (Freestone et al. 2007).

Later, during the medieval period, churches' glass were found to be stained with yellow and red colors by the incorporation of colloidal AgNPs and AuNPs, respectively (Schaming and Remita 2015). Heiligtag and Niederberger (2013) reported that Mesopotamians began to use glazed ceramics for decoration purposes during the ninth century. These decorations were distinctive because they exhibited unique optical features because of the existence of Ag and/or CuNPs which were isolated within the outer ceramic layers. These decorations represented a perfect example of incorporating metal nanoparticles which would display exciting coloration, i.e., bright blue and green coloration, when exposed to reflection. A double layer of AgNPs with an approximate range of 5–10 nm was found in the outer layers, and larger particles of a size of 5–20 nm were found in the inner layers. Far ahead, the red glass was manufactured using that process worldwide.

In the middle of the nineteenth century, a similar technique was employed to generate the popular Satsuma glass in Japan (Nakai et al. 2004). The Satsuma glass was colored by the absorption properties of CuNPs that were helpful in producing the ruby color. It was reported by Rytwo (2008) that in 5000 BC, cloths and wools were bleached in Cyprus by clay. In 1857, Michael Faraday described the process of synthesizing colloidal AuNPs, and from that time the scientific arena of nanomaterial synthesis was initiated. Moreover, he postulated that there was a major difference between the optical features of Au colloids and their bulk counterparts. This might be considered as one of the earliest reports illustrating the effect of quantum size. Later, the reason behind the color variation of metal colloids was further described and explained by Mie (1908). During 1940s, silicon oxide nanoparticles (SiO_2 NPs) were produced to be used as an alternative for carbon black for rubber reinforcement purposes (Rittner and Abraham 1998).

Today, manufactured nanomaterials possess significant advantages over their bulk counterparts involving improvement of conductivity strength, durability, lightness, as well as providing beneficial properties (e.g., self-healing, self-cleaning, anti-freezing, and antibacterial). They can also reinforce materials for construction purposes or for safety issues (Jeevanandam et al. 2018). In 2003, Samsung company started to release electronic devices with the advantage of employing antibacterial technology which was given the trade name of Silver Nano™. Such devices involved air conditioners, air purifiers, refrigerators, washing machines, and vacuum cleaners, which used ionic AgNPs (Jeevanandam et al. 2018). Additionally, nanostructured materials are widely employed in auto production, for instance, as fillers in tires to enhance adhesion to the road, as fillers in the car body to increase the stiffness, and also as transparent layers used for constructing heated, mist- and ice-free window panes (Jeevanandam et al. 2018).

At the end of 2003, Mercedes-Benz managed to exploit metal and nonmetal NPs for paint finishing for their car series. This coating had the advantage of increasing scratch resistance and enhancing the appearance by increasing the car gloss. Kreuter (2007) reported that in 2005, a human serum albumin material composed of paclitaxel was synthesized, commercialized, and released in the market and was given the trade name of Abraxane™. Vance et al. (2015) reported that in 2014, approximately 1814 nanotechnology-based consumer products were commercially available in markets of over than 20 countries around the world. Yet, this number is in huge increase.

1.4 Classification of Nanomaterials

Nanomaterials are characterized by their extraordinary properties which can be tailored to be applied in different disciplines. Among these characteristic features are the reduction in spatial dimensions as well as their small surface-area-to-volume ratio. Basically, nanomaterials are classified into three main categories including zero-dimensional, one-dimensional, and two- dimensional structures as described in Table 1.1.

1.5 Types of Nanomaterials

Recently, most nanomaterials produced are categorized into four categories, namely, carbon-based nanomaterials, inorganic-based nanomaterials, organic-based nanomaterials, and composite-based nanomaterials as listed in Table 1.2 (Jeevanandam et al. 2018).

Table 1.1 Classification of nanomaterials

Structure of nanomaterial	Definition	Examples	Description	Application	References
Zero-dimensional structures	Structures that are confined in three dimensions	Nanoparticles, nanograins, nanoshells, nanocapsules, nanorings, fullerenes, colloidal particles	Particles which range in size from a few tens to a few hundreds of nanometers	Aid in fabricating novel composite materials, for example, synthesis of plastics with high resistance to UV light, synthesis of more effective drugs, and catalysts	Tiwari et al. (2018)
One-dimensional structures	Structures that are confined in two dimensions	Carbon nanotubes (CNTs)	Nanotubes are referred to as hollow cylinders with either single or multilayered walls with a diameter in the range of few nanometers. Nanotubes are characterized by high tensile strength, elasticity, high electrical conductivity, flexibility, thermal conductivity, good electron field emitting properties, low thermal expansion coefficient, and high aspect ratio	Nanotubes are used to construct enormously strong materials with different applications including vehicle manufacturing, nanocircuits, mechanical gears for nanomachines, and display screens	Chen et al. (2006)
		Nanowires (quantum wires)	They differ from the nanotubes in that they are not hollow	Nanowires are more favorable in making electronic chips and devices	
Two-dimensional structures	Structures that are confined in one dimension	Nano-membranes, nanofilms, nanolayers, graphene	They possess a variety of optical, chemical, mechanical, biodegradable, and biocompatible properties	Tissue engineering and in drug delivery	Lalwani et al. (2013), Wan et al. (2016)

Table 1.2 Categories of nanomaterials

Classification	Examples	Description
Carbon-based nanomaterials	Fullerenes (C60), carbon nanotubes (CNTs), carbon nanofibers, carbon black, graphene, and carbon onions	Nanomaterials which contain carbon exhibit different morphological structures like hollow tubes, spheres, and ellipsoids
Inorganic-based nanomaterials	Metal, metal oxide nanoparticles such as AuNPs, AgNPs, ZnO NPs, TiO_2 NPs	They exhibit versatile shapes, i.e., rod, spherical, hexagonal, cuboid
Organic-based nanomaterials	Dendrimers, micelles, liposomes and polymer NPs.	Nanomaterials made up of organic matter
Composite-based nanomaterials	Metal organic frameworks (MOFs)	They are multiphase NPs in which NPs combine with other NPs or combine with larger or with bulk-type materials. The composites are constructed by the combination of carbon-based, metal-based, or organic-based NMs with any type of metal, ceramic, or polymer bulk materials

1.6 Characteristic Features of Nanomaterials

Nanomaterials possess unique physical and chemical features including shape, size, aspect ratio, surface area, size distribution, crystallinity, agglomeration state, and surface morphology/topography. Discussed below are some of the exceptional properties of nanomaterials.

1.6.1 Optical Properties

Nanomaterials are characterized by unique optical properties which can be easily identified via different spectroscopic techniques. Electron transition between the two states of emission and adsorption results in determining the optical properties of the prepared nanomaterial (Mohan Bhagyaraj and Oluwafemi 2018). Different metal nanoparticles exhibit a great difference in their optical features based on their particle size. For instance, the color of the dispersed nanoparticles can be totally variable based on the prepared size and shape; for example, silver nanoparticles (AgNPs) might exhibit a blue and a yellow color at sizes 40 and 100 nm, respectively. Moreover, if the prepared AgNPs possess a prism shape, the visible color becomes red. Raza et al. (2016) performed a study in 2016, in which he and his coauthors managed to synthesize AgNPs with different sizes and shapes by solution-based chemical reduction methodology. Silver nitrate ($AgNO_3$) was used as a precursor for silver ions, sodium borohydride and tri-sodium citrate (TSC) were used as reductants, while polyvinyl pyrrolidone (PVP) was employed as a stabilizing agent. Different colors were obtained including faint yellow, dark yellow, faint brown, dark brown, and a greenish color as well. Additionally, it has been noticed that as the size of gold nanoparticles (AuNPs) increases, the color of the colloidal suspension of gold turns from red to yellow. For instance, AuNPs exhibit a yellow, red, and purple colors at sizes 2–5, 10–20, and > 20 nm, respectively (Mohan Bhagyaraj and Oluwafemi 2018).

The manifestation of nanoparticles' color is mainly dependent upon the surface plasmon resonance (SPR) effect. SPR effect is a resonance which takes place because of the interaction between the outer electron bands of the prepared nanoparticles along with the light wavelength. Light photons cause excitation to the particles' outer electrons, and consequently the outer electrons on the metal particles vibrate at a certain wavelength and absorb light which corresponds to that resonance. The scattering mathematic theory of both Mies and Rayleigh scattering clarified the link between particle size and color (González et al. 2014).

1.6.2 Magnetic Properties

The interaction between the magnetic spins of the constituent material with the electron charges results in the confinement of the unique magnetic properties of a nanomaterial (Mohan Bhagyaraj and Oluwafemi 2018). The fabricated nanomaterials exhibit different magnetic features owing to the large surface-area-to-volume ratio which characterize the majority of nanostructures. Moreover, the constituent atoms undergo different magnetic interactions with the neighboring atoms, resulting in various magnetic features. Super-paramagnetism may result because of the fluctuation in the energy of magnetic anisotropy. Super-paramagnetic materials are free of coercivity and remanence. They can be very useful in a wide range of disciplines, for instance, bioprocessing, magnetic resonance imaging, and bioprocessing in addition to refrigeration. Interestingly, both bulk platinum and gold display no magnetic properties, while at the nano-sized scale, they possess magnetic features.

1.6.3 Electronic Properties

The increase in the wave-like quantum mechanical properties and the lack of scattering centers majorly affect the electronic properties of nanoscaled materials. The discrete nature of the energy states becomes apparent when the size of the system becomes comparable with the de Broglie wavelength of the electrons. But a fully discrete energy spectrum is only observed in three-dimensional structures. In some cases, below a critical length scale, the conducting materials become insulators as the energy bands cease to overlap. The intrinsic wave-like nature of the nanostructures makes the electrons tunnel quantum mechanically between two closely adjacent nanostructures. Applying a voltage between the aligned discrete energy levels of two nanostructures, causes the occurrence of resonant tunneling, which results in an abrupt increase in tunneling current. Highly confined structures like quantum dots are very sensitive to the presence of other charge carriers and the charge of the dot. These Coulomb blockade effects result in conduction processes involving single electrons, and as a result, they require only a small amount of energy to operate a switch, transistor, or memory element. All these phenomena can be utilized to produce radically different types of components for electronic, optoelectronic, and information- processing applications, such as resonant tunneling transistors and single-electron transistors.

1.6.4 Mechanical Properties

The mechanical features of nanoparticles such as hardness and elasticity contribute to the proper employment of the prepared nanoparticles and aid in evaluating their roles and mechanism of action. Hardness can be measured by applying a technique which is called nano-indentation technique. Additionally, it can be applied to determine the elastic properties of particles. However, different factors might affect the measurement of the mechanical properties of nanomaterials such as the uniform dispersion of nanoparticles on a hard substrate, the appropriate use of loads onto the particle, the accurate position of the particle, the measurement of particle deformation, etc. (Wagner et al. 2011). Also, instrumental errors with the instrument calibration and the calculation models might be also taken into account while measuring the mechanical features of the prepared nanomaterials.

1.6.5 Catalytic Properties

Nano-catalysts possess unique catalytic properties such as high selectivity and reactivity when compared with their counterparts. Interestingly, several experimental studies on nano-catalysts showed the relationship between particle size and the catalytic activity of the prepared nano-catalyst. Furthermore, other factors play an important role in determining the catalytic reactivity and selectivity of nanostructures such as geometry, oxidation state, composition, and the surrounding chemical/physical environment. For instance, nanoparticles synthesized from different transition element oxides possess awesome catalytic properties because of the large surface area. Mori and Shitara (1994) suggested that there is a relation between materials' catalytic activity and the roughness of its surface. Moreover, in a study conducted by Häkkinen et al. (2003), another factor affects the catalytic activity of nanomaterials which was the ability of small clusters to jiggle between favorable isomers with different energies within a chemical reaction.

1.7 Synthesis Approaches of Nanomaterials

In order to expand the application of nanomaterials in different disciplines, it is vital that such materials become available with the preferred shape, size, morphology, crystallinity, and chemical composition. To date, numerous techniques are available for synthesizing nanomaterials. However, they can be fabricated via two main approaches referred to as "top-down" and "bottom-up" approaches (Fig. 1.1) (Birnbaum and Pique 2011). The first approach is the top to down in which bigger particles are broken down into smaller and smaller dimensions with different shapes (Balasooriya et al. 2017). This approach involves different techniques such as mechanical milling, lithography, laser ablation, electrochemical explosion, attrition, chemical etching, etc. However, the top-down approach has certain disadvantages like the formation of imperfect surface structure. Sometimes the prepared nanomaterials exhibit somehow broad size distribution. Furthermore, in the prepared nanomaterials, certain impurities might exist.

The second approach is the bottom-up approach in which nanoparticles are built up via the smaller building blocks (i.e., atoms, monomers, molecules, etc.). In the bottom-up approach, the individual building blocks of atoms and molecules fit together and become self-assembled accurately and placed where they are desired. Different techniques such as pyrolysis, spinning, plasma or fire sparing, sol-gel process, biological method, chemical vapor deposition, atomic condensation, etc. follow the bottom-up approach. Bottom-up approach possesses more advantages than the top-down approach including decrease in the dimension of the prepared nanoparticles, large surface area, large surface energy, spatial confinement, and reduced imperfections.

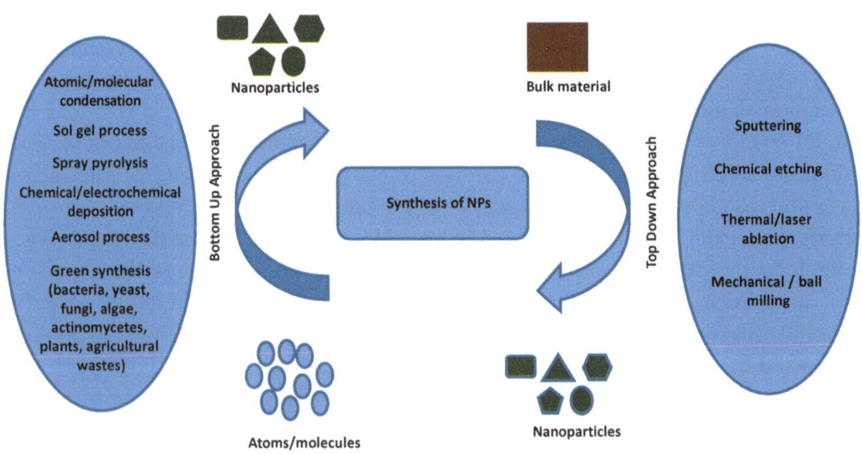

Fig. 1.1 Nanoparticle synthesis approaches

1.8 Synthesis Routes of Nanomaterials

1.8.1 Physical Routes for Synthesis of Nanomaterials

1.8.1.1 Physical Vapor Deposition (PVD)

Physical vapor deposition (PVD) is a vaporization coating procedure which involves the transfer of material to atomic levels (Tulinski and Jurczyk 2017). PVD is somehow like the chemical vapor deposition (CVD) method. They differ only in the state of the precursor starting material, as it has to be in a solid form in the case of PVD and in a gaseous form in the case of CVD. PVD processes involve the techniques of both pulsed laser deposition (PLD) and sputtering. PVD is performed under vacuum conditions and involves the four following stages which are evaporation, transportation, reaction, and deposition. During the first stage, i.e., evaporation, the starting material is targeted by a high-energy source, for instance, a beam of ions or electrons. This results in vaporizing the materials' surface atoms. The following stage involves the transportation of vaporized atoms from the targeted material to the substrate in order to be coated. This coating is comprised of metal oxides, carbides, nitrides, etc. Additionally, during the transportation stage, the vaporized atoms react with the suitable type of gas. Such gases may be nitrogen, oxygen, or methane. PVD coatings are essential as they provide better hardness, wear resistance, and minimized friction and are enriched with oxidation resistance. Hence, nanomaterials prepared by PVD are substantially applied in optical, magnetic, optoelectronical, and microelectronic devices. Other applications may include disciplines of tribology, corrosion resistance, decorative coatings, and thermal protection.

Thermal evaporation is one of the most applied techniques of PVD. In this technique, the substance is first heated in order to be evaporated by the effect of energy either in the form of laser, electrical current, arc discharger, electron beam, etc. till reaching the required temperature. It is worthy to note that in order to obtain definite film properties, the temperature must be higher than 100 °C. The thermally unconfined molecules or atoms attach around the coating substrate (Vossen et al. 1991). The main difference between thermal evaporation and sputtering techniques is that evaporation relies on thermal excitation in order to transfer the coating material to the gas phase, while sputtering technique transfers the coating material into the gas phase via impulse transfer.

1.8.1.2 Pulsed Laser Deposition (PLD)

Pulsed laser deposition (PLD) is considered one of the easiest physical routes for depositing thin films like metals, oxides, semiconductors, nitrides, organic compounds, etc. This technique is derived from PVD processes which imply a high-power with short-pulsed laser radiation on solid substrates (Tulinski and Jurczyk 2017). In this technique, a laser beam passes through the vacuum chamber and strikes the target material. Atoms are cleaved from the bulk material by vaporizing the bulk material at the non-equilibrium state. PLD has versatile applications, for instance, synthesizing thin films of nanoscaled metal oxides and manufacturing multilayered superconducting materials, nitride films, as well as ceramic oxides (Zhang et al. 2014).

1.8.1.3 Ion Beam Implantation or Ion Implantation Technique

Ion implantation is a physical technique in which ions with high energy strike into the surface of a solid sample. An ion source is responsible for generating these ions. Additionally, these ions can be produced via a secondary energy source which aids in ionizing the existing gases, e.g., nitrogen or methane. The device is comprised of six major components involving vacuum system, ion source, accelerator, magnetic analyzer, beam scanning, as well as target chamber (Tulinski and Jurczyk 2017). An important portion is the ultrahigh vacuum (UHV) which usually accompanies the apparatus of ion beam implantation in order to prevent ion scattering as well as ion beam contamination.

For ions to be selectively chosen, a magnetic analyzer is employed to speed up ion selection via a flight tube. At the flight tube, ions are divided based on their mass ratio as well as their ionic charge. Afterwards, the mass-selected ions are speeded up under the effect of high energy in an acceleration tube with an acceleration voltage ranging between 30 and 300 Kv. Then, they become concentrated into a beam and directed to the target surface. The capability of the ion implantation technique to deal with the microstructures at the atomic levels is attributed to its high quenching rate. Ion implantation is characterized by its accuracy, its elasticity, and its ability to manufacture nanocrystalline materials.

1.8.1.4 Plasma Synthesis

One of the physical routes which are employed in order to synthesize, activate, and functionalize materials is the low-temperature plasma. It is employed for synthesizing nanomaterials through a thermal reaction in which material is heated till being evaporated within induction plasma (Tulinski and Jurczyk 2017). Then, the resultant vapors are exposed to a very rapid quenching reaction zone. The quenching gases can be either inert gases, for instance, nitrogen (N) or argon (Ar), or reactive gases as ammonia (NH_3) or methane (CH_4) based on the type of nanomaterial being fabricated (Tulinski and Jurczyk 2017). The produced nanopowder is collected via a porous filter located far from the plasma reaction. The manufactured nanoparticles are in the range of 20 to 100 nm based on the employed quenching atmosphere. It is worth noting that the production rate may reach few hundred grams per hour to 3–4 kg per hour, but this differs from one material to another depending upon its physical properties. Different methods such as alloying of surface plasma help to upgrade the surface properties of the targeted materials (Miklaszewski et al. 2011).

1.8.1.5 Laser Evaporation Method

Laser evaporation is considered a very promising bottom-up technique which facilitates the synthesis of magnetic nanopowders. The employed laser is responsible for evaporating the precursor metal oxides which are the main sources for the required metals. The produced nanoparticles are generated by the effect of rapid condensation and nucleation because of the applied temperature gradient (Kurland et al. 2007, 2009). The particle size as well as its magnetic phase can be easily controlled by adjusting the laser power as well as atmosphere composition (Stötzel et al. 2013).

1.8.2 Chemical Methodologies for Synthesis of Nanomaterials

Nanomaterials synthesized via chemical-based methodologies depend on the interaction between different ion solutions in specific quantities under controlled circumstances such as pressure, temperature, pH, and stirring rate. Chemical-based methodologies offer a high confinement over the particle structure, shape, and size. They include chemical vapor deposition, epitaxial growth, colloidal dispersion, sol-gel, hydrothermal route, microemulsions, polymer route, precipitation processes, etc. (Tulinski and Jurczyk 2017).

1.8.2.1 Chemical Vapor Deposition (CVD)

Chemical vapor deposition (CVD) is a chemical procedure which can be applied to synthesis of solid thin film coatings, production of highly pure powders and bulk materials, as well as fabrication of composite materials (Dobkin and Zuraw 2003). CVD depends on flowing gases within a chamber containing the heated objects which need to be coated. Chemical reactions take place close to the hot surface leading to the manufacture of thin films on the surface. Usually, chemical byproducts are released out of the chamber along with the remaining unreacted gases. There are many kinds of CVD. They may be designed in either hot wall or cold wall reactors. Also, some upgraded CVD processes have been introduced by the involvement of photons, lasers, plasma, ions, and hot filaments.

CVD possess several advantages in depositing thin films: (i) CVD films are accurate, and (ii) the deposited materials are highly pure and are deposited fairly rapid. On the contrary, the disadvantage of CVD involves the hazardous materials that are employed during deposition as some precursors may be hazardous, corrosive, and flammable.

1.8.2.2 Epitaxial Growth

Epitaxy refers to the organized growth of single film material above a certain substrate (Foord et al. 1997). Epitaxy is a linking procedure that combines the scientific art of crystal growth and manufacturing of devices and circuits with applications in microelectronics, optoelectronics, and photonics. Versatile techniques of epitaxy exist including liquid phase epitaxy (LPE), hydride vapor phase epitaxy (HVPE), ultrahigh vacuum chemical vapor deposition (UHV-CVD), chemical beam epitaxy, metal organic vapor phase epitaxy (MOVPE), and molecular beam epitaxy (MBE) (Tulinski and Jurczyk 2017). Homoepitaxy describes the process when the material grown has the same composition of the substrate, while heteroepitaxy describes the process when the material's structure differs from that of the substrate.

1.8.2.3 Sol-Gel Technique

In this technique, first, the concentrated suspension of a metallic oxide or hydroxide is formed (sol). The sol is then dehydrated by evaporation or solvent extraction, resulting in the formation of colloidal suspension (gel) (Liu et al. 2018). Controlled heating in autoclave or in vacuum converts the gel to finely divided powder, with particle size in the range of 0.03–0.1 μm. This method can produce extremely homogeneous mixtures of two or more components because the mixing of ingredients takes place at the atomic level in a liquid rather than in the solid state. A wide range of pure and mixed oxides can be produced with controlled size and composition, high surface area and high purity due to the absence of grinding and pressing steps. However, the sol-gel method possesses some disadvantages also such as relatively high cost of precursors, long processing time, large shrinkage during processing, and the possibility of the formation of hard agglomerates (Ullattil et al. 2017).

1.8.2.4 Coprecipitation Technique

The most commonly used method for the synthesis of multicomponent oxide is coprecipitation method, which produces a "mixed" precipitate comprising two or more insoluble species that are simultaneously removed from the solution. The precursors used in this method are mostly inorganic salts (nitrate, chloride, sulfate, etc.) that are dissolved in water or any other suitable medium to form a homogeneous solution with clusters of ions (Shinde et al. 2017). The crystal growth and their aggregation are influenced by the concentration of salt, temperature, the actual pH, and the rate of pH change that force the dissolved inorganic salts to precipitate as hydroxides, hydrous oxides, or oxalates. Generally, a calcination step is necessary to transform the hydroxide into crystalline oxides. In most of the binary, ternary, and quaternary systems, a crystallization step is necessary, which is generally achieved by thermal treatment or, more elegantly, by a hydrothermal procedure in high-pressure autoclaves. A large number of reports are available on the synthesis of ultrafine oxide powders (Bumb et al. 2008), oxide-oxide composites, and biomaterial by coprecipitation reactions (Liang et al. 2011; Kalam et al. 2018). The advantages of coprecipitation reactions are (i) the homogeneity of component distribution, (ii) the relatively low reaction temperature, (iii) the fine and uniform particle size with weakly agglomerated particles, and (iv) the low cost. However, these reactions are highly susceptible to the reaction conditions, and because of incomplete precipitation of the metal ions, control over the stoichiometry of the precursors is rather difficult to achieve. In addition, the coprecipitation reactions are not suited for certain oxides/hydroxides, for instance, in the case of amphoteric systems.

1.8.2.5 Flame Spray Pyrolysis Technique (FSP)

FSP is a process of a gas phase combustion synthesis method which enables the production of a broad range of materials in the form of nanostructured powders with high specific surface area and primary particle size in the range of nanometers. There are two main routes for injection of the precursor in this method including; (1) in homogenous phase, through an evaporation system at controlled temperature, or (2) through a spray system, in liquid phase. This last methodology offers a greater flexibility in terms of type of precursors to be used and flow control to determine the amount of produced material (Solero 2017). The advantage of FSP is the use of a wide variety of possible low-cost precursors (mainly in the field of metal oxides such as TiO_2, Al_2O_3), obtaining a final product with high purity and relatively narrow size distribution (Tok et al. 2006). FSP seems to be a versatile process allowing a strict control of the produced nanomaterial; particularly product size and morphology which strictly depend on precursor concentration and dispersion gas flow rate (Height et al. 2006).

1.8.2.6 CO_2 Laser Pyrolysis Technique

Laser pyrolysis method is a technique used to synthesize ultrafine powders by heating a mixture of reactant vapor and inert gas with a laser. Laser pyrolysis is a very suitable gas phase process for the synthesis of a wide range of nanoparticles at laboratory scale. The principle of the method is based on the decomposition of gaseous or liquid reactants by a high-power CO_2 laser, followed by a quenching effect. The literature reports the possibility to produce carbides, nitrides, oxides, metals, and composite nanoparticles by this process (Borsella et al. 2011).

1.8.3 Mechanical Methodologies for Synthesis of Nanomaterials

Various mechanical methodologies are usually employed to obtain small-sized nanostructures. The main working principle of such mechanical methods is dependent upon the generation of very fine-grained powders (at the nanometer scale) and then consolidated through hot isostatic pressing (HIP) (Groza 1999). Mechanical processes include mechanical grinding (Prasad Yadav et al. 2012), high-energy ball milling (Arulmani et al. 2018; Rane et al. 2018), mechanical alloying (MA) (Wang et al. 2019), and reactive milling (Kirti et al. 2018). The mechanical techniques possess several advantages as they are cost-effective, simple procedures and a large feedstock powder of the desired material can be generated. However, certain disadvantages may arise as aggregation of the prepared nanopowders, and contamination may take place from the process equipment (Tulinski and Jurczyk 2017). It is worth mentioning that these mechanical synthesis methods are employed to synthesize inorganic compounds as well as metals or alloys.

1.8.3.1 Milling Processes

Milling processes have the ability to manufacture nanomaterials, bio-nanomaterials, and/or bio-nanocomposites (Niespodziana et al. 2010; Tulinski and Jurczyk 2012; Balcerzak et al. 2015). Different types of ball mills have been developed involving tumbler, attrition, shaker, vibratory, and planetary mills. Dense materials such as tungsten carbide or steel are much preferred since the kinetic energy of the balls is a function of both their velocity and mass (Tulinski and Jurczyk 2017). Moreover, by applying vibration with small amplitude and high frequency, high-energy milling forces can be attained.

High-energy ball milling (HEBM) is a top-down approach. HEBM is an effective, simple type of a ball milling process in which the powder mixture is placed in a ball mill, and then it becomes exposed to collisions with high energy from the balls. High-energy ball milling is sometimes referred to as mechanical alloying (Rane et al. 2018). In 1970, Benjamin proposed a pioneering development on the ball milling technique (Benjamin 1970). Then, it was successfully applied to manufacture oxide dispersion strengthened (ODS) iron- and nickel-based alloys which were heavily employed in aerospace industry (Bhadeshia 1997). In 1983, Koch managed for the first time to prepare successfully by applying the mechanical alloying technique (Koch et al. 1983). Till now, mechanical alloying has been used to synthesize ODS materials like highly thermal-resistant materials, magnetic materials, hydrogen storage materials, amorphous materials, superconducting materials, and nanocrystalline materials (Bidabadi et al. 2013). During the process of HBEM, coarse grained structures are subjected to disassociation due to the intensive cyclic deformation which is done by the milling of the stiff balls within a high-energy mill shaker (Koch et al. 1983). HBEM can produce nanoparticles within the range of 4–26 nm. Three main types of HBEM exist including the shaker mill, the planetary mill, and the attritor mill (Wang et al. 2019). Another advantage of this process is that it is characterized by its simplicity and can manufacture massive amounts of the desired materials reaching to tons (Rane et al. 2018). Yet, contamination from the surrounding environment may occur (e.g., balls and vials) and from the surrounding atmosphere. Consequently, a number of actions have been performed such as usage of surfactants, alloy-coated milling media, and protective atmospheres to lessen the contamination problem.

Cryomilling is also referred to as low-temperature mechanical alloying. It is somehow similar to HEBM, but it differs from HEBM in that the metallic powders become severely milled in conditions containing a cryogen (e.g., liquid argon or liquid nitrogen) or under low temperature (Wang et al. 2019). During the 1980s, cryomilling was extensively applied to manufacture ODS materials, intermetallic compounds, nanocrystalline materials, etc. (Kumar and Biswas 2019). Cryomilling mainly depends upon solid-state reactions as well as atomic diffusion to produce alloys. But it has been noted that extreme cold welding of particles may lead to particles' aggregation. Hence, a balance between cold welding and fracturing should be taken into account during cryomilling.

1.8.3.2 Severe Plastic Deformation (SPD)

Sever plastic deformation (SPD) is a metal forming process in which ultrafine metals are generated by the introduction of an ultra-large plastic strain to a bulk metal (Rosochowski 2005). The basic purpose a SPD process is the manufacturing of highly strengthened parts. The traditional metal forming processes like forging, rolling, and extrusion form structures which have low thickness with a very thin diameter thus becoming not preferable during structure manufacturing. Hence, various processes have been explored like equal-channel angular pressing (ECAP) (Lapovok 2005), accumulative roll bonding (ARB) (Toth and Gu 2014), high-pressure torsion (HPT), repetitive corrugation and straightening (RCS) (Sunil and Sunil 2015; Langdon 2016), and cyclic extrusion compression (CEC) (Pardis et al. 2011).

1.8.3.3 Lithography

Lithography is the most applicable mechanical route for synthesizing a number of ordered nanoparticles in an array (Gentili et al. 2005). In 1796, the Bavarian researcher Alois Senefelder invented the process of lithography where the desired materials were patterned onto a base substrate using masks (Nayfeh 2018). Lithography is the process in which one- and two-dimensional structures are fabricated and at least one of the lateral dimensions is in the nanoscale. Lithography is subcategorized into electron beam lithography, photolithography, X- ray and extreme UV lithography, soft lithography, colloidal lithography, nanoimprint lithography, focused ion beam and neutral atomic beam lithography, scanning probe lithography, and atomic force microscope nanolithography (Kumar et al. 2018). As mentioned, different lithographic processes are widely employed in the industry, among which is the optical lithography. Optical lithography is a top-down approach which is an effective technique in fabricating materials for microelectronic and nanoelectronic devices. Additionally, it is further applied to synthesize integrated circuits (Nalamalpu et al. 2015).

1.8.3.4 Etching

Etching is used in either nanofabrication or microfabrication in which chemical layers are removed from the wafer's surface during manufacturing (Kolasinski 2005). In each etching step, parts of the tested wafer are protected from the etchant by the help of a "masking" material which governs resistance against etching. These masking materials may be photoresist materials which are formulated using photolithography. In some cases, these masking materials may be composed of silicon nitride which is characterized by its long-term validity.

1.9 Techniques Employed for Nanomaterials' Characterization

For further exploration of the unique and extraordinary features of nanomaterials, accurate and detailed characterization of the material has to be performed. This can be done by the help of different characterization techniques which provide optical, morphological, electrical, and magnetic information about the prepared nanomaterial. A great progress concerning the technologies and the characterization instruments has been devoted. Such techniques possess the advantage of characterizing materials without causing any significant modification or any damage to the nanomaterials under study (Omran et al. 2019a, b).

1.9.1 Spectroscopic Characterization Techniques

The interaction between the matter and the electromagnetic radiation is referred to as spectroscopy (Bumbrah and Sharma 2016). The main principles of the spectroscopic techniques are dependent upon absorption, emission, fluorescence, or scattering (Skoog et al. 2017). Different spectroscopic techniques are employed to determine the optical properties of nanostructures. Some of the main optical characterization techniques include UV/visible spectroscopy (UV/Vis), photoluminescence spectroscopy (PL), surface-enhanced Raman spectroscopy (SERS), Brillouin spectroscopy, etc.

1.9.1.1 Ultraviolet-Visible Spectroscopy (UV/Vis)

UV/Vis spectroscopy is a kind of a spectroscopic technique in which light in the UV range (200–400 nm) is absorbed by the molecule. Electrons become excited from the ground state to a higher energy state (Chirayil et al. 2017). The energy difference between both that of the ground state and that of the higher energy state becomes equal to the amount of energy being absorbed. The apparatus of the UV/Vis spectrophotometer includes the following parts: reference and sample beams and deuterium or tungsten lamp for the wavelengths of the ultraviolet and the visible regions, respectively, monochromator and a detector. The UV spectrum is the spectrum which appears when a sample is exposed to UV light (Chirayil et al. 2017). Samples are placed inside glass, quartz, silica, or plastic cuvettes within the apparatus in front of the light path to be ready for measurement. It is worth noting that both the plastic and the glass cuvettes absorb wavelengths at 310 nm, so they should not be used for studying absorbance lower than that wavelength. Hence, quartz cuvettes are much preferred as they can be used for measuring absorption in the ultraviolet range (above 180 nm). Two types of beams exist; one is the reference beam, and the other one is the sample beam. The reference beam is emitted from the light source to the detector with no interactions with the sample. On the contrary, the sample beam interacts with the sample. Usually, the ratio between the intensities of the reference and the sample beams is determined by the help of the detector (Marvin et al. 2003). When a difference is found in the intensities, that means the reference beam's intensity is higher than that of the sample. Another important term is transmittance which takes place when the light beam passes through a solution, part of the light becomes absorbed, and the rest becomes transmitted via the solution. Therefore, the ratio between the light entering the sample to the light which exits the sample at a certain wavelength is referred to as "transmittance." Absorbance is referred to as the negative logarithm of transmittance.

UV/Vis spectroscopy follows the Beer-Lambert law. The Beer-Lambert law states that when a beam of a monochromatic light passes through a solution of an absorbing material, the rate of reduction in the intensity of the radiation in addition to the thickness of the absorbing solution is directly proportional to the incident radiation and to the concentration of the solution (Chirayil et al. 2017; Titus et al. 2019). The expression of the Beer-Lambert law is

$$A = \log\left(\frac{Io}{I}\right) = ECL$$

where A is the absorbance, Io is the light intensity that leaves the sample cell, I is the light intensity that is incident upon the sample cell, E is the molar absorptivity, C is the molar concentration of the solute, and L is the length of the sample cell (cm). From the Beer-Lambert law, it can be concluded that as the number of molecules which are capable of absorbing light becomes greater at a certain wavelength, the greater will be the magnitude of light absorption, and this is the basic working principle of UV/Vis spectrophotometer. UV/Vis spectrophotometer is majorly employed in the discipline of analytical chemistry to provide quantitative determination of analytes like transition metal ions, biological macromolecules, and highly conjugated organic compounds. Additionally, it helps to determine the size and concentration of the prepared nanoparticles. It also gives an indication concerning nanoparticles' aggregation/agglomeration. Moreover, it is a quite simple and effective technique to predict the stability of the prepared nanoparticles. Simply, as particles become destabilized, the original absorption peak decreases in intensity (because of the reduction of stable nanoparticles), and frequently the peak becomes broadened, and sometimes a secondary peak will occur at longer wavelengths (due to the formation of aggregates) (Chirayil et al. 2017).

1.9.1.2 Raman Spectroscopy (RS)

In 1928, C.V. Raman along with K.S. Krishnan published together the first article applying the Raman technique (Raman and Krishnan 1928). Raman spectroscopy was named after its inventor C.V. Raman. Raman spectroscopy (RS) is a multipurpose technique in which a variety of samples can be analyzed. It provides qualitative and quantitative analyses. By evaluating the frequency of the scattered radiation, qualitative analysis can be carried out, while quantitative analysis is performed via the measurement of the intensity of the scattered radiation (Bumbrah and Sharma 2016).

In RS, a beam of monochromatic light illuminates the molecules of a tested sample and originates a scattered light. The scattered light has a frequency which is different from that of the incident or the inelastic light collision and is used to display the Raman spectrum. When a monochromatic radiation hits the sample molecules, it scatters in every direction. Much of this scattered radiation has a frequency which is equal to the frequency of incident radiation and makes up the Rayleigh scattering. However, only a small fraction of the scattered radiation possesses a frequency that is different from that of the incident radiation and constitutes the Raman scattering. Raman Stokes in Raman spectrum appear when the frequency of the incident radiation is higher than that of the scattered radiation. Contrary, the Raman anti-stokes lines appear when the frequency of the incident radiation is lower than that of the scattered radiation (Dent and Smith 2005). Raman Stokes bands are more intensive than Raman anti-Stokes and thus are easily to be measured in conventional Raman spectroscopy, while anti-Stokes bands are measured with fluorescing samples.

Raman spectra are recorded over a range of 4000–10 cm^{-1}. Nonetheless, Raman active modes for organic molecules vibrations occur in the range of 4000–400 cm^{-1}. Raman spectrophotometers can be either dispersive or nondispersive. A prism or grating are usually used in dispersive Raman spectrophotometer, while nondispersive Raman spectrophotometer depends on an interferometer like Michelson interferometer (Bumbrah and Sharma 2016). Raman spectroscopy is a spectroscopic technique which helps in studying the vibrational and rotational modes of a specimen by the help of Raman scattering or by the scattering of monochromatic laser light (Bumbrah and Sharma 2016; Pilot et al. 2019). Typically, the sample is hit by a laser beam. A lens is responsible for collecting the electromagnetic radiation from the laser hit and then passes through a collimator. Stokes of Raman scattering are initiated because of the molecular excitation from the ground state to a vibrational excited state. Changes in polarizability affect Raman scattering, whereas the Raman shift is dependent upon the involved vibrational level. Advanced versions of Raman spectroscopy involve stimulated Raman spectroscopy, surface-enhanced Raman spectroscopy (SERS), and resonance Raman spectroscopy (Titus et al. 2019).

1.9.1.3 Photoelectron Spectroscopy (PES)

Photoelectron spectroscopy (PES) is a technique in which electron energy from solids, liquids, and gases is evaluated by the photoelectric effect (Ghosh 1983). The photoelectric effect is the main principle of physics behind the PES technique. PES is categorized into two main types based upon the source of exciting radiation, namely, ultraviolet PES (UPS) and X-ray PES (XPS). In UPS, the photon energy ranges from 10 to 50 eV which is greater than the traditional energy range (i.e., 2–5). Consequently, electrons are released from the sample surface under the photoelectric effect. UPS is a helpful technique which aids in studying chemical bonding, valence energy levels, and particularly, the bonding of molecular orbitals. Furthermore, UPS facilitates the observation of fine structures with high resolution due to the vibrational motion of molecular ions. On the other hand, XPS is a quantitative spectroscopic technique which determines the elemental composition of a sample at parts per thousand (Matsushima and Yamauchi 2019). Irradiation of a material results in obtaining the X-ray photoelectron spectra. XPS is also referred to as electron spectroscopy. XPS gives information about the local bonding of atoms.

1.9.1.4 Infrared Spectroscopy (IR)

Infrared spectroscopy (IR) spectroscopy is a technique which provides data by the interaction of infrared radiation with the tested sample. The resultant scanned spectrum ranges from 4000 to 400 cm^{-1}. Changes in the electric dipole moment of a specific molecular functional group are the keys behind generation of an IR spectrum (Dominguez et al. 2014; Chaber et al. 2017). Infrared active molecules are the molecules whose dipole moments are altered during vibration. IR spectroscopy basically aids in identifying the type of chemical bonds in a sample which are revealed by absorbing the characteristic wavelength of infrared radiation due to the presence of particular functional groups (Kumar et al. 2019). Any form of samples can be analyzed via IR spectroscopy, whether it is liquid, solid, powder, or film.

IR spectroscopy provides qualitative in addition to quantitative data. By employing IR spectroscopy, getting deeper insight of materials' structural analysis and interactions at molecular level has touched the ground. In short, IR spectroscopy is a very promising analyzing technique in getting more information regarding the chemical structure of small molecules, nanomaterials, natural products, and other biomolecules. Also, it provides easiness in identifying the functional groups in a tested sample. Additionally, determination and characterization of supramolecular interactions is attainable (Sondhi et al. 2009; Ellis and Goodacre 2006). Hence, IR spectroscopy offers several applications in different disciplines like organic chemistry, drug discovery, and drug design and provides precious information regarding the morphological transitions within the phase(s) of metal/metal oxide nanoparticles, carbon nanoparticles, and graphene quantum dots (GQDs) and their interactions with biomolecules (Bayda et al. 2017).

Fourier transform infrared spectroscopy (FTIR) is a very common technique employing the working principle of IR. It has been used for more than 50 years. FTIR provides information about the molecular interactions as well as the molecular structure of a specific material via absorbing the infrared light. Absorption takes place because of the excitation from the ground energy level to a higher energy level. FTIR spectroscopy can be used for the characterization of nanomaterials, biomolecules, and organic molecules and has the capability to clarify the structure of proteins. Each band which appears in the FTIR spectrum represents a fingerprint to a certain functional group (Hambardzumyan et al. 2011). FTIR introduces a rapid, accurate, and a non-destroying technique to confine the different ligands which are attached to NPs depending on their vibrational signature (López-Lorente and Mizaikoff 2016). Another apparatus applying the working principles of IR spectroscopy is called attenuated total reflection (ATR)-FTIR spectroscopy. It has the ability to determine the adsorbed molecular species structure. Additional advantages of ATR-FTIR spectroscopy are the minimization of the required steps needed for sample preparation as well as the absence of spectral reproducibility which are present in traditional IR mode (Beasley et al. 2014).

1.9.1.5 Mass Spectrometry (MS)

Mass spectrometry (MS) provides data regarding the elemental/ molecular composition and structure. It is very helpful in providing data concerning the physiochemical features of nanomaterials like mass, composition, and structure. Moreover, it is a very important characterization technique especially in toxicological, environmental, and nanotechnological studies. The main principle of MS upon which it can differentiate between charged particles is their mass-to-charge ratio. The combination of MS and inductively coupled plasma (ICP) aids in atomizing and ionizing the elements with high accuracy as well a provision of elemental chemical quantification analysis. Yet, the equipment is somehow costly and lacks databases while identifying certain species (Lavigne et al. 2013).

1.9.2 Morphological Characterization Techniques

Different microscopic techniques are employed to observe the morphological structure of nanomaterials including scanning electron microscope (SEM), field emission scanning electron microscopy (FESEM), scanning tunneling microscope (STM), high-resolution transmission electron microscope (HRTEM), atomic force microscope (AFM), etc.

1.9.2.1 Scanning Electron Microscope (SEM)

In scanning electron microscope (SEM), an electron beam is usually directed towards the specimen rather than a light beam, as in the case of optical microscopes (Joshi et al. 2008). SEM differs from the conventional light microscopes as the images are magnified by light in the case of the light microscope. An electron gun is responsible for emitting the electron beam, and it is located at the top of the microscope. Generally, there are two main kinds of electron guns: (i) field emission guns, which release a strong electric field that splits electrons from the atom, and (ii) thermionic guns in which the filament is heated till the electrons get away. The tested samples are subjected to scanning by high-energy electron beams (Titus et al. 2019). When the emitted beam of electrons hits the specimen surface, three kinds of electrons are released, and they are the backscattered (or primary) electrons, secondary electrons, and Auger electrons. Both of the secondary and the backscattered electrons are very important for SEM. High-resolution images generated by SEM demonstrate details of approximately 1–5 nm via the secondary electrons. Additionally, the energy dispersive X-ray part is usually equipped with SEM in order to identify the elemental composition of the tested sample. The backscattered electrons are also used to form the image in this technique.

The electron beam is passed through a scanning coil and then to a final lens. This results in beam deflection in horizontal and vertical directions so that the performance becomes more efficient and faster. Then, signals become detected and enlarged by the aid of electronic devices, which display the signals into images on a cathode ray tube. The resultant displayed image is considered a distribution map of the signal intensity which is emitted from the specimen scanned area. It is important to note that proper sample preparation is very essential as this analysis will not work properly if the samples are not well prepared. Nonmetal samples need prepping with a sputter coater (Nixon 1971). Specimens are usually conducted with gold as a thin conducting layer. Coating the specimen with gold takes place via argon gas and an electric field. Electrons are removed from the argon by exerting an electrical field, thus releasing positively charged ions. These positively charged ions become attracted to the negatively charged gold. The argon ions separate gold atoms, which in turn fall onto the specimen which then becomes covered with a thin conductive coating. Water removal is important in traditional SEM, because of the vacuum vaporization of water molecules which causes reduction in the clarity of the resultant image. Interestingly, new versions of SEM do no longer need a full vacuum (Titus et al. 2019).

1.9.2.2 Transmission Electron Microscopy (TEM)

A transmission electron microscope (TEM) gives details concerning the morphological structure, crystallinity, size, shape, and the compositional data of the tested sample by the help of an electron beam. TEM can reveal the finest details of the specimen under study, and in some cases it can reveal internal structures like individual atoms. TEM differs from the light microscope in that it uses electrons rather than light. Additionally, the glass lenses in the light microscope are replaced with electromagnetic lenses, and images are displayed on a screen instead of being viewed via an eye piece. TEM uses electrons instead of light. As electrons' wavelength is far smaller than that of the light, the image resolution obtained from TEM images is way better than that of the light microscope (Chirayil et al. 2017). Specimens are introduced onto grids in order to be imaged. However, certain samples are cut into very thin films by an ultramichrotome using a diamond knife under freezing conditions (Cheville and Stasko 2014). The images are displayed in black and white by TEM. Operating the imaging process requires special training, and samples must have the ability to withstand vacuum. High-resolution transmission electron microscope (HRTEM) is a very powerful microscopic technique to study the atomic features of materials (Chirayil et al. 2017). It is very helpful in imaging semiconductors, metals, sp^2-bonded carbon (e.g., graphene, carbon nanotubes), etc., and particularly it is very useful in the field of nanotechnology.

1.9.2.3 Scanning Transmission Electron Microscopy (STEM)

Scanning transmission electron microscope (STEM) combines the main principles of both TEM and SEM. It helps in imaging nanostructures with different modes and gives data concerning the elemental composition of the tested specimen. Samples must be very thin in order to be imaged by STEM by the help of an electron beam. Numerous types of electrons and electromagnetic signals are generated by the interaction between the electron nanoprobe and the specimen. The prementioned signals are then gathered together to produce the required images or to obtain diffraction patterns which can in turn be analyzed to reveal some spectroscopic information. For instance, high-angle annular dark-field (HAADF) or Z-contrast images provide information concerning the structural composition of the sample by collecting high-angle scattered electrons (Chirayil et al. 2017). The presence of subangstrom or subnanometer electron probes helps in providing details concerning shape, size, surface and crystal structures, and any emerging defects.

1.9.2.4 Scanning Probe Microscope (SPM)

Scanning probe microscopes (SPM) are extremely beneficial for imaging and characterizing materials at the nanoscale levels as well as imaging atoms. SPM depends on light waves to image the samples. A very fine probe referred to as a "tip" is responsible for sample scanning. The surface is scanned via an atomically sharp probe. It provides a three-dimensional topographic image of the specimen being imaged at the atomic scale. SPM is a very powerful microscope which has a resolution of less than 1 nm. Generally, two types of modes are usually employed by researchers, and they are the contact mode and non-contact mode. In contact mode, the force is remained constant between the tip and the surface, thus giving the ability for quick imaging of the specimen surface. In tapping non-contact mode, the cantilever oscillates by touching the specimen surface. And it is very beneficial in imaging a soft surface.

There exist numerous types of scanning probe microscopes involving atomic force microscope (AFM) which measures the electrostatic forces between the tested sample and the cantilever tip, magnetic force microscopes which measure the magnetic forces, and scanning tunneling microscopes (STM) which measure the electrical current that flows between the sample and the cantilever tip (Chirayil et al. 2017). Thus, electrical, magnetic properties as well as the sample topography can be easily attained.

1.9.2.5 Atomic Force Microscopy (AFM)

The atomic force microscope (AFM) is a powerful microscope that aids in studying materials at the nanoscale levels (Binnig et al. 1987). It displays three-dimensional images with plenty of surface measurements, and at an atomic resolution, it meets the requirements of researchers in different scientific fields. Sample preparation is not hard when it is being imaged by AFM. Moreover, it helps in estimating the surface roughness of polymer nanocomposites and has the ability to define the surface texture of several types of materials. It is worth noting that this microscopic technique does not cause any damage to the tested sample. The sample surface is scanned via a sharp tip. When the sample surface approaches the tip, a small deflection takes place due to the attractive forces between the tip and surface which can be detected by a laser beam. A photosensitive diode is an important part of AFM which helps in tracking the reflected beam's direction which might occur due to the cantilever deflection. AFM produces a morphological image of the scanned surface by maintaining the laser position constant via controlling the tip height (Titus et al. 2019).

Three kinds of scanning modes for AFM exist, and they are the contact mode, noncontact mode, and the tapping mode. In contact mode, the tip scans the sample's surface, and then by pushing the cantilever against the tested sample by a piezoelectric positioning element, the repulsive forces on the tip is generated. Afterwards, evaluation of the cantilever deflection takes place and the images are displayed. In noncontact mode, the tip flutters over the sample surface, and then the attraction forces between the tip and the specimen are estimated, and hence the images are produced. In the tapping mode, a piezoelectric crystal is responsible for generating a high-resolution image. The cantilever oscillates because of the piezo motion. Oscillation is reduced as the cantilever begins to touch the sample surface, and this reduction aids in measuring the different characteristics of the tested specimen.

1.9.2.6 Confocal Laser Scanning Microscopy (CLSM)

Confocal laser scanning microscopy (CLSM) is a potent technique to generate sharp images of a sample that might be blurred when examined under a conventional microscope. This technique became very popular between the scientific and industrial communities. Different scientific fields favor the use of CLSM like life, biological, and materials science disciplines, cell biology, microbiology, genetics as well as in semiconductor examination. Furthermore, it is very useful in quantum optics and imaging of nanocrystals (Fellers and Davidson 2007). This technique enables taking a high number of images with different depths, a process referred to as optical sectioning. The light beam becomes focused by the help of an objective lens, and afterwards the object is subjected to scanning by a computerized scanning device. The light points emitted from the tested specimen are then detected by a photomultiplier tube (PMT), and the output from PMT is turned into an image and demonstrated by the computer screen (Fellers and Davidson 2007).

1.9.2.7 Scanning Tunneling Microscopy (STM)

Scanning tunneling microscope (STM) produces images with an atomic scale. The main principle of AFM depends upon quantum tunneling (Titus et al. 2019). The sample surface is imaged by a fine probe equipped with a tip with the aid of a piezoelectric crystal, and the resultant tunneling current is measured. When the conducting tip gets into contact with the specimen surface, electrons flow through the vacuum. The piezoelectric crystal creates images by adjusting the sample surface with the tip distance, and hence the current between them maintains constant. Topography of the sample surface is determined by plotting the tip height as a function with its lateral position over the specimen (Titus et al. 2019). STM can be used at different conditions either in vacuum or in air, liquid, or gas and at a wide range of temperatures. It is worth noting that the tip should be entirely clean and sharp in order to work properly. Carbon nanotube tips are used in STM (Pasquini et al. 2005). It is a potential tool that provides facility to characterize materials.

1.9.2.8 Scanning Near-Field Optical Microscopy (SNOM)

Scanning near-field optical microscopy (SNOM) is a microscopic technique which provides a description of the topographic structure as well as the optical fluorescence features of the studied material (Chirayil et al. 2017). Images are created by SNOM through sample scanning with a small aperture at a certain distance, which is illuminated from the back side and is responsible for recording optical information and expressing the transmitted fluorescent light into an image. The obtained optical image is a result of scanning each point and each line in the sample. The reflected or transmitted light becomes restricted when sample scanning is done from a small distance. Among the advantages of SNOM is that no sample preparation is required and it is used in imaging different kinds of samples, i.e., transparent, conductive, and non-conductive ones. It can be applied in different types of research such as nanotechnology, nano-optics and nano- photonics.

1.9.3 Other Important Characterization Techniques

1.9.3.1 Dynamic Light Scattering (DLS) and Zeta Potential

Dynamic light scattering (DLS) is a powerful technique to tailor the average size particle of nanoparticles. In DLS, light is scattered from laser source and goes through the tested colloidal solution. The intensity of the scattered light is then analyzed as a function of time, and hence the average particle size can be estimated. It is worth noting that nanoparticles' diffusion rate strongly influences the time delay (Meulendijks et al. 2018). Brownian motion is the main working principle behind DLS (Mailer et al. 2015). Small-sized particles move faster than the large-sized ones and hence less light is scattered. An important term is the hydrodynamic diameter which is defined as the diameter of a solid sphere which would be the same as the hydrodynamic friction of the molecule of interest. Additionally, DLS provides information about the state of agglomeration of nanoparticles.

When particles are charged with opposite charges, a thin layer is created which is referred to as "stem layer," and the outer layer becomes diffusive and contains the weekly associated ions. Electric double layers are generated because of both the stem and the outer layers (Sapsford et al. 2011). Zeta potential refers to the shear surface electric potential. It is estimated by effect of charged species velocity towards the electrode under the effect of an external electric field within the sample solution (Clogston and Patri 2011). Usually, nanoparticles possess a negative or positive charge on their surface. For instance, particles with a negative zeta potential value bind to the positively charged surfaces and vice versa. When an electric field is applied under the control of DLS, particles start moving due to the interaction between the charged particle and the electric field. Zeta potential measures the effective electric charge on the tested nanoparticle's surface and provides a calculated charge stability of the colloidal particles. The magnitude of the zeta potential provides information about the particle stability. High stability of the prepared nanoparticles can be detected via the higher magnitude of zeta potential. For more clarification, particles tend to agglomerate in zeta potential range of 0–5 mV; minimally, moderately, and highly stable particles are usually determined at the ranges of 5–20, 20–40, and >40 mV, respectively (Titus et al. 2019).

1.9.3.2 Energy Dispersive X-Ray Analysis (EDX)

Energy dispersive X-ray analysis (EDX) is an elemental analysis which is equipped along with electron microscopes. It basically depends on the emission of X-rays that demonstrates the type of elements present in the tested specimen. EDX is a powerful analytical technique to detect nanoparticles. It provides quantitative and qualitative information. EDX has the advantage of sample overall mapping by the accurate analysis of the near-surface elements and also has the capability to evaluate the elemental percentage in the tested sample. EDX is used in conjunction with SEM or TEM. Usually, an electron beam with an energy of 10–20 keV hits the sample's surface which in turn leads to emission of X-rays from the tested material (Joshi et al. 2008). The composition and quantity of nanoparticles can be evaluated using EDX. For example, nanoparticles like silver, gold, palladium, etc. can be easily identified using EDX. It is worth mentioning that elements with low atomic number are hard to be detected by EDX (Titus et al. 2019).

1.9.3.3 X-Ray Diffraction (XRD)

To obtain information regarding the phase identification, quantification, average size particle, lattice distortion, deviation of a specific element from the ideal composition, and nanocrystalline orientation, X-ray diffraction is always employed (Titus et al. 2019). The desired material that is being analyzed is mounted on the goniometer and then bombarded via the emitted X-rays with gradual rotation. The atomic planes interact with the X-ray beam, and then the rest of the beam is absorbed, scattered, refracted, or diffracted by the tested sample. Usually, X-rays are diffracted by each element in a totally different way, relying on the type and the arrangement of the atoms. X-rays are known to be short-wavelength electromagnetic radiation. A high voltage is kept steady between the electrodes; thus attraction between the electrons and the metal target occurs. At the striking point, X-rays are generated and radiated in every path. Afterwards, the generated X-rays are collimated and concentrated towards the tested sample which is finely grounded. X- rays are detected by a detector, and signal processing takes place via a microprocessor or electronically performed. X-ray scan or a spectrograph is obtained by the applied angle variation between the source, sample, and the detector. Scattering takes place when the X-ray impacts the crystal lattice. Diffraction occurs when scattering comes in phase with other plane scatterings. It is worth noting that each crystalline material has its unique atomic structure and hence diffracts the X-rays in a totally distinctive pattern (Titus et al. 2019).

In 1913, Bragg formulated Bragg's equation (Bragg 1913) which measures the diffraction angle. The Bragg equation is

$$2d\sin\theta = n\lambda$$

where d denotes to the spacing between the planes, θ denotes to incidence angle, n denotes to an integer, and λ denotes to beam wavelength.

1.9.3.4 Thermogravimetric Technique

Thermogravimetric technique enables to track the mass changes of a nanomaterial. This technique enables to provide information like desorption, absorption, adsorption, sublimation, decomposition, oxidation, and reduction in cases of nanomaterials. Different materials can be analyzed using this technique such as paints, films, and so on. The equipment is made up of different parts involving electronic microbalance, sample holder, furnace, temperature programmer, and recorder. The main task of the microbalance is to record any changes of sample mass. Two types of microbalance exist, the deflection type and the null-point type. The sample holder is the place where the sample is located and is referred to as a crucible. It is attached to the weighing arm of the microbalance. Different types of crucible are available including deep, retort cup, loosely covered, and shallow pan types. Furnaces are responsible in providing a specific heat rate. Temperature measurement is achieved via thermocouples, and the temperature programmer controls the temperature rate during analysis.

1.9.3.5 Nuclear Magnetic Resonance (NMR)

Nuclear magnetic resonance (NMR) is a characterization technique which handles atomic nuclei containing both angular momentum and magnetic moments when exposed to an external magnetic field (MacArthur 2016). NMR provides detailed information concerning the chemical environment structure, possible reactions, and expected dynamics of tested atomic nuclei (Kumar et al. 2019). NMR detects each spectral line of the tested nuclei with high sensitivity. The main principle of NMR is to get the most use of the magnetic properties of specific nuclei and translate them into chemical data output. NMR provides data regarding the physiochemical features involving structure, functionality, and purity. Recently, a pulsed field gradient NMR has been developed to estimate the diffusion capability of nanomaterials, and under which the species interaction and the size can be analyzed (Deshayes et al. 2010). Among the main advantages of using NMR is that small amounts of the samples are required, but the main drawbacks are represented in time consumption as well as the low detection sensitivity (Mullen et al. 2010).

1.10 Arising Era of Nanobiotechnology

In nanobiotechnology, the term begins with the prefix "nano" and the word "biotechnology" which refers to the employment and self-assembly of biologically oriented molecules towards different technological applications (El-Gendy and Omran 2019). Nanobiotechnology represents the fusion point between the disciplines of biotechnology and nanotechnology (Manju and Sreenivasan 2010). It is a multidisciplinary field of science which deals with the biomimetic synthesis, manipulation, and controlled bio-functionalization of structures at the nanoscale level (Mogoşanu et al. 2016; Omran et al. 2018a). It covers a wide range of scientific disciplines including chemistry, physics, engineering, biology, etc. Nanobiotechnology was initiated by the development of the atomic force microscope (AFM) which facilitated atomic imaging in the 1980s. Hess and Jaeger (2010) demonstrated that nanobiotechnology was fully involved as an important scientific discipline since the year 2000. Christopher Lowe was the pioneer scientist in introducing nanobiotechnology by his publication in the journal *Current Opinion in Structural Biology* in 2000 in which he introduced this emerging discipline to the scientific communities (Lowe 2000). Afterwards, in 2004, two books were edited by Christof Niemeyer and Chad Mirkin. These two books provided a reference to various researchers in the community of nanobiotechnology worldwide (Hess and Jaeger 2010). Briefly, nanobiotechnology continues to be the science of the hour. As a result of the outstanding properties of nanomaterials, the number of engineers and scientists who keep working on and exploring materials at the nanoscale is enormously increasing (Guo 2013). As mentioned earlier, when materials are in the nanoscale level, their fundamental chemical, physical, and mechanical properties differ from their bulk counterparts.

Since many years ago, green-based technologies have been the focus point of the whole world to protect human health as well as the surrounding environment from any hazardous issues that may arise. The beginning of green technology took place in purifying the water system; afterwards other green techniques were introduced involving the provision of catalytic converters for cars, treatment plants recycling, solar panels, and many others (Guo 2013). Green chemistry and engineering arise as an important branch that would help the science of nanotechnology to imitate natural processes to lessen any hazardous effect which may result from the conventional chemical and physical synthesis methodologies. The main target of green chemistry/engineering is to enhance industries to act somehow similar to ecosystems or like living cells, in which benign constituents are employed, biogenic wastes are recycled, and energy can be effectively monitored.

Not surprisingly, cells may act as green nano-factories. Natural components are utilized by cells at room temperature to assemble nanostructures. These reactions are carried out without using any toxic solvents, hence the combination between nanotechnology and green chemistry facilitates the manufacture of environmentally sustainable nanomaterials. Such combination is worthy to be kept on and be developed for many reasons, among which is that this combination helps to synthesize clean nanomaterials from the beginning without causing any environmental problems. Applying a green nano-approach to nanotechnology ultimately helps the environment to boost environmental performance by waste detoxification. Additionally, more green products are manufactured to replace the old wasteful and hazardous ones.

Using biological components to produce nanoscaled materials is the basic simple principle of nanobiotechnology (Omran et al. 2018b; El-Aswar et al. 2019; Omran et al. 2019a). Therefore, green nanotechnology helps to achieve the main tasks of nanotechnology as they may be applied in electronic, computing, biological and clinical applications. The target of the upcoming chapters is to show by several selected examples how these biological approaches tend to introduce a great contribution to nanoscience.

References

Artioli G, Angelini I, Polla A (2008) Crystals and phase transitions in protohistoric glass materials. Phase Transit 81:233–252

Arulmani S, Anandan S, Ashokkumar M (2018) Introduction to advanced nanomaterials. In: Nanomaterials for green energy. Elsevier, Cambridge, MA, pp 1–53

Balasooriya ER, Jayasinghe CD, Jayawardena UA et al (2017) Honey mediated green synthesis of nanoparticles: new era of safe nanotechnology. J Nanomater 2017, Article ID 5919836:1–10. https://doi.org/10.1155/2017/5919836

Balcerzak M, Jakubowicz J, Kachlicki T, Jurczyk M (2015) Hydrogenation properties of nanostructured Ti_2Ni-based alloys and nanocomposites. J Power Sources 280:435–445

Ball AS, Patil S, Soni S (2019) Introduction into nanotechnology and microbiology. Methods Microbiol 46:1–18

Balzani V (2005) Nanoscience and nanotechnology: a personal view of a chemist. Small 1:278–283

Basu PK (1997) Theory of optical processes in semiconductors: bulk and microstructures. Clarendon press, Oxford

Bayda S, Hadla M, Palazzolo S et al (2017) Bottom-up synthesis of carbon nanoparticles with higher doxorubicin efficacy. J Control Release 248:144–152

Beasley MM, Bartelink EJ, Taylor L, Miller RM (2014) Comparison of transmission FTIR, ATR, and DRIFT spectra: implications for assessment of bone bioapatite diagenesis. J Archaeol Sci 46:16–22

Benjamin JS (1970) Dispersion strengthened superalloys by mechanical alloying. Metall Trans A 1:2943–2951

Bhadeshia H (1997) Recrystallisation of practical mechanically alloyed iron-base and nickel- base superalloys. Mater Sci Eng A 223:64–77

Bidabadi ARS, Enayati MH, Dastanpoor E et al (2013) Nanocrystalline intermetallic compounds in the Ni–Al–Cr system synthesized by mechanical alloying and their thermodynamic analysis. J Alloys Compd 581:91–100

Binnig G, Gerber C, Stoll E et al (1987) Atomic resolution with atomic force microscope. Europhys Lett 3:1281–1286

Birnbaum AJ, Pique A (2011) Laser induced extraplanar propulsion for three-dimensional microfabrication. Appl Phys Lett 98:134101

Borsella E, D'Amato R, Terranova G et al (2011) Synthesis of nanoparticles by laser pyrolysis: from research to applications. Energia Ambiente Innov 4:54–64

Bragg WL (1913) The structure of some crystals as indicated by their diffraction of X-rays. Proc R Soc London Ser A 89:248–277

Brun N, Mazerolles L, Pernot M (1991) Microstructure of opaque red glass containing copper. J Mater Sci Lett 10:1418–1420

Bumb A, Brechbiel MW, Choyke PL et al (2008) Synthesis and characterization of ultra-small superparamagnetic iron oxide nanoparticles thinly coated with silica. Nanotechnology 19:335601–335608

Bumbrah GS, Sharma RM (2016) Raman spectroscopy – basic principle, instrumentation and selected applications for the characterization of drugs of abuse. Egypt J Forensic Sci 6:209–215

Chaber R, Łach K, Szmuc K et al (2017) Application of infrared spectroscopy in the identification of Ewing sarcoma: a preliminary report. Infrared Phys Technol 83:200–205

Chen Y, Li CP, Chen H, Chen Y (2006) One-dimensional nanomaterials synthesized using high-energy ball milling and annealing process. Sci Technol Adv Mater 7:839–846
Cheville NF, Stasko J (2014) Techniques in electron microscopy of animal tissue. Vet Pathol 51:28–41
Chirayil CJ, Abraham J, Mishra RK et al (2017) Instrumental techniques for the characterization of nanoparticles. Elsevier Inc, Amsterdam, pp 1–36
Clogston JD, Patri AK (2011) Zeta potential measurement. In: Characterization of nanoparticles intended for drug delivery. Springer, New York, pp 63–70
Dent G, Smith E (2005) Modern Raman spectroscopy: a practical approach. Wiley, London
Deshayes S, Maurizot V, Clochard M-C et al (2010) Synthesis of specific nanoparticles for targeting tumor angiogenesis using electron-beam irradiation. Radiat Phys Chem 79:208–213
Dobkin D, Zuraw MK (2003) Principles of chemical vapor deposition. Springer Science & Business Media, Dordrecht
Dominguez G, Mcleod AS, Gainsforth Z et al (2014) Nanoscale infrared spectroscopy as a non-destructive probe of extraterrestrial samples. Nat Commun 5:5445–5446
Drexler KE (1981) Molecular engineering: an approach to the development of general capabilities for molecular manipulation. Proc Natl Acad Sci 78:5275–5278
Dykman L, Khlebtsov N (2012) Gold nanoparticles in biomedical applications: recent advances and perspectives. Chem Soc Rev 41:2256–2282
El-Aswar EI, Zahran MM, El-Kemary M (2019) Optical and electrochemical studies of silver nanoparticles biosynthesized by *Haplophyllum tuberculatum* extract and their antibacterial activity in wastewater treatment. Mater Res Express 6:105016–105032
El-Gendy NS, Omran BA (2019) Green synthesis of nanoparticles for water treatment. In: Nano and bio-based technologies for wastewater treatment. Wiley, Hoboken, pp 205–263
Ellis DI, Goodacre R (2006) Metabolic fingerprinting in disease diagnosis: biomedical applications of infrared and Raman spectroscopy. Analyst 131:875–885
Fellers TJ, Davidson MW (2007) Introduction to confocal microscopy. J Investig Dermatol 132:1–5
Feynman RP (1960) There's plenty of room at the bottom. California Institute of Technology Eng Sci Mag 23:22–36
Foord JS, Davies GJ, Tsang WT (1997) Chemical beam epitaxy and related techniques. Wiley, Chichester; New York
Freestone I, Meeks N, Sax M, Higgitt C (2007) The Lycurgus cup—a roman nanotechnology. Gold Bull 40:270–277
Gentili E, Tabaglio L, Aggogeri F (2005) Review on micromachining techniques. In: AMST'05 advanced manufacturing systems and technology. Springer, Vienna, pp 387–396
Ghosh PK (1983) Introduction to photoelectron spectroscopy. Wiley, New York
González AL, Noguez C, Beránek J, Barnard AS (2014) Size, shape, stability, and color of plasmonic silver nanoparticles. J Phys Chem C 118:9128–9136
Groza JR (1999) Nanosintering. Nanostruct Mater 12:987–992
Guo KW (2013) An overview of green nanotechnology. In: Bio-nanotechnology: a revolution in food, biomedical and health sciences. Blackwell Publishing Ltd, Oxford, pp 311–354
Häkkinen H, Abbet S, Sanchez A et al (2003) Structural, electronic, and impurity-doping effects in nanoscale chemistry: supported gold nanoclusters. Angew Chem Int Ed 42:1297–1300
Hambardzumyan A, Molinari M, Dumelie N et al (2011) Structure and optical properties of plant cell wall bio-inspired materials: cellulose–lignin multilayer nanocomposites. C R Biol 334:839–850
Height MJ, Mädler L, Pratsinis SE, Krumeich F (2006) Nanorods of ZnO made by flame spray pyrolysis. Chem Mater 18:572–578
Heiligtag FJ, Niederberger M (2013) The fascinating world of nanoparticle research. Mater Today 16:262–271
Hess H, Jaeger L (2010) Nanobiotechnology editorial overview. Curr Opin Biotechnol 21:373–375
Jeevanandam J, Barhoum A, Chan YS et al (2018) Review on nanoparticles and nanostructured materials: history, sources, toxicity and regulations. Beilstein J Nanotechnol 9:1050–1074

Johnson-McDaniel D, Barrett CA, Sharafi A et al (2013) Nanoscience of an ancient pigment. J Am Chem Soc 135:1677–1679

Joshi M, Vandana P (2008) Nanostructured lipid carrier (NLC) based gel of celecoxib. Int J Pharm 346:124–132

Kalam A, Al-Sehemi AG, Assiri M et al (2018) Modified solvothermal synthesis of cobalt ferrite ($CoFe_2O_4$) magnetic nanoparticles photocatalysts for degradation of methylene blue with H_2O_2/visible light. Results Phys 8:1046–1053

Kirti S, Bhandari VM, Jena J et al (2018) Exploiting functionalities of biomass in nanocomposite development: application in dye removal and disinfection along with process intensification. Clean Technol Environ Policy 20:981–994

Koch CC, Cavin OB, McKamey CG et al (1983) Preparation of "amorphous" $Ni_{60}Nb_{40}$ by mechanical alloying. Appl Phys Lett 43:1017–1019

Kolasinski R (2005) Oblique angle sputtering yield measurements for ion thruster grid materials. In 41st AIAA/ASME/SAE/ASEE Joint Propulsion Conference & Exhibit (p. 3526)

Kreuter J (2007) Nanoparticles—a historical perspective. Int J Pharm 331:1–10

Kumar U, Panda D, Biswas KG (2018) Non-lithographic copper-wire based fabrication of microfluidic reactors for biphasic flow applications. Chem Eng J 344:221–227

Kumar S, Biswas A (2019) A unified TOPSIS approach to MADM problems in interval-valued intuitionistic fuzzy environment. In Computational Intelligence: Theories, Applications and Future Directions-Volume II, Springer, Singapore, pp 435–447

Kumar A, Khandelwal M, Gupta SK et al (2019) Fourier transform infrared spectroscopy: Data interpretation and applications in structure elucidation and analysis of small molecules and nanostructures. In Data Processing Handbook for Complex Biological Data Sources (pp. 77–96). Academic Press

Kurland HD, Grabow J, Staupendahl G et al (2007) Magnetic iron oxide nanopowders produced by CO_2 laser evaporation. J Magn Magn Mater 311:73–77

Kurland H-D, Grabow J, Staupendahl G et al (2009) Magnetic iron oxide nanopowders produced by CO_2 laser evaporation—'in situ' coating and particle embedding in a ceramic matrix. J Magn Magn Mater 321:1381–1385

Lalwani G, Henslee AM, Farshid B et al (2013) Two-dimensional nanostructure-reinforced biodegradable polymeric nanocomposites for bone tissue engineering. Biomacromolecules 14:900–909

Langdon TG (2016) A comparison of repetitive corrugation and straightening and high-pressure torsion using an Al-Mg-Sc alloy. Integr Med Res 5:353–359

Lapovok RYE (2005) The role of back-pressure in equal channel. J Mater Sci 40:341–346

Lavigne J-P, Espinal P, Dunyach-Remy C et al (2013) Mass spectrometry: a revolution in clinical microbiology? Clin Chem Lab Med 51:257–270

Leonhardt U (2007) Optical metamaterials: invisibility cup. Nat Photonics 1:207–208

Liang Y, Li Y, Wang H et al (2011) Co_3O_4 nanocrystals on graphene as a synergistic catalyst for oxygen reduction reaction. Nat Mater 10:780–786

Liu G, Li Z, Xu L et al (2018) The dynamics and adsorption of Cd (II) onto hydroxyapatite attapulgite composites from aqueous solution. J Sol-Gel Sci Technol 87:269–284

López-Lorente ÁI, Mizaikoff B (2016) Recent advances on the characterization of nanoparticles using infrared spectroscopy. TrAC Trends Anal Chem 84:97–106

Lowe CR (2000) Nanobiotechnology: the fabrication and applications of chemical and biological nanostructures. Curr Opin Struct Biol 10:428–434

MacArthur BKE (2016) The use of annular dark-field scanning transmission electron microscopy for quantitative characterisation. Johnson Matthey Technol Rev 60:117–131

Mailer AG, Clegg PS, Pusey PN (2015) Particle sizing by dynamic light scattering: non-linear cumulant analysis. J Phys Condens Matter 27:145102–114511

Manju S, Sreenivasan K (2010) Functionalised nanoparticles for targeted drug delivery. In: Biointegration of medical implant materials. Elsevier, Duxford, pp 267–297

Marvin LF, Roberts MA, Fay LB (2003) Matrix-assisted laser desorption/ionization time-of- flight mass spectrometry in clinical chemistry. Clin Chim Acta 337:11–21

Matsushima N, Yamauchi J (2019) First-principles X-ray photoelectron spectroscopy binding energy shift calculation for boron and aluminum defects in 3C-silicon carbide. Jpn J Appl Phys 58:031001–031012

Meulendijks N, van Ee R, Stevens R et al (2018) Flow cell coupled dynamic light scattering for real-time monitoring of nanoparticle size during liquid phase bottom-up synthesis. Appl Sci 8:108–118

Mie G (1908) Beiträge zur Optik trüber Medien, speziell kolloidaler Metallösungen. Ann Phys 330:377–445

Miklaszewski A, Jurczyk MU, Jurczyk K, Jurczyk M (2011) Plasma surface modification of titanium by TiB precipitation for biomedical applications. Surf Coat Technol 206:330–337

Mogoşanu GD, Grumezescu AM, Bejenaru LE, Bejenaru C (2016) Natural and synthetic polymers for drug delivery and targeting. In: Nanobiomaterials in drug delivery. Elsevier, San Diego, pp 229–284

Mohan Bhagyaraj S, Oluwafemi OS (2018) Nanotechnology: the science of the Invisible. Elsevier Ltd, India, pp 1–18

Mori S, Shitara Y (1994) Tribochemical activation of gold surface by scratching. Appl Surf Sci 78:269–273

Mullen DG, Fang M, Desai A et al (2010) A quantitative assessment of nanoparticle– ligand distributions: implications for targeted drug and imaging delivery in dendrimer conjugates. ACS Nano 4:657–670

Nakai I, Numako C, Hosono H, Yamasaki K (2004) Origin of the red color of Satsuma copper-ruby glass as determined by EXAFS and optical absorption spectroscopy. J Am Ceram Soc 82:689–695

Nalamalpu A, Kurd N, Deval A, et al (2015) Broadwell: a family of IA 14nm processors. In: 2015 Symposium on VLSI Circuits (VLSI Circuits). IEEE, Intel Corporation, Hillsboro, OR, USA, pp C314–C315

Nayfeh MH (2018) Fundamentals and applications of nano silicon in plasmonics and fullerenes: current and future trends. Elsevier, Amsterdam

Nelson DJ, Strano M (2006) Richard Smalley: saving the world with nanotechnology. Nat Nanotechnol 1:96–98

Niespodziana K, Jurczyk K, Jakubowicz J, Jurczyk M (2010) Fabrication and properties of titanium–hydroxyapatite nanocomposites. Mater Chem Phys 123:160–165

Nixon WC (1971) The general principles of scanning electron microscopy. Philos Trans R Soc Lond Ser B Biol Sci 261:45–50

Omran BA, Nassar HN, Fatthallah NA et al (2018a) Characterization and antimicrobial activity of silver nanoparticles mycosynthesized by *Aspergillus brasiliensis*. J Appl Microbiol 125:370–382

Omran BA, Nassar HN, Fatthallah NA et al (2018b) Waste upcycling of *Citrus sinensis* peels as a green route for the synthesis of silver nanoparticles. Energy Sources, Part A Recover Utilization and Environmental Effects 40:227–236

Omran BA, Nassar HN, Younis SA et al (2019a) Physiochemical properties of *Trichoderma longibrachiatum* DSMZ 16517-synthesized silver nanoparticles for the mitigation of halotolerant sulphate-reducing bacteria. J Appl Microbiol 126:138–154

Omran BA, Nassar HN, Younis SA et al (2019b) Novel mycosynthesis of cobalt oxide nanoparticles using *Aspergillus brasiliensis* ATCC 16404: optimization, characterization and antimicrobial activity. J Appl Microbiol 128:438–457

Pardis N, Talebanpour B, Ebrahimi R, Zomorodian S (2011) Cyclic expansion-extrusion (CEE): a modified counterpart of cyclic extrusion-compression (CEC). Mater Sci Eng A 528:7537–7540

Pasquini A, Picotto GB, Pisani M (2005) STM carbon nanotube tips fabrication for critical dimension measurements. Sensors Actuators A Phys 123:655–659

Paul S, Chugh A (2011) Assessing the role of Ayurvedic 'bhasms' as ethno-nanomedicine in the metal based nanomedicine patent regime. J Intellectual Property Rights 16:509–515

Pilot R, Signorini R, Durante C et al (2019) A review on surface-enhanced Raman scattering. Biosensors 9:57–156

Prasad Yadav T, Manohar Yadav R, Pratap Singh D (2012) Mechanical milling: a top down approach for the synthesis of nanomaterials and nanocomposites. Nanosci Nanotechnol 2:22–48

Raman CV, Krishnan KS (1928) A new type of secondary radiation. Nature 121:501–502

Rane AV, Kanny K, Abitha VK, Thomas S (2018) Methods for synthesis of nanoparticles and fabrication of nanocomposites. In: Synthesis of inorganic nanomaterials. Elsevier, SPi Global, India, pp 121–139

Rastogi A, Singh P, Haraz FA, Barhoum A (2018) Biological synthesis of nanoparticles: an environmentally benign approach. In: Fundamentals of Nanoparticles. Elsevier Inc, Typeset by Thomson Digital, India, pp 571–604

Raza M, Kanwal Z, Rauf A et al (2016) Size- and shape-dependent antibacterial studies of silver nanoparticles synthesized by wet chemical routes. Nanomaterials 6:74–89

Rittner MN, Abraham T (1998) Nanostructured materials: an overview and commercial analysis. JOM J Miner Met Mater Soc 50:37–38

Rosochowski A (2005) Processing metals by severe plastic deformation. In: Solid state phenomena. Trans Tech Publications, Stäfa, pp 13–22

Rytwo G (2008) Clay minerals as an ancient nanotechnology: historical uses of clay organic interactions, and future possible perspectives. Macla 9:15–17

Sapsford KE, Tyner KM, Dair BJ et al (2011) Analyzing nanomaterial bioconjugates: a review of current and emerging purification and characterization techniques. Anal Chem 83:4453–4488

Schaming D, Remita H (2015) Nanotechnology: from the ancient time to nowadays. Found Chem 17:187–205

Shinde M, Qureshi N, Rane S et al (2017) Instantaneous synthesis of faceted iron oxide nanostructures using microwave solvothermal assisted combustion technique. J Nanosci Nanotechnol 17:5024–5030

Skoog DA, Holler FJ, Crouch SR (2017) Principles of instrumental analysis. Cengage Learning Asia Pte Ltd, Singapore

Solero GAG (2017) Synthesis of nanoparticles through flame spray pyrolysis: experimental apparatus and preliminary results. Nanosci Nanotechnol 7:21–25

Sondhi SM, Rani R, Gupta PP et al (2009) Synthesis, anticancer, and anti-inflammatory activity evaluation of methanesulfonamide and amidine derivatives of 3, 4-diaryl-2-imino-4- thiazolines. Mol Divers 13:357–366

Stötzel C, Kurland H-D, Grabow J et al (2013) Control of the crystal phase composition of Fe x O y nanopowders prepared by CO2 laser vaporization. Cryst Growth Des 13:4868–4876

Sunil BR, Sunil BR (2015) Repetitive corrugation and straightening of sheet metals. Lasers Mater Manuf Process 30:1261–1270

Taniguchi N, Arakawa C, Kobayashi T (1974) On the basic concept of 'nano- technology'. In: Proceedings of the international conference on production engineering. Japan Society of Precision Engineering, Tokyo, pp 18–23

Titus D, Samuel JJE, Roopan SM (2019) Nanoparticle characterization techniques. In: Green synthesis characterization and applications of nanoparticles, Micro and nano technologies. Elsevier Inc, Amsterdam, pp 303–319

Tiwari P, Srivastava M, Mishra R et al (2018) Economic use of waste Musa paradisiaca peels for effective control of mild steel loss in aggressive acid solutions. J Environ Chem Eng 6:4773–4783

Tok AIY, Boey FYC, Zhao XL (2006) Novel synthesis of Al2O3 nano-particles by flame spray pyrolysis. J Mater Process Technol 178:270–273

Toth LS, Gu C (2014) ScienceDirect tutorial review ultrafine-grain metals by severe plastic deformation. Mater Charact 92:1–14

Tulinski M, Jurczyk M (2012) Nanostructured nickel-free austenitic stainless steel composites with different content of hydroxyapatite. Appl Surf Sci 260:80–83
Tulinski M, Jurczyk M (2017) Nanomaterials synthesis methods. In: Metrology and standardization of nanotechnology, pp 75–98
Ullattil SG, Thelappurath AV, Tadka SN et al (2017) A sol-solvothermal processed 'black TiO_2'as photoanode material in dye sensitized solar cells. Sol Energy 155:490–495
Vance ME, Kuiken T, Vejerano EP et al (2015) Nanotechnology in the real world: redeveloping the nanomaterial consumer products inventory. Beilstein J Nanotechnol 6:1769–1780
Visweswara Rao P, Hua Gan S (2015) Recent advances in nanotechnology-based diagnosis and treatments of diabetes. Curr Drug Metab 16:371–375
Vossen JL, Kern W, Kern W (1991) Thin film processes II. Gulf Professional Publishing, Oxford
Wagner R, Moon R, Pratt J et al (2011) Uncertainty quantification in nanomechanical measurements using the atomic force microscope. Nanotechnology 22:455703–455713
Walter P, Welcomme E, Hallégot P et al (2006) Early use of PbS nanotechnology for an ancient hair dyeing formula. Nano Lett 6:2215–2219
Wan J, Lacey SD, Dai J et al (2016) Tuning two-dimensional nanomaterials by intercalation: materials, properties and applications. Chem Soc Rev 45:6742–6765
Wang J, Wu S, Suo X-K, Liao H (2019) The processes for fabricating nanopowders. In: Advanced nanomaterials and coatings by thermal spray, Micro and nano technologies. Elsevier Inc, San Diego
Zhang H, Yang J, Wang S et al (2014) Film thickness dependence of microstructure and superconductive property of PLD prepared YBCO layers. Physica C 499:54–56
Zsigmondy R (1914) Influence of the quantity of peptising agent on the properties of hydrosols. Hydrosols of stannic acid. Z Anorg Chem 89:210

Chapter 2
Prokaryotic Microbial Synthesis of Nanomaterials (The World of Unseen)

2.1 Introduction

Several research studies clearly demonstrated that green biologically based methodologies are non-hazardous, low-cost and are considered as an environment-friendly alternative to the conventional physical and chemical synthesis methods (Makarov et al. 2014; Gowramma et al. 2015; El-Gendy and Omran 2019). Both of microorganisms and plants have been proved since a long time to have the capability to absorb and accumulate metal ions from the surrounding environment (Shah et al. 2015). Such biological entities possess attractive properties which candidate them as potent biological bio-nano-factories to produce green-synthesized nanomaterials. By playing this role, they help in decreasing environmental pollution as well as retrieving metals from industrial wastes. They majorly rely upon inherent biochemical routes to convert inorganic metallic ions to their nanoparticle state (Baker et al. 2013). Different discipline can get benefits behind the capability of microorganisms to interact, extract, and accumulate metallic ions from the surrounding environment including biotechnological, medical, pharmaceutical, bioleaching, and bioremediation applications (Shah et al. 2015).

One of the main reasons behind the interaction of microorganisms with their surrounding environment is the lipid-based amphipathic membrane structure which enables different oxidation-reduction reactions to happen and thereby promoting plenty of biochemical conversions (Mandal et al. 2006). Several reported investigations demonstrated the extracellular and intracellular synthesis of nanomaterials using either unicellular or multicellular microorganisms (Fernández et al. 2016; Kumaresan et al. 2018; Baygar et al. 2019; Sreedharan et al. 2019). Still much research regarding the significant oxidation-reduction reactions, nucleation, and consequent nanoparticle growth and formation have to be spotted on (Shah et al. 2015). Additionally, designing biologically synthesized nanomaterials with definite shape and size needs more investigations. Since culturing methods are critically

B. A. Omran, *Nanobiotechnology: A Multidisciplinary Field of Science*,
Nanotechnology in the Life Sciences, https://doi.org/10.1007/978-3-030-46071-6_2

important in case of microorganisms, a full investigation to elucidate the effect of different parameters that may affect the size and morphology of the prepared nanoparticles is a very important aspect. Optimization of culturing parameters like temperature, nutrients, medium pH, light, shaking speed, metal salt and biomass concentrations, and buffer strength can considerably raise the enzyme activity (Iravani 2011). The biological synthesis of nanomaterials by microorganisms is a bottom-up approach at which nanoparticles are biosynthesized via reduction/oxidation reactions. Different biological molecules secreted by microorganisms usually mediate these reactions like proteins, enzymes, and sugars (Prabhu and Poulose 2012). This chapter discusses in details the ability of prokaryotic microorganisms to biosynthesize nanomaterials.

2.2 The Unseen World of Prokaryotes

The word *prokaryote* is originated from the Greek language which combines the word *pro*, "before" with karyon, "nut or kernel." Prokaryotic microorganisms are single-celled microorganisms. Prokaryotes refer to microorganisms in which the micro-compartments are not separated from cytoplasm by membranes and the nuclear division takes place by fission (Killham and Prosser 2015). They do not possess either a definite shape of nucleus or any other organelles. The intracellular water-soluble constituents such as deoxyribonucleic acid (DNA), proteins, and metabolites are organized together within the cytoplasm and enclosed by an outer cellular membrane, rather than being in distinct cellular compartments. The prokaryotic life started over 4 billion years ago, feeding off different nutrients like atmospheric carbon dioxide, carbon monoxide, hydrogen, steam, nitrogen, and ammonia (Whitman et al. 1998). It is worth noting that prokaryotes play a very important role in several biogeochemical cycles and other essential ecosystem functions.

2.2.1 Classification of Prokaryotes

Prokaryotes are divided into two major domains including eubacteria/bacteria and archaea (Woese et al. 1990). They can be easily distinguished from each other depending upon their unique genetic markers. Both are single-celled microorganisms. They possess simple structure of cell membrane which is found just after the cell margin. The plasma membrane represents a semipermeable barrier that regulates the entrance and the existence of components inside and outside the cell. The plasma membrane is made up of a bilayer of lipids and proteins bonded together with non-covalent linkages. Hence, it is not considered as a firm structure. The cell wall provides the plasma membrane with the needed rigidity as the plasma membrane is surrounded by the cell wall. The bacterial cell wall is composed of peptidoglycan which is usually made up of proteins and oligosaccharides with no lipids. The reason behind the thickness and rigidity of the cell wall comes from the peptidoglycan which gives the cell its own shape. Bacteria are divided into two major groups based on their wall structures involving Gram-positive and Gram-negative bacteria. In 1884, the Danish bacteriologist managed to develop the Gram staining technique and was named after him (Gram 1884). The main difference between Gram-negative bacteria and Gram-positive ones is that they do not retain the crystal violet dye (Jurat-Fuentes and Jackson 2012). This mainly relies on the thickness of the peptidoglycan layer as well as the absence or presence of an outer lipid membrane. Gram-positive bacteria are characterized by the presence of a thick layer of peptidoglycan with no outer lipid membrane. Contrary, Gram-negative bacteria have a thin peptidoglycan layer and possess an outer lipid membrane. Bacteria also possess an outer membrane which completely surrounds the cell wall in addition to the plasma membrane. The region between both the outer membrane and the plasma membrane is referred to as periplasmic space. The outer membrane has a totally different molecular composition from that of the plasma membrane. The outer membrane consists of unique lipids mainly glycolipids, for example, lipopolysaccharides (LPS). Porins are among the proteins which are found in the outer membrane.

Archaea are dominant microorganisms in soil and marine environments, and they represent a substantial portion of Earth's microbial diversity as described by (DeLong 1992; Karner et al. 2001; Bintrim et al. 1997). Archaea are prokaryotes which usually inhabit extreme environments with severe conditions such as extreme salinity, temperature, and pH and have strange metabolism (Clark et al. 2016). They are usually found in hot springs and halophilic and acidic environments. Archaea is organized into three main kingdoms: *Crenarchaeota*, *Euryarchaeota*, and *Korarchaeota* (Canfield 2005). Various electron donors and acceptors can be utilized by archaea. They play a vital role in global biochemical cycles (Itävaara et al. 2016).

2.3 Biomimetic Synthesis of Nanoparticles via Prokaryotic Microorganisms

The major advances in technology through the last decades resulted in increasing the basic standards to guarantee the safety of human health and the surrounding ecosystem. The increase in industrial activities led to several environmental regulations regarding their actions. As mentioned in the previous chapter, nanotechnology is one of the important science disciplines that deal with materials at the nanoscale level (Jeyaraj et al. 2015). Yet, their synthesis is of a major concern, as chemical methods involve the use of hazardous solvents, generation of unsafe reaction products which affect the working personnel. Additionally, physical methodologies are somehow limited because of the low yield, the requirement of extreme conditions like high temperature and pressure (Mallick et al. 2004). Both are also complicated, expensive and require the presence of reducing, capping, and stabilizing agents. On the other hand, biogenesis of NPs is a possible alternative to the chemical and physical synthesis methodologies which depends only on biological agents for synthesizing NPs. It is less complicated, cost-effective, and most importantly non-toxic, and usually there is no need to use reducing, capping, and stabilizing agents as they are naturally existed within the biological entities (El-Gendy and Omran 2019; Omran et al. 2019). Nanobiotechnology is considered an enormously emerging scientific field which influences different disciplines including chemistry, engineering, biology, medicine, pharmaceutical fields, and many others. Different biological entities can be employed in the biomimetic synthesis of nanomaterials like bacteria, actinomycetes, yeast, fungi, algae (micro- and macro-), plant materials, agro-industrial wastes, diatoms, etc. Microorganisms proved to be important nano-bio-factories that have high potential as environment-friendly and cost-effective entities that are devoid from toxic, harsh chemicals with no demand for high energy. Microorganisms are characterized by their capability to accumulate heavy metals and detoxify them by the help of different reductase enzymes. These enzymes have the potential of reducing metal salts to their nanoparticle state.

2.4 Common Synthesis Mechanisms for Synthesizing Nanomaterials Using Microbes

Usually metabolic activities of microbial cells result in producing enzymes which help microorganisms to grab target ions from the surrounding environment and convert them to their nanoparticle state. Synthesis mechanisms can be classified into extracellular and intracellular mechanisms. The extracellular synthesis of NPs mainly depends on trapping the ions of the targeted metal salt on the cell surface and reducing the ions by the help of the existed enzymes (Li et al. 2011). The extracellular synthesis of NPs using microbes is mainly dependent upon the activity of nitrate reductase (Hulkoti and Taranath 2014). Several publications assured the role played by nitrate reductase in the extracellular synthesis of NPs (Durán et al. 2005; He et al. 2007; Ingle et al. 2008). According to Manivasagan et al. (2016), NPs produced extracellularly have wider applications involving bio-imaging, electronics, optoelectronics, and in sensor technology as when compared with the intracellularly prepared NPs.

Whilst, the intracellular synthesis of NPs depends on the transportation of metal ions inside the microbial cell and then subsequent bio-reduction to NPs takes place in presence of enzymes. Several biological agents other than enzymes react with metal ions leading to the formation of NPs (Hulkoti and Taranath 2014). The cell wall plays a vital role in the intracellular synthesis of NPs. An electrostatic interaction takes place between the positively charged metal ions with negatively the charged cell wall. The intracellular enzymes existed within the cell wall reduce metal ions to NPs, and then these NPs diffuse off through the cell wall (Hulkoti and

Taranath 2014). To initiate the intracellular synthesis of nanomaterials, an important extra processing step is required which is ultrasonication or the use of a suitable detergent. Ultrasonication refers to the process in which sound energy is applied to aid in particles agitation within the desired sample. It helps in extracting different cellular constituents from microbial cells, algae, plants, and seaweeds (Akpan et al. 2019; Cheaburu-Yilmaz et al. 2019). The usually employed frequencies of an ultrasonic are higher than 20 Kilohertz (kHz). Ahmad et al. (2003b) reported the intracellular synthesis of 5–15 nm sized gold nanoparticles (AuNPs) by *Rhodococcus* sp. Electron microscopy (EM) analysis revealed the presence of thin sections of AuNPs with good monodispersity on the cell wall and the cytoplasmic membrane as well. It was noticed that the synthesized particles were more concentrated on the cytoplasmic membrane than on the cell wall. Mukherjee et al. (2001) explained the intracellular synthesis mechanism of AuNPs by *Verticillium* sp. In this study, the intracellular mechanism was explained based on the trapping of the targeted metal ions, e.g., $AuCl_4^-$ ions, bioreduction, and capping with stabilizing agents. The study demonstrated that the fungal cell surface gets in contact with metal ions and an electrostatic interaction occurs resulting in ion trapping and the enzymes existed within the cell wall reduce the metal ions to their nanoparticle state.

2.5 Biosynthesis of Nanomaterials Using Bacteria

Nanobiotechnology is a scientific discipline that combines nanotechnology with biotechnology. The main target of nanobiotechnology is the development of environmentally benign technology for synthesizing safe, green, and low-cost nanomaterials (Velmurugan and Iydroose 2014). Basically, chemical synthetic methodologies depend mainly on using toxic chemicals and non-polar solvents, and they usually generate hazardous chemical species which limit the use of NPs in clinical and pharmaceutical applications (Rajeshkumar and Naik 2018; Puja and Kumar 2019). Therefore, there was an urgent need to develop biocompatible, clean, non-hazardous, and eco-friendly technology to synthesize nanomaterials. In the meantime, biological techniques are considered green, non-toxic, sustainable, cost-effective, and eco-friendly processes (Saratale et al. 2018). For these reasons, scientists started to focus on biological entities to biosynthesize nanomaterials (Thakkar et al. 2010). Several investigations have been carried out to test the capability of microorganisms like bacteria, actinomycetes, and fungi to mediate the synthesis of nanomaterials (Omran et al. 2018; Afzal et al. 2019; Hassan et al. 2019; Omran et al. 2019). It is worth noting that extracellular synthesis of nanomaterials has more commercial applications in various fields. Additionally, in order to the formation of polydispersed nanomaterials, it is essential to perform optimization for the parameters that may enhance the synthesis of monodispersed nanomaterials. According to Narayanan and Sakthivel (2010), intracellular production results in synthesizing nanomaterials of specific dimensions and with less polydispersity. Some of the studies that investigated the extracellular and intracellular biosynthesis of nanomaterials using bacteria are discussed below with the highlight upon the newest studies.

Usually, bacteria live and survive even in harsh and toxic environmental circumstances, for instance, the high concentrations of heavy metals (Faramarzi and Sadighi 2013). Yet, different natural defense mechanisms have been evolved by bacteria to overcome such extreme conditions including efflux pumps, intracellular sequestration, and extracellular precipitation in order to cope up with such stress conditions (Iravani 2014). In 1980, Beveridge and Murray reported for the first time the extracellular deposition of AuNPs on *Bacillus subtilis* cell wall by reacting with gold chloride solution (Beveridge and Murray 1980). In another report introduced by Klaus-Joerger et al. (2001), *Pseudomonas stutzeri* AG259 biosynthesized AgNPs intracellularly. This bacterium was characterized by its high tolerance to silver ions. The prepared AgNPs ranged in size from few nm to 200 nm by the influence of NADH-dependent reductase enzyme which released electrons and became oxidized to NAD^+. As a result, electron transfer led to the bioreduction of silver ions to AgNPs. Parikh and co-authors demonstrated the capability of *Morganella* sp. to biosynthesize AgNPs extracellularly (Parikh et al. 2008). The bacterium was basically isolated from an insect gut and was identified via by 16S rRNA gene sequencing as *Morganella* sp. It belongs to the family Enterobacteriaceae. It was deposited in GenBank with the accession number EF525539. The supernatant color changed from yellow to clear brown suspension of AgNPs and was further assessed via UV/Vis spectroscopy which depicted a characteristic SPR peak at 415 nm. Optimization of $AgNO_3$ concentration was tested in the range of 1–10 mM. SPR characteristic peak increased at 5 mM concentration, and then the intensity started to decrease at higher concentrations. The obtained results showed that 5 mM concentration of Ag^+ ions was the most appropriate. TEM images of AgNPs biosynthesized by *Morganella* sp. were spherical in shape with a size of 20 nm. Two absorption bands centered at approximately 1550 and 1650 cm^{-1} were depicted via FTIR spectrum and corresponded to amide II and amide I bands, respectively, which indicated the existence of proteins around the surface of AgNPs which might act as capping and stabilizing agents. Tropical and subtropical countries suffer from vector and vector-borne diseases which represent a huge threat to public health (Klempner et al. 2007). Mosquitoes are among the main vectors of different diseases that affect humans as well as animals. According to James (1992), they are the major vectors in causing diseases like yellow fever, malaria, dengue fever, filariasis, etc. which result in high mortalities each year. Due to resistance emergence towards chemical insecticides, new alternatives are now the main focus of many researchers (Headrick and Goeden 2001). *Anopheles subpictus* is recognized as a vector of malaria as reported by Chandra et al. (2010). It breeds in several habitats including rain water, rice field accumulations (Dhanda and Kaul 1980), waste water disposal systems, and irrigated sites (Mukhtar et al. 2003). *Aedes aegypti* is an insect which is severely anthropophilic, and it is endemic in tropical and subtropical regions in various countries. It is widely adapted to urban habitats, invading cities and human settlements. It is the major vector behind the transmission of urban yellow fever, dengue, and other arboviroses. Its main occurrence usually takes place in periods of raining summer till early winter (Barreto and Teixeira 2008). According to the World Health Organization (WHO), approximately five billion people are at risk by infections caused by *A. aegypti* in more than 100 countries (World Health Organization 2005). *Bacillus thuringiensis* is usually employed as a biological pesticide due to its high activity against insects (Becker et al. 2010). According to several publications, the Gram-positive endospore-forming bacterium, *B. thuringiensis* releases parasporal crystalline inclusions in which polypeptides (δ-endotoxin) are present. These toxins exert a toxic effect towards mosquito larvae (Singh et al. 1996). Srivastava and Constanti (2012) managed to intracellularly biosynthesize different types of metal NPs including Pd, Ag, Rh, Ni, Fe, Co, Pt, and Li via *Pseudomonas aeruginosa*. Neither external capping nor stabilizing agents were involved in this study. In a study carried out by Marimuthu et al. (2013), the larvicidal activity of *B. thuringiensis*-derived cobalt nanoparticles (CoNPs) was tested against malaria vector, *A. subpictus*, and dengue vector, *A. aegypti* (Diptera: Culicidae). *B. thuringiensis* (MTCC-6941) was provided by the Institute of Microbial Technology (IMTECH), Chandigarh, India. The synthesized CoNPs using *B. thuringiensis* culture supernatant was analyzed by XRD analysis. XRD revealed three distinct diffraction peaks at 27.03°, 31.00°, and 45.58° which corresponded to 102, 122, and 024 planes, respectively. Based on Scherer's formula, the average particles' size was found to be 85.3 nm. Crystallinity and purity of the biosynthesized CoNPs were assured by the presence of sharp peaks and absence of unidentified ones. Spherical- and oval-shaped particles were detected via FESEM as well as TEM with an approximate size of 84.81 nm. FTIR spectra implicated role of the peak at 3436 cm^{-1} for O–H hydroxyl group and 2924 cm^{-1} for methylene C–H stretch in the formation of CoNPs. The LC50 values of the larvicidal activity of biosynthesized CoNPs against *A. subpictus* and *A. aegypti* were 3.59 and 2.87 mg/L, respectively. Different bacterial isolates were isolated from municipal wastes, Pollachi, Tamil Nadu, India. Serial dilution was employed to isolate heterotrophic bacteria followed by spread plate technique in nutrient agar. The pure isolated bacterial strains were inoculated in nutrient broth and then incubated at 37 °C for 24 h and were placed on an orbital shaker at 220 round per minute (rpm). After the incubation period, the cultures were tested for nitrate reductase assay. Approximately 56 isolates were obtained and identified, amongst *B. subtilis* EWP-46 which is known as a potent nitrate reducer and was further selected for this study. The capability of nitrate reducing bacterium *Bacillus subtilis* EWP-46 cell-free extract to reduce silver ions into their nanoparticle state (AgNPs) was conducted by Velmurugan and Iydroose (2014). Optimization of the different physicochemical factors that would affect the synthesis of AgNPs was performed. Among these factors are hydrogen ion concentration, temperature, silver ion concentration, and reaction time. After optimization, the highest concentration of AgNPs was attained at pH 10.0, temperature 60 °C, 1 mM $AgNO_3$, and reaction time 12 h. A characteristic SPR peak at 420 nm was detected via the UV/Vis spectrophotometer. Presence of elemental silver signaling peaks was confirmed by SEM-EDX spectra. Size of the prepared AgNPs was approximately in the range of 10–20 nm as observed by atomic force microscope (AFM) and TEM. The evidence for the existence of biomolecules which might be involved in bioreduction of silver ions was provided by FTIR. The crystalline nature of AgNPs was attained by XRD. Ammonium precipitation methodology was carried out to extract the proteins from *B. subtilis* EWP-46 cell-free extract, and the molecular weight of the protein was further identified via sodium dodecyl sulfate-polyacrylamide gel electrophoresis (SDS-PAGE). An intense band revealed the presence of a purified protein with a molecular weight of 43 kDa which might be attributed to nitrate reductase enzyme as demonstrated by Morozkina and Zvyagilskaya (2007). 96-well microtiter plate was employed to determine the minimum inhibitory concentration (MIC) of biosynthesized AgNPs against *P. fluorescens* and *S. aureus*. The MIC values against *P. fluorescens* and *S. aureus* were 129 ± 10.8 and 116 ± 12.6 μg/ml, respectively. Extremophilic microorganisms are distinctive microorganisms as they possess unique abilities in growing under extreme surrounding conditions such as high temperature, extreme salinity, and pH. Additionally, they have the capability to be potential producers of different valuable biomaterials. Moreover, according to Murphy et al. (2008) and Bhattacharya et al. (2013), they can be cultivated under non-sterile conditions due to their high resistance to extreme conditions, so they are economically recommended. Das and co-authors investigated the capability of a strain of *Bacillus cereus*; isolated from a soil contaminated with heavy metals to produce AgNPs extracellularly at ambient temperature within 24 h. The synthesized AgNPs showed surface plasmon resonance properties which could be useful in various applications (Das et al. 2014). Radiation-resistant bacterial strain *Deinococcus radiodurans* had the potential to produce AgNPs extracellularly via the bioreduction of AgCl solution (Kulkarni et al. 2015). The biosynthesized AgNPs exhibited a broad-spectrum antibacterial and anti-biofilm activities against both Gram-negative and Gram-positive bacteria. Moreover, AgNPs displayed outstanding anticancerous potential against human breast cancer cell lines. It was shown that AgNPs inhibited proliferation of cancer cell lines as was revealed from cytotoxicity and cell viability assays.

In a study introduced by Juibari and co-authors, a simple and environment-friendly process for the biosynthesis of AuNPs extracellularly was achieved by the thermophilic strain *Ureibacillus thermosphaericus* (Juibari et al. 2015). The isolated strain was molecularly identified via 16S rDNA which in turn showed 96% similarity with *U. thermosphaericus* and was deposited in the National Center for Biotechnology Information (NCBI) data bank and was given the accession no.: HM6261192. *U. thermosphaericus* supernatant had the ability to bio-reduce aqueous $AgNO_3$ after 24 h, and the color of the colloidal solution changed from pale yellow to red or purple. AuNPs characteristic absorption peak appeared between 500 and 530 nm via UV/Vis spectrophotometer. The prepared AuNPs exhibited different sizes ranging from 30 to 200 nm. TEM revealed that the particles were polydispersed spheres. XRD demonstrated the crystallinity of the biosynthesized AuNPs. Two strong peaks were revealed at 2θ value ranging from 0 to 80 (44.5° and 77.62°), and they corresponded to the 200 and 311 of the *fcc* structure of gold particles, respectively (Shankar et al. 2004). FTIR showed the presence of a band at 3285 cm^{-1} which is assigned to amide stretching vibrations. This highlighted the linkage between amide groups and gold. Additionally, two peaks appeared at 1649 and 1528 cm^{-1} corresponding to amide I and II bonds of proteins, respectively (Dong et al. 1995). According to Song et al. (2009) and Rai et al. (2006), proteins adhering to AuNPs via either free amine groups or carboxylate ions of their amino acid residues result in subsequent stabilization of the biosynthesized AuNPs. Authors focused the spot on the capability of the supernatant at the stationary phase to achieve the highest biosynthesis of AuNPs when compared with the other growth phases (lag phase and exponential phase). This observation was attributed to the high activity of the enzymes which might be included within the reduction reaction during the stationary phase than in the other phases (Juibari et al. 2015).

Zinc nanoparticles (ZnNPs) were successfully biosynthesized by *Pseudomonas hibiscicola* which was isolated from the effluents of electroplating industry in Mumbai (Punjabi et al. 2018). Formation of ZnNPs was observed via the formation of white precipitate. Formation of ZnNPs was confirmed by the UV/Vis spectrophotometer through the appearance of a sharp peak at 340 nm. The average particle size distribution was measured by DLS and was found to be 110 nm with a polydispersity index of −0.2. The zeta potential of the ZnNPs was found to be 24.64 mV. FTIR revealed the presence of spectral bands corresponding to amine, alkane, and secondary alcohols which might have acted as stabilizing and capping agents. XRD assured the crystallinity of ZnNPs. Nanoparticle tracking analysis (NTA) helped to measure the particle size distribution which was in the range of 5–90 nm, and the mean size was 62 nm. Additionally, the estimated concentration was 4.72×10^8 particles/ml. Spherical-shaped ZnNPs with a size of 60 nm were observed via TEM. Cytotoxicity was carried out by evaluating nanoparticle concentration in the range of 0.3–10 mg/ml. Generally, compound toxicity is determined in terms of IC50 value. IC50 value refers to the concentration at which cell growth is inhibited by 50%. Based on the effect of NPs on Vero cell line, the IC50 value of ZnNPs were found to be 6.24 mg/ml. Gram-positive bacteria were the most sensitive towards ZnNPs such as *Staphylococcus aureus* and its drug-resistant variant Methicillin-resistant *Staphylococcus aureus* (MRSA). ZnNPs were found to be effective against *Mycobacterium tuberculosis* and its MDR strain with MIC of 1.25 mg/ml. Interesting results were obtained by the synergistic activity between ZnNPs and gentamicin (590μg/mg) in case of MRSA. The resultant data proved that ZnNPs were effective against drug-resistant strains and against hospital-acquired infections. Copper nanoparticles (CuNPs) were successfully prepared via the metal-reducing bacterium *Shewanella oneidensis* as demonstrated by Kimber et al. (2018). This is the first study to investigate the bioreduction of Cu (II) ions to CuNPs via an anaerobic metal-reducing bacterium. According to Lloyd et al. (2011), this bacterial strain possesses a high potential for producing a wide range of catalytically active metallic NPs. The CuNPs were biogenically produced by washing the bacterial cells of *S. oneidensis* MR1 and the addition of 0.2×10^{-6} M Cu (II) ions ($CuSO_4$) in combination with 100×10^{-3} M lactate which acted as an electron donor. The color of the colloidal solution changed from colorless to pink after 3 h of reaction time. The color change was considered a primarily observation for the biological synthesis of CuNPs. Thin sections taken by the transmission electron microscope revealed the intracellular synthesis of CuNPs with a size range of 20–40 nm. Additionally, certain precipitates were noticed in the extracellular matrix, forming large agglomerates, and were associated with the outer cell membrane. Interestingly, the prepared NPs were observed as intracellular precipitates within both the cytoplasm and periplasm. EDX spectrum showed the presence of Cu signaling peaks. The size distribution of 88% of the prepared particles was in the range of 20–50 nm, while around 5% were found in large agglomerates with a size of 200 nm. A novel microscopic technique referred to as serial block-face scanning electron microscopy (SBFSEM) was further employed to confirm and image biomass which supported the prepared particles. The 3D electron images along with the data obtained from TEM–EDX and STEM–EDX data confirmed that the CuNPs were synthesized intracellularly and that the estimated size falls within the range of 20–50 nm. X-ray absorption near-edge spectroscopy (XANES) and extended X-ray absorption fine structure spectroscopy demonstrated that the prepared CuNPs were cubic. Yet, atomic resolution images as well as electron energy loss spectroscopy suggested the occurrence of partial oxidation of the surface layers to cuprous oxide (Cu_2O) upon exposure to air. This study introduced an innovative, facile, biological biosynthesis method for producing efficient CuNPs catalysts. *S. loihica* PV-4 was obtained from the Deutsche Sammlung von Mikroorganismen und Zellkulturen (DSMZ) bacterium collection under the number DSM 17748. Ultra-small nanoparticles (USNPs) like palladium, platinum, and gold nanoparticles were attained by *Shewanella loihica* PV-4 electrochemically active biofilms (EABs) as demonstrated by Ahmed et al. (2018). USNPs refer to the class of nanoparticles whose size range between 1 and 10 nm. USNPs can be employed in diverse fields involving catalysis (Wang et al. 2012), photocatalysis (Khan et al. 2015a, b), energy conversion and storage, chemical manufacturing, biological applications, biomedicines (Costo et al. 2011; Huang et al. 2012; Huo et al. 2014), semiconductors (Dawson and Kamat 2001), and environmental treatments (Drexler 2004). According to Babauta et al. (2012) and Khan et al. (2013), microbial biofilms that are grown on artificial electrodes are usually referred to as electrochemically active biofilm (EAB). EAB has been extensively utilized during the last decades particularly in microbial fuel cells (MFCs) systems to yield energy from wastewater (Ahmed et al. 2018). Furthermore, EABs have emerged as a platform for nanoparticle synthesis (Kalathil et al. 2011). *Rhodobacter sphaeroides* is a Gram-negative, non-sulfur purple bacterium which has the capability to grow under several nutritional and environmental conditions. It has the potential to sequester cobalt and nickel (Calvano et al. 2014; Volpicella et al. 2014) and reduce oxyanions, such as chromate, selenite, and tellurite (Italiano et al. 2012). *R. sphaeroides* is considered an important microbial nano-bio-factory as large quantity of bacterial biomass can be obtained since it can easily grow in easily available low-cost growth medium. Most importantly, it is non-pathogenic and does not need any special precautions during handling. Cells of *R. sphaeroides* strain R26 were provided from the German Collection of Microorganisms and Cell Cultures (DSMZ number 2340) (Italiano et al. 2018). In this study, authors made a trial to investigate the influence of Au (III) on growth of *R. sphaeroides*. Hence, growth curves were obtained in the presence of increasing rates of Au (III) concentrations. It was found that the growth rate was reduced by increasing Au (III) concentrations. Moreover, it was noticed that exposure to gold (i.e., at a concentration of 40 mM) did not cause any substantial decrease in population size at the stationary phase. The duration of lag phase was increased upon exposure to gold ions at concentrations higher than 10 mM. This was interpreted due to the occurrence of a resistance mechanism which was activated by Au (III). Purple precipitates were formed in cultures with metabolically active cells, suggesting the extracellular reduction process of Au (III) to Au (0) in the growth medium as defense mechanism against gold exposure (Edwards and Thomas 2007). At concentrations higher than 10 μM, no increase in population size was noticed even after 16 days, which might be due to the detoxification mechanism(s) that became not sufficient to conserve the cellular integrity. TEM micrographs helped to gain more insight concerning the exposure of *R. sphaeroides* cells to gold ions. The extracellular formation of AuNPs was clearly indicated via the TEM images which revealed the presence of large spherical aggregates and other irregular shapes. For more confirmation, bacterial cells were observed via SEM-EDS to detect the presence of Au in the extracellular medium. Beside the presence of gold signals, signals from C, O, and Na atoms were observed which probably might belong to biomolecules surrounding the nanoparticle surface. Biogenic AuNPs derived from *R. sphaeroides* exhibited a high catalytic activity in degrading nitro-aromatic compounds. Hence, this study paved the route towards developing green-synthesized NPs for the biocatalytic reduction processes of different nitro-aromatic contaminants. In a study performed by Quinteros et al. (2019), the cell-free supernatant of *Pseudomonas aeruginosa* mediated the biosynthesis of AgNPs. TEM images of AgNPs revealed homogenous distribution of spheroidal particles with a size of 25 ± 8 nm. The particles were surrounded by corona. Different characteristic peaks corresponding to functional groups of carbohydrates and proteins were illustrated by FTIR. SDS-PAGE and mass spectroscopy (MS) were carried out to identify the type of proteins existed within the supernatant of *P. aeruginosa*. The identified protein profile revealed the presence of different proteins such as outer membrane structural proteins (e.g., OprG and glycine zipper 2 TM domain-containing protein). Other proteins were found like phospholipid-binding protein MlaC, PhoP/Q, low Mg^{2+} inducible outer membrane protein H1, uncharacterized protein PA1579, azurin, cold-shock proteins and bacteriohemerythrin. Moreover, proteins essential to induce response against oxidative and environmental stresses were also detected such as alkyl hydroperoxide reductase subunit C and RNA-binding protein Hfq. Proteins involved in metabolic reactions were determined involving inorganic pyrophosphatase, lipid A deacylase PagL, pterin-4-alphacarbinolamine dehydratase, glycine cleavage system H protein 1 and YgdI/YgdR family lipoprotein (Quinteros et al. 2019). These biomolecules were suggested to act as capping agents for the biosynthesized AgNPs. Extracellular biosynthesis of AgNPs was attained by the bacterial strain *Bacillus subtilis* KMS2–2 as demonstrated by (Mathivanan et al. 2019). The strain *B. subtilis* KMS2–2 was isolated from a sediment sample located at Uppanar estuary, Southeast coast, India. Later, it was identified via the traditional staining biochemical techniques, followed by molecular identification via 16 rDNA sequencing. In this study, the effects of calcination (200 °C for 30 min) upon the structural characteristics as well as AgNPs antibacterial potential were investigated. The synthesized AgNPs before and after calcination were rounded in shape and crystalline in nature with a size ranging from 18–100 nm to 49–153 nm according to the observations of SEM and XRD, respectively. Presence of functional groups related to bioactive compounds in the crude AgNPs sample assured their role as capping materials. Agar well diffusion assay was employed to investigate the antibacterial activity of the crude and calcined AgNPs against several pathogenic bacterial strains. Among which are *Staphylococcus aureus* MTCC 2940, *Pseudomonas fluorescens* MTCC 1749, *Escherichia coli* MTCC 1610, *Proteus mirabilis* MTCC 425, and *Bacillus cereus. Pseudomonas fluorescens* MTCC 1749 was the most sensitive as the zone of inhibition (ZOI) reached 16 mm, while the lowest inhibition zone (8 mm) was observed against *Proteus mirabilis* MTCC 425. Besides, this study illustrated that the calcined AgNPs did not exert any inhibition effect towards the mentioned bacterial pathogens. Eventually, authors recommended in the follow-up studies to focus upon the inhibitory mechanism of action of biosynthesized AgNPs against pathogenic bacteria. Raja et al. (2019) reported the biological synthesis of AgNPs and AuNPs from a bacterial strain. Symbiotic bacterial strain KPR-8B (*P. luminescens*) was isolated from *Heterorhabditis indica*. Molecular identification of the KPR-8B symbiotic bacterial strain was carried out by 16S rDNA sequencing method. The preliminary indication for the extracellular synthesis of nanoparticles was observed via the color change from pale yellow to purple in case of AuNPs and from yellow to brown in case of AgNPs. UV/Vis spectroscopy displayed a strong SPR at 560 nm and 440 nm for AuNPs and AgNPs, respectively. XRD analysis of *P. luminescens* KPR-8B of AuNPs showed the 2Θ values of 31.03, 43.83, 64.14, and 77.25 which corresponded to (111), (200), (220), and (311), respectively. Similarly, XRD for the synthesized AgNPs depicted 2Θ values at 31.74, 45.76, 57.03, and 76.35 which were attributed to (111), (200), (220), and (311), respectively. Bragg's reflections showed that the prepared NPs were face-centered cubic and crystalline in nature. FTIR peaks pointed out the presence of different biomolecules like aromatics, alcohols, alkynes, alkenes, carboxylic acids, and alkyl halides. Micrographs of high-resolution transmission electron microscope (HRTEM) of KPR-8B-derived AuNPs and AgNPs clearly showed rounded particles with size ranging from 14–46 nm for AuNPs and 17–40 nm for AgNPs. The selected area electron diffraction (SAED) pattern demonstrated the crystalline nature of the prepared particles. Zeta potential showed the presence of negative charge on AuNPs and AgNPs with zeta potential values of – 23.1 and – 23.4 mV, respectively. Average size particle distribution was 179.1 and 504.2 for AuNPs and AgNPs, respectively, as revealed by DLS. According to Zhang et al. (2005b), Selenium (Se) is an essential trace element in every living organism and is usually recommended to be supplemented with various proteins as it assists the immune system to perform its functions appropriately in order to inhibit cell damage and help to control the functions of thyroid gland. Moreover, Se exerts a powerful anticancer and an antimicrobial effects (Wadhwani et al. 2016). Selenium nanoparticles (SeNPs) can be prepared via physical methodologies like laser ablation, UV radiation, hydrothermal techniques, etc. (Iranifam et al. 2013). Besides, SeNPs can be synthesized via chemical methodologies such as acid decomposition, precipitation method, reduction using ascorbic acid, sodium dodecyl sulfate (SDS), sulfur dioxide and glucose, etc. Yet, despite the design of nanoparticles with a definite shape and size via these methods, they require the use of chemicals, harsh conditions like acidic pH, and high temperature which make them unsuitable for applications in the medical field (Wadhwani et al. 2016). Therefore, researchers have focused their efforts towards the preparation of biologically synthesized SeNPs as depicted by Afzal et al. (2019). In this study, 20 cyanobacterial strains were purchased from CFTRI, Mysore, New Delhi, and Tiruchirappalli. The present study investigated the capability of 20 cyanobacterial strains to biosynthesize SeNPs, and they involve:

Anabaena variabilis NCCU-441, *Arthrospira indica* SOSA-4, *Arthrospira maxima* SAE-4988, *Arthrospira indica* SAE-84, *Calothrix brevissema* NCCU-65, *Chroococcus* NCCU-207, *Gloeocapsa gelatinosa* NCCU-430, *Lyngbya* NCCU-102, *Microchaete* sp. NCCU-342, *Nostoc muscorum* NCCU-442, *Nostoc punctiforme*, *Nostoc sphericum*, *Oscillatoria* sp. NCCU-369, *Phormidium* sp. NCCU-104, *Plectonema* sp. NCCU-204, *Scytonema* sp. NCCU-126, *Spirulina* CPCC-695, *Spirulina platensis* NCCU-S5, *Synechocystis* NCCU-370, *Westiellopsis prolifica* NCCU-331.

Monitoring of color change from faint blue to different degrees of orange color was used to assess the biosynthesis of SeNPs. *Anabaena variabilis* NCCU-441, *Arthrospira indica* SOSA-4, *Gloeocapsa gelatinosa* NCCU-430, *Oscillatoria* sp. NCCU-369, and *Phormidium* sp. NCCU-104 were found to be the best five tested cyanobacterial strains. SEM images proved the presence of spherical and polydispersed NPs. The smallest size of SeNPs was found to be biosynthesized by *Arthrospira indica* SOSA-4 (11.8 nm). While, the largest size of SeNPs was observed by *Arthrospira indica* SAE-84 with a size of 60 nm. EDX analysis revealed the presence of SeNPs signaling at 1.4 keV. Further, the antioxidant activities of best five cyanobacterial strains (*Gloeocapsa gelatinosa* NCCU-430, *Anabaena variabilis* NCCU-441, *Arthrospira indica* SOSA-4, *Oscillatoria* sp. NCCU-369, *Phormidium* sp. NCCU-104) were tested by the 2,2-diphenyl-1-picrylhydrazyl (DPPH) and SOR scavenging assays. The best antioxidant activity was attained by *Arthrospira indica* SOSA-4 with IC_{50} 81.67 ± 0.77 μg/ml (SOR) and 73.94 ± 1.53 μg/ml (DPPH). Zirconium dioxide (ZrO_2) is an interesting material with multiple technological applications including oxygen sensors, ceramic making, fuel solid electrolytes in electrochromic devices, fuel cells, dental frameworks, ornaments, and water purification (Puigdollers et al. 2016). One pot green synthesis of ZrO_2 NPs with an average particle size of 44 ± 7 nm was obtained via the extremophilic *Acinetobacter* sp. KCSI1 as demonstrated by Suriyaraj et al. (2019). The tested bacterium *Acinetobacter* sp. (GenBank accession number: KX881380) was isolated from ilmenite mine ore, Kerala, India. The crystallinity of ZrO_2 NPs was confirmed by XRD and Raman spectroscopy. Well-aligned crystals of ZrO_2 NPs were observed via HRTEM and SAED images. The zeta potential value was 36.5 ± 5.46 mV. FTIR spectrum revealed the occurrence of characteristic bands at 855 and 510 cm^{-1} corresponding to the bending vibration of Zr-O-Zr, which in turn confirmed the synthesis of ZrO_2 NPs (Singh and Nakate 2014). AFM revealed data concerning the mechanical behavior of the biosynthesized NPs including hardness, and Young's modulus was found to be 9.206 ± 2.22 and 0.285 ± 0.13 GPa, respectively. No detectable cytotoxicity of ZrO_2 NPs was observed upon fibroblast cells (L929). Thus, results suggested that the synthesized ZrO_2 NPs is biocompatible and not harmful for environmental applications. Table 2.1 represents a list of different metal and metal oxide NPs biosynthesized by bacteria.

Table 2.1 A representative list of bacterial-mediated biosynthesis of metal and metal oxide NPs

Bacteria	Produced NPs	Size (nm)/shape	Synthesis location	Characterization techniques	Application	References
***Corynebacterium strain* SH09**	AgNPs	10–15, spherical	Intracellular	UV/Vis spectrophotometry, TEM, XRD, FTIR, EDX	–	Zhang et al. (2005a)
Bacillus licheniformis	AgNPs	50	Intracellular	UV/Vis spectrophotometry, SEM, EDX, XRD	–	Kalimuthu et al. (2008)
***Morganella* sp.**	AgNPs	20 ± 5, spherical	Extracellular	UV/Vis spectrophotometry, XRD	–	Parikh et al. (2008)
Bacillus licheniformis	AgNPs	50, spherical	Intracellular	UV/Vis spectrophotometry, SEM, EDX, XRD	–	Kalimuthu et al. (2008)
Bacillus cereus	AgNPs	4–5, spherical	Intracellular	UV/Vis spectrophotometry, HRTEM, SAED, XRD, FTIR	–	Ganesh Babu and Gunasekaran (2009)
Escherichia coli	AgNPs	50, spherical	Extracellular	UV/Vis spectrophotometry, TEM, XRD, fluorescence microscopy, DLS	–	Gurunathan et al. (2009)
Corynebacterium glutamicum	AgNPs	5–50, irregular shapes	Intracellular	XRD, EDX, TEM	–	Sneha et al. (2010)
Rhodobacter sphaeroides	AgNPs	7, spherical	Extracellular	UV/Vis spectrophotometry, XRD, HRTEM, SAED	–	Bai et al. (2011)
Pseudomonas aeruginosa	AgNPs	20–50, spherical	Extracellular	UV/Vis spectrophotometry, XRD, TEM	Antimicrobial activity	Oza et al. (2012b)
Bacillus flexus	AgNPs	12–61, spherical, triangular	Extracellular	UV/Vis spectrophotometry, AFM, FESEM, XRD, EDX, TEM, FTIR	Antibacterial activity	Priyadarshini et al. (2013)
Serratia nematodiphila	AgNPs	10–31, spherical	Extracellular	UV/Vis spectrophotometry, TEM, XRD	Antibacterial activity	Malarkodi et al. (2013)

Stenotrophomonas	AgNPs, AuNPs	AgNPs (40–60), AuNPs (10–50)	Extracellular	UV/Vis spectrophotometry, TEM, DLS, EDX	–	Malhotra et al. (2013)
***Bacillus strain* CS 11**	AgNPs	42–92,	Extracellular	UV/Vis spectrophotometry, TEM	–	Das et al. (2014)
Bacillus thuringiensis	AgNPs	43.52–142.97	Extracellular	UV/Vis spectrophotometry, SEM, EDX	Larvicidal activity	Najitha Banu et al. (2014)
***Bacillus brevis* (NCIM 2533)**	AgNPs	41–68, spherical	Extracellular	UV/Vis spectrophotometry, AFM, SEM, FTIR	Antibacterial activity	Saravanan et al. (2018)
Pseudomonas mandelii	AgNPs	1.9–14.1, spherical	Extracellular	UV/Vis spectrophotometry, XRD, HRTEM, FESEM, EDX, FTIR, AFM	Larvicidal activity	Mageswari et al. (2015)
***Bacillus licheniformis* Dahb1**	AgNPs	18.69–63.42, spherical	Extracellular	UV/Vis spectrophotometry, XRD, EDX, FTIR, TEM and AFM	Antibacterial activity	Shanthi et al. (2016)
***Pseudomonas fluorescens* CA 417**	AgNPs	5–50, spherical, near spherical, triangular, hexagonal	Extracellular	UV/Vis spectrophotometry, FTIR, EDX, XRD, TEM, DLS	Antibacterial activity	Syed et al. (2016)
***Pseudomonas aeruginosa* strain SN5**	AgNPs	35–60,	Extracellular	UV/Vis spectrophotometry, XRD, TEM,	Antibacterial activity	Naik et al. (2017)
***Pseudomonas aeruginosa* ATCC 27853**	AgNPs	33–300, spherical	Extracellular	UV/Vis spectrophotometry, FTIR, SEM	Antimicrobial activity	Peiris et al. (2017)
Pseudomonas aeruginosa	AuNPs	15–30, spherical	Extracellular	UV/Vis spectrophotometry, TEM, fluorescence microscopy	–	Husseiny et al. (2007)
Rhodopseudomonas capsulata	AuNPs	10–20, nanowires, nanospheres at high and low concentrations of gold ions, respectively	Extracellular	UV/Vis spectrophotometry, TEM, EDX, FTIR, SDS-PAGE, ICP-MS	–	He et al. (2008)

(continued)

Table 2.1 (continued)

Bacteria	Produced NPs	Size (nm)/shape	Synthesis location	Characterization techniques	Application	References
Stenotrophomonas maltophilia	AuNPs	40, spherical	Extracellular	UV/Vis spectrophotometry, electrophoresis, zeta potential, FTIR. EDX, TEM	–	Nangia et al. (2009)
Shewanella oneidensis	AuNPs	12 ± 5	Extracellular	UV/Vis spectrophotometry, FTIR, DLS, XRD	Antibacterial activity	Suresh et al. (2011)
***Pseudomonas veronii* AS41G**	AuNPs	5–25, spherical	Extracellular	UV/Vis spectrophotometry, FTIR, XRD, TEM	Antibacterial activity	Baker and Satish (2015)
***Lactobacillus kimchicus* DCY51**	AuNPs	5–30, spherical	Intracellular	UV/Vis spectrophotometry, FETEM, XRD, DLS, FTIR	Antioxidant activity	Markus et al. (2016)
Marine *Shewanella* sp. CNZ-1	AuNPs	15	Intracellular	UV/Vis spectrophotometry, TEM, XRD, XPS	–	Zhang and Hu (2019)
Rhodopseudomonas palustris	CdS NPs	8.01 ± 0.25	Intracellular	UV/Vis spectrophotometry, XRD, TEM, HRTEM, SAED, FTIR	–	Bai et al. (2009)
Gluconacetobacter xylinus	CdS NPs	30	–	UV/Vis spectrophotometry, SEM, XRD, FTIR, TGA, PL	–	Li et al. (2009)
Bacillus licheniformis	CdS NPs	2–10, cubic	Extracellular	UV/Vis spectrophotometry, XRD, TEM, FTIR, EDX, TGA, photoluminescence spectroscopy	–	Bakhshi and Hosseini (2016)
***Lactobacillus* sp.**	Ti NPs	40–60, spherical	Extracellular	TEM, XRD	–	Prasad et al. (2007)
Rhodobacter sphaeroides	ZnS NPs	8	Intracellular	UV/Vis spectrophotometry, photoluminescence spectra, XRD, TEM, EDX	–	Bai et al. (2006)

Halococcus salifodinae **BK18**	Se NPs	28, rod	Intracellular	XRD, SAED, TEM	Anticancer activity against HeLa cell lines	Srivastava et al. (2014)
Bacillus cereus strain **HMH1**	Iron oxide NPs	18.8–28.3, spherical	Extracellular	UV/Vis spectrophotometry, FESEM, DLS, FTIR, EDX, VSM	Cytotoxic activity	Fatemi et al. (2018)
Escherichia **sp.**	CuNPs	22.33–39, spherical	Extracellular	UV/Vis spectrophotometry, SEM, TEM, XRD, FTIR	Photocatalytic degradation and treatment of textile effluents	Noman et al. (2019)

2.5.1 *Elucidation of Bacterial-Mediated Mechanism for Synthesis of Nanoparticles*

The exact mechanism of biologically produced NPs varies from organism to organism. Yet, the biosynthesis of NPs follows a general scheme wherein metal ions either penetrate into the microbial cell or remain on the microbial surface, and by the presence of enzymes, they get reduced into NPs (Yin et al. 2016). Metal ions are reduced by enzymes to form nuclei of noble metals, which later on grow and accumulate NPs (Mukherjee et al. 2002).

AgNPs are the most comprehensively studied type of noble metal NPs which may be due to their massive applications. In this context, in 2008, Kalimuthu and co-authors illustrated the role of nitrate reductase enzyme in the biosynthesis of AgNPs using *Bacillus licheniformis* (Kalimuthu et al. 2008). *B. licheniformis* is known to secrete the cofactors NADH and NADH-dependent enzymes, particularly nitrate reductase. It is worth noting that nitrate ions help to induce enzyme secretion. Cofactors like NADH in NADH-dependent nitrate reductases enzymes are required for generating metal NPs. Sintubin and colleagues evaluated the ability of different bacterial strains to biologically fabricate AgNPs including *Lactobacillus* spp., *Enterococcus faecium*, *Pediococcus pentosaceus*, *Lactococcus garvieae*, and *Staphylococcus aureus* and the Gram-negative bacteria *Escherichia coli* and *Pseudomonas aeruginosa* (Sintubin et al. 2009). *Lactobacillus* sp. was found to have the potential to biosynthesize AgNPs. Authors hypothesized another mechanism for AgNPs synthesis by *Lactobacilli* in which the increase in pH leads to a competition between the metal ions and the protons upon the negatively charged binding sites. Additionally, high pH induces the opening of monosaccharide rings like glucose and converts it to open-chain aldehydes. The aldehydes confer the reducing power. In presence of metal ions, the aldehydes become oxidized to carboxylic acids and the metal ions are reduced at the same time.

2.6 Biomimetic Synthesis of Nanomaterials Using Actinomycetes

Actinomycetes are also referred to as actinobacteria. Actinomycetes are classified as bacteria, but they are somehow distinctive; thus they represent an individual group (Pepper and Gentry 2015). They are unicellular Gram-positive prokaryotes, filamentous, and aerobic bacteria possessing DNA with high G + C composition and are able to produce various kinds of antibiotics (Barka et al. 2016). Actinomycetes are prokaryotic microorganisms whose characteristics and general functions are intermediated between both of bacteria and fungi. The number of actinomycetes that has been identified is way less than the bacterial numbers. In a morphological point of view, actinomycetes are somehow similar to fungi as they exist as elongated cells which branch into hyphae or filaments (Pepper and Gentry 2015). It is worth noting that the fungal hyphae are distinguished from actinomycetes hyphae on basis of size as actinomycete hyphae are much smaller than the fungal hyphae. They flourish in aerobic and warm soils. They are heterotrophs which depend on decomposable organic matter as nutrients. They proliferate in soils which have plenty of plant and animal residues. One of the distinctive features of actinomycetes that make them differ from bacteria is their ability to consume several soil constituents such as chitin, cellulose, and hemicellulose. Interestingly, actinomycetes produce the characteristic earthy aroma which is often smelled during the spread of soil compost.

Despite of being identified as soil microorganisms, marine actinomycetes started recently to gain popularity. Various secondary metabolites were identified from marine actinomycetes which added a new whole dimension to microbially bioactive natural metabolites (Zotchev 2012). Importantly, actinomycetes possess high extent of proteins which significantly increases the productivity of the biosynthesis approach. Actinomycetes have been well studied for the synthesis of antibiotics and hydrolytic enzymes (Saratale et al. 2012; Saratale et al. 2018). They represent an important source of effective antibiotics which provide treatment against several infectious human pathogens. Several publications demonstrated that actinomycetes isolated from Ethiopia and Kenya soils exerted an antibacterial activity against *E. coli* (Bizuye et al. 2017; Kibret et al. 2018; Rotich 2018). Kibret et al. (2018) and Tabrizi et al. (2013) reported that *Streptomyces* species isolated from Ethiopian and Iranian soil produce bioactive natural metabolite which had an antimicrobial effect against *Salmonella bodii* and *S. typhi*.

According to what was mentioned earlier, actinobacteriology is one of the interesting vital emerging research disciplines. The group of streptomycetes is an extremely important group as 50–55% of existed antibiotics are produced by *Streptomyces* sp. (Manivasagan et al. 2014). Different species of *Streptomyces* are employed in several enzymatic and pharmaceutical industries (Alani et al. 2012). Recently, different actinobacteria species have been isolated and are considered potential synthesizers of metal nanoparticles. Among these genera are *Thermomonospora* sp. (Ahmad et al. 2003a), *Rhodococcus* sp. (Ahmad et al. 2003b), *Streptomyces avidinii* (Park et al. 2006), *Streptomyces hygroscopicus* (Sadhasivam et al. 2012), *Nocardia farcinica* (Oza et al. 2012a), *Streptomyces naganishii* (MA7) (Shanmugasundaram et al. 2013), *Nocardiopsis* sp. MBRC-1 (Manivasagan et al. 2013), and *Streptomyces* sp. (Karthik et al. 2014).

In 2003, the alkalotolerant actinomycete (*Rhodococcus* sp.) was reported by Ahmad et al. (2003b) to synthesize AuNPs intracellularly with a size of approximately 5–15 nm. A vivid purple color of the actinomycete cells was observed after the addition of 10^{-3} M aqueous solution of tetrachloroauric acid ($HAuCl_4$) for 24 h. Micrographs of TEM revealed the presence of thin sections of monodispersed AuNPs located on the cell wall in addition to the cytoplasmic membrane. This indicated that mainly the reduction of gold ions under the effect of enzymes existed in both the cell wall and the cytoplasmic membrane. Interestingly, no toxic effect was exerted by these metal ions and that was proven by cell multiplication. The capability of the alkalothermophylic actinomycetes strain, *Thermomonospora* sp., to bioreduce $HAuCl_4$ was investigated by Ahmad et al. (2003c). The bioreduction of the aqueous chloroaurate ions after being exposed to the biomass of *Thermomonospora* sp. was easily tracked by UV/Vis spectroscopy. A gold surface plasmon band appeared at 520 nm and steadily increased in intensity until complete reduction of $AuCl_4^-$ ions. The prepared AuNPs exhibited a pink-ruby red color which was formed due to the excitation of surface plasmon vibrations of the synthesized AuNPs. The recorded FTIR spectrum showed the existence of two bands at 1660 and 1530 cm^{-1} which could be attributed to the amide I and II bands of proteins, respectively. As a result, stability of the as-prepared AuNPs was attributed to the linkages between AuNPs along with either the free amine groups or the cysteine residues. TEM image clearly demonstrated the presence of dense uniformly sized AuNPs. XRD revealed that the synthesized particles were face-centered cubic (*fcc*) and nanocrystalline in nature. Ahmad and his coworkers recommended the identification of proteins which might be involved in the reduction of gold ions and the subsequent capping and stabilization of the prepared AuNPs. Additionally, an actinomycetes strain, *Thermomonospora* sp., was reported by Sastry et al. (2003) to have the potential to synthesize AuNPs. *Thermomonospora* sp. was isolated from self-heating compost in the Barabanki district, Uttar Pradesh, India. Formation of AuNPs was observed via the formation of an intense red color. The recorded UV/Vis spectra showed a characteristic absorption peak of AuNPs at 520 nm. The peaks steadily increased in intensity as a function with time. Complete reduction of the $AuCl_4^-$ ions was attained at approximately 120 h. The prepared AuNPs were tested for their stability, and they remained stable for nearly 6 months. Sapkal and Deshmukh (2008) tested the capability of ten actinomycetes strains isolated from Himalayan Mountain to microbially synthesize AuNPs. Out of the ten isolated strains, four strains coded with D10, HM10, ANS2, and MSU had the potential to synthesize AuNPs intracellularly. The strain HM10 exerted the highest potency towards synthesizing AuNPs. The prepared AuNPs exhibited different shapes involving circular- and rod-shaped morphological structures. TEM and X-ray diffraction (XRD) analysis demonstrated the average size particle which was in the range of 18–20 nm. UV/Vis spectrophotometer assured the bioreduction of $HAuCl_4$ to AuNPs within 24 h. It is worth noting that the strain HM10 expressed an observable growth at different concentrations of $HAuCl_4$ (1–10 M). The synthesized AuNPs by the selected strain HM10 exhibited a good antibacterial action against *Staphylococcus aureus* and *Escherichia coli* as noted via the agar well diffusion assay. The isolated HM10 strain was identified as *Streptomyces viridogens*. Streptomyces *hygroscopicus* is an actinomycetes strain that is commonly known to produce plenty of antibiotics and enzymes which are of commercial importance. The extracellular biological synthesis of AgNPs by the supernatant of *S. hygroscopicus* was investigated by Sadhasivam et al. (2010). A brown color appeared by the addition of $AgNO_3$ to *S. hygroscopicus* supernatant. It was noticed that the intensity of the brown color increased till reaching 24 h and then remained steady. The synthesized particles showed a SPR peak at 420 nm and 425 nm, thus assured the biosynthesis of AgNPs. Field emission scanning electron microscopy (FESEM) as well as TEM revealed the spherical shape of the AgNPs with a size ranging from 20 to 30 nm. EDX was employed to confirm the purity of the synthesized AgNPs.

Strong signals corresponding to silver were observed via the EDX spectrum. It is worth mentioning that other peaks corresponding to silicon (Si), oxygen (O), carbon (C), and chloride (Cl) were also detected. This was most likely due to the borosilicate glass on which the tested sample was coated as a technical preparation of FESEM. Besides, the biosynthesized AgNPs significantly prevented the growth of some medically important pathogenic Gram-positive bacteria, e.g., *Bacillus subtilis* and *Enterococcus faecalis*; Gram-negative bacteria, e.g., *E. coli* and *Salmonella typhimurium*; and yeast *Candida albicans*. Waghmare et al. (2011) demonstrated the potential of *Streptomyces* sp. HBUM171191 to synthesize manganese and zinc NPs intracellularly. This was carried out by the reduction of manganese sulfate ($MnSO_4$) and zinc sulfate ($ZnSO_4$) with the help of HBUM171191. In a study carried out by Sivalingam et al. (2012), soil samples were collected from mangrove sediment of Pichavaram, Tamil Nadu, India, and placed in sterile air-free polyethylene bags and stored at 4 °C. Morphological and biochemical characterization was performed to identify the isolated strain. Partial gene sequence (855 nucleotides) of the isolated strain was deposited at GenBank database at NCBI and took the accession no. JQ231271.1. It was identified as *Streptomyces* sp. BDUKAS10 based significantly upon biochemical, physiological characterization and 16S rDNA molecular sequence analysis. The potency of *Streptomyces* sp. BDUKAS10 to biosynthesize AgNPs was evaluated. A yellowish brown coloration developed after 16 h of reaction period. Maximum color intensity was observed at 36 h. A maximum absorption SPR peak appeared at 441 nm as observed by UV/Vis spectrophotometer. FTIR spectra provided an evidence for the presence of proteins which might be the possible reason behind the reduction capability of *Streptomyces* sp. BDUKAS10. EDX revealed the presence of major peaks of silver. Spherical-shaped particles were observed via TEM in a size range of 21–48 nm. The antibacterial activity of the biosynthesized AgNPs was evaluated against *Bacillus cereus* (MTCC 1272), *Pseudomonas aeruginosa* (MTCC 1688), and *Staphylococcus aureus* (MTCC 96). Maximum ZOI was observed against *B. cereus*. A facile and an environment-friendly method was carried out to biologically synthesize AuNPs. Several soil samples were collected from Songon copper mine in northwest of Iran. Soil samples were serially diluted and spread on starch casein agar plates. First, the isolates were morphologically identified as *Streptomyces* sp. Afterwards, a molecular characterization based on 16SrRNA was performed. The isolated pure colonies were identified as *Streptomyces* sp. ERI-3. The ability of the cell-free supernatant of *Streptomyces* sp. ERI-3 to produce AuNPs was studied by Zonooz et al. (2012). The color of the cell-free supernatant changed from whitish yellow to reddish purple color. A sharp characteristic SPR peak appeared at 540 nm which indicated the formation of AuNPs. The Bragg reflections corresponded to the lattice planes of gold and confirmed its crystalline nature. Cylindrical- and spherical-shaped AuNPs were demonstrated via the TEM images. The influence of different parameters that would affect the biosynthesis reaction was investigated. Among these parameters are reaction time, temperature, and concentration of $HAuCl_4$. The highest production of AuNPs was attained at 12 h of reaction period at temperature of 30 °C, with 3 mM of $HAuCl_4$ concentration at pH 6. Alani et al. (2012) compared between one fungal and one actinomycete isolated thermo-alkalo-tolerant strains. They were further identified as *Aspergillus fumigatus* and *Streptomyces* sp. Both were tested for their ability to synthesize AgNPs. The change in the color of the autolyzed cell-free cultures from colorless to light brown and then to dark brown after 24 h was a preliminary observation for the synthesis of AgNPs. A characteristic sharp peak appeared at 420 nm as was confirmed by UV/Vis spectrophotometer. The initial formation kinetics were faster with *A. fumigatus*, but the formation continued for a longer period with *Streptomyces* sp., resulting in higher concentrations after 48 h. TEM images revealed the high dispersity of the prepared AgNPs by both microorganisms with a size ranging from 15 to 45 nm in case of *A. fumigatus*, while in case of *Streptomyces* sp., the size was in the range of 15–25 nm. The high productivity and the narrower size distribution of *Streptomyces* sp. made it more preferable than *A. fumigatus*. Khadivi Derakhshan et al. (2012) collected ten soil samples from Songon copper mine in Northwest Iran. A *Streptomyces* strain was isolated and was further identified as *Streptomyces griseus*. The isolated *S. griseus* from Songon copper was able to synthesize AuNPs extracellularly for the first time within 48 h of reaction period as revealed by UV/Vis spectrophotometer. Such limited reaction time is advantageous for industrial downstream production. Additionally, the extracellular synthesis of AuNPs is of a major advantage, as there is no need for an extra step (i.e., ultrasonication) to release the synthesized AuNPs. TEM micrographs revealed the formation of well-dispersed and spherical-shaped particles. Five strong peaks were revealed by XRD spectrum including (38.269), (44.600), (64.678), (77.549), and (82.352), which were in agreement with Bragg's reflection of AuNPs. AgNPs were successfully synthesized using the actinobacterium *Rhodococcus* NCIM 2891 as investigated by Otari et al. (2012). A change in color from colorless to brown took place after 18 h of reaction time which indicated the formation of AgNPs. A distinctive SPR peak of AgNPs appeared at 420 nm as detected by UV/Vis spectrophotometer. Scanning electron microscope (SEM) revealed the presence of circular-shaped particles with a size of approximately 100 nm. Characteristic silver signals were detected via EDX at 3 keV. TEM images showed that the average size of the prepared AgNPs was 10 nm and assured the spherical shape of the prepared particles. Four different planes corresponding to (111), (200), (220), and (311) planes were identified via SAED and confirmed the *fcc* structure of elemental silver. The resultant XRD data matched the ones obtained from SAED and assured the crystalline nature of the biosynthesized AgNPs. Oza et al. (2012a) investigated the ability of *Nocardia farcinica* to biologically fabricate AuNPs. The tested strain was obtained from the National Collection of Industrial Microorganisms (NCIM), Pune. The synthesized AuNPs were spherical in shape, with a size of 15–20 nm. Rapid change in color in less than 5 s took place after adding the gold ions to *N. farcinica* cell filtrate. The color changed from pale yellow to wine red color at 100 °C by using100 ppm gold perchlorate salt. The impact of different pH on biosynthesis of AuNPs was investigated. Surprisingly, at pH 4, the UV/Vis spectral analysis revealed the occurrence of a sharp peak of AuNPs centered at 540 nm. Contrary, at alkaline pH (8–10), flat absorption peaks appeared; hence, this confirmed that at alkaline pH, the gold ions were less reduced and the efficacy of capping proteins decreased. Furthermore, at pH 6, a broad hump appeared at 600 nm which illustrated the formation of large polydispersed NPs. The larger size might be attributed to the coalescence of smaller nuclei. The above mentioned results assured the high impact of pH in forming thermodynamically stable AuNPs. Different actinomycetes strains were isolated from several soil samples as reported by Subashini and Kannabiran (2013). One potential isolate exhibited the capability to synthesize AgNPs. Molecular taxonomic characterization was employed to identify the isolated actinomycetes strain and was identified as *Streptomyces* sp. VITBT7 (accession number JX188053). The synthesized AgNPs showed a distinctive surface plasma resonance (SPR) peak at 420 nm. The biologically synthesized 20–70 nm sized AgNPs were spherical in shape. *Streptomyces* sp. VITBT7 derived AgNPs exhibited antimicrobial activity against different bacterial and fungal pathogens. Various soil samples were collected from Kodaikanal hill station, Tamil Nadu, India. Soil samples were serially diluted and spread on starch casein nitrate agar plates. The plates were then incubated at 30 °C for 7d as demonstrated by Chauhan et al. (2013). Pure colonies were obtained and were identified as *Streptomyces* sp. JAR1 through the molecular taxonomic characterization based upon 16S rRNA sequencing technique. Synthesis of AgNPs was attained by using the extracellular culture filtrate of *Streptomyces* sp. JAR1. Change in coloration took place from pale yellow to yellowish brown color. UV/Vis spectral analysis was employed to monitor and track the formation and stability of the fabricated AgNPs in the colloidal solution. A sharp peak appeared between 420 and 425 nm. The influence of different time intervals 24, 48, and 72 h was investigated. It was found that by increasing the reaction time, the peaks got more intense. Chauhan and co-authors illustrated that the mechanism behind the bioreduction of silver ions to their nanoparticle state which was due to the reduction process of electron enzymatic shuttle. Furthermore, authors pointed out the role of microbially produced NADH and NADH-dependent enzymes such as nitrate reductase which might be responsible for reducing the silver ions leading in the subsequent formation of AgNPs. FTIR assured the presence of proteins which could be also involved in the bioreduction process. Silver characteristic peaks were revealed from EDX. Moreover, AFM was performed to determine the topological appearance involving information regarding roughness, porosity, and fractal behavior of the annealed and deposited films. Additionally, the estimated size of the biosynthesized AgNPs was found to be 68.13 nm. Different bacterial, yeast, and fungal strains such as *Staphylococcus aureus*, *Ganoderma* sp. JAS4, *Scedosporium* sp. JAS1, *Fusarium* sp., and *Candida tropicalis* were efficiently inhibited by the prepared AgNPs. For the first time, a study conducted by Verma and co-authors successfully managed to biologically fabricate nanotriangular AuNPs from the endophytic actinomycetes strain *Saccharomonospora* sp. (Verma et al. 2013). Endophytes are microorganisms like bacteria, fungi, or actinomycetes that live within plant materials in a symbiotic relationship (Hassan et al. 2018). The tested strain was isolated from sterilized surface root tissues of *Azadirachta indica.* The addition of 1 mM of $HAuCl_4$ to the biomass of *Saccharomonospora* sp. resulted in yielding prismatic gold nanotriangles as confirmed by TEM. The SDS-PAGE profiling confirmed the existence of proteins at 42 and 50 kilodalton (kD), and hence, they were involved in biosynthesis in addition to the stabilization of the prepared NPs. The characterization of the biosynthesized particles was monitored by UV/Vis spectroscopy, AFM, and XRD. Yet, the mechanism of shape orientation was not revealed. Eventually, this study showed the possibility of shape control which needs further investigations. In a study performed by Karthik et al. (2014), the marine actinobacteria *Streptomyces* sp. LK3 mediated the biological synthesis of AgNPs. The synthesized AgNPs exhibited a characteristic absorption peak at 420 nm. Different XRD intense peaks were revealed at 2Θ values of 27.51, 31.87, 45.57, 56.56, 66.26, and 75.25 which corresponded to (210), (113), (124), (240), (226), and (300) Bragg's bands of AgNPs. The FTIR spectra displayed prominent peaks at 3417 cm^{-1} (OH stretching due to alcoholic group) and 1578 cm^{-1} (C=C ring stretching) which confirmed the presence of biomolecules that might have been involved in the bioreduction process. The biosynthesized AgNPs were spherical in shape with a size of 5 nm as demonstrated by TEM images. Topographical and surface morphological structures of the synthesized AgNPs were studied via AFM. Furthermore, in order to confirm the presence of the enzyme, nitrate reductase, *Streptomyces* sp. LK3 was inoculated in nitrate. After the incubation period, two reagents were added including; reagent A (sulfanilic acid) and reagent B (N, N-dimethyl-1-naphthylamine). The changes in color from pale yellow to the characteristic red or pink color indicated a positive result. It is worth noting that if the suspension color changed to pink-red before adding zinc powder, the reaction is positive, and thus the presence of nitrate reductase was confirmed. If no change in color took place after the addition of the two reagents A and B, small amount of zinc powder is added, and the tube is shaken strongly and then is allowed to stand at room temperature for 10–15 min to observe color change. In this case, no color change was noticed after the addition of sulfanilic acid and alpha naphthylamine. Thus, a small amount of zinc dust was added, and also no color change took place. The synthesized AgNPs displayed significant acaricidal activity against *Haemaphysalis bispinosa* and *Rhipicephalus microplus*. The resultant LC50 value was 16.10 and 16.45 mg/L against *R. microplus* and *H. bispinosa*, respectively. Thus, *Streptomyces* sp. LK3 derived AgNPs could be recommended as a safe alternative to replace the traditional acaricidal (anti-parasitic) agents as it can be provided as an anti-parasitic formulation. Additionally, it can be easily scaled up for industrial applications. Saravana Kumar et al. (2015) reported the capability of an actinomycetes strain isolated from a soil sample in an agriculture field in Tamil Nadu, India, to effectively mediate the extracellular synthesis of AgNPs. The isolated strain was genetically identified as *Streptomyces* sp. 09 PBT 005. A primarily observation for detecting the synthesis of AgNPs was the color change from colorless to dark brown. The change in color took only few minutes to reveal. The UV/visible spectrum demonstrated the presence of a sharp peak at 440 nm. According to Shankar et al. (2003), the appearance of a specific absorption peak appears due to the excitation of longitudinal plasmon vibrations and formation of quasi-linear super structures. The prepared particles were roughly spherical, and they were big in size with a diameter average ranging between 198 and 595 nm.

An actinobacterium strain CGG11n was isolated from the mineral horizon of a pure stand of *Picea sitchensis* Carriere (Sitka spruce) from the Southern end of

Hamsterley Forest, County Durham, UK, as demonstrated by Buszewski et al. (2016). The isolated strain was identified by 16S rRNA gene sequence and was referred as *Streptacidiphilus albus* DSMZ 41753^{T}. In this study, the biosynthesized AgNPs were perfectly prepared by *Streptacidiphilus* sp. when the cell-free extract was combined with a concentration of 3 mmol^{-1} $AgNO_3$ after 7 days of incubation. The formation of biosynthesized AgNPs was confirmed by the change in color from yellow to dark brown via a green single step. UV/Vis spectrophotometer displayed an absorption peak in the range 420–430 nm. Elemental analyzer revealed the elemental analysis. Different elements were detected C, H, N: 30.45 ± 001%; 6.58 ± 002%; 4.25 ± 001%, respectively. The content of silver in the culture medium was estimated by inductively coupled plasma-mass spectroscopy (ICP-MS) and was found to be around 9.03 ± 0.01 mg l^{-1}. Interestingly, this study showed the variation in the zeta potential value along with changes in pH. Additionally, the average size distribution varied with the different pH ranges. The experimental data showed that at pH 4.5–8.5, the size of biosynthesized AgNPs was 100 ± 15 nm. The nanoparticles' average size distribution decreased from 1500 to 200 nm in a pH range of 2–4. Above pH 10, the size of the prepared particles started to increase. TEM illustrated the shape of the prepared particles as spherical ones. XRD confirmed the crystallinity of the biosynthesized AgNPs. EDX depicted silver signals at 3 keV. The antimicrobial potential of AgNPs alone and in combination with antibiotics against some clinical strains was carried out. These strains were *Staphylococcus aureus*, *Bacillus subtilis*, *Escherichia coli*, *Pseudomonas aeruginosa*, *Klebsiella pneumoniae*, *Proteus mirabilis*, and *Salmonella infantis*. The prepared AgNPs mitigated the growth of most bacterial strains. The most sensitive strain was *P. aeruginosa* (10 mm), followed by *S. aureus*, *B. subtilis*, and *P. mirabilis* (all 8 mm). Both of *K. pneumoniae* and *E. coli* were more resistant recording 6 and 2 mm, respectively. Additionally, the synergistic effect of both the biosynthesized AgNPs in combination with different commercially available antibiotics was also estimated. The combination between bio (AgNPs) along with tetracycline, ampicillin, kanamycin, and neomycin recorded the best results then followed by streptomycin and gentamycin against *E. coli*, *K. pneumoniae*, and *S. infantis*. Based on these results, AgNPs biosynthesized from *Streptacidiphilus* sp. strain CGG11n offer valued contribution to pharmaceutical industries

Składanowski et al. (2017) isolated an actinomycete strain which was then identified as *Streptomyces* sp. strain NH21. It was isolated from a humic layer of acidic pine in forest soil. This strain had the potential to biologically synthesize AgNPs and AuNPs. The physicochemical characteristic features of the obtained particles were studied via different spectroscopic and microscopic analyses. A broad peak of AgNPs appeared between 404 and 424 nm while for AuNPs was 564 nm. The size of AgNPs and AuNPs was 44 and 10 nm, respectively as revealed by TEM and nanoparticle tracking analyzer (NTA). XRD assured that the prepared particles were in the nanocrystal form. The zeta potential values of AgNPs and AuNPs were − 9.95 mV and − 14.5 mV, respectively. The obtained data revealed that the prepared particles were stable and monodispersed which might be attributed to the electrostatic repulsion forces between particles within the colloidal solution. *Streptomyces xinghaiensis* OF1 strain successfully mediated the synthesis of AgNPs as demonstrated by Wypij and co-colleagues (Wypij et al. 2018). The actinomycetes OF1 strain was isolated from sediment samples of Lonar Crater of Maharashtra, India. Afterwards, the isolated pure strain was characterized by a nearly complete 16S rRNA gene sequence (1414 nucleotides) and was mostly closely related to *S. xinghaiensis* S187 and was deposited in GenBank under the accession number of KY 523106. A dense brown color of the cell filtrate of *S. xinghaiensis* OF1 was observed after being treated with 1 mM aqueous solution of $AgNO_3$. The biosynthesized AgNPs was further characterized via UV/Vis spectrophotometer in which a SPR peak appeared at 420 nm. TEM images of *S. xinghaiensis* OF1 derived AgNPs showed the presence of spherical-shaped polydispersed particles with a size ranging from 5 to 20 nm. The average size particle as well as their concentration was depicted by NTA. It was revealed that the average size particle of the prepared AgNPs was approximately 64 ± 49 nm and their concentration was 2.7 × 107 particles ml^{-1}. Zetasizer showed that the particles were negatively charged with a polydispersed index value of −15.7 Mv. FTIR demonstrated the presence of five absorbance bands at 3432, 2925, 1631, 1385, and 1033 cm^{-1}. The biologically prepared AgNPs from *S. xinghaiensis* OF1 strain displayed the highest inhibitory effect against *Pseudomonas aeruginosa* (ATCC 10145), followed by *Candida albicans* (ATCC 10231) and *Malassezia furfur* (DSMZ 6170). But each of *Bacillus subtilis* (PCM 2021), *Escherichia coli* (ATCC 8739), *Klebsiella pneumoniae* (ATCC 700603), and *Staphylococcus aureus* (ATCC 6538) were much less sensitive to the biosynthesized AgNPs. The minimum inhibitory concentration (MIC) values were recorded as follows: 16, 32, 32, 64, 64, and 256 μg ml^{-1}, *P. aeruginosa, M. furfur, E. coli, B. subtilis, K. pneumoniae*, and *S. aureus* respectively. The minimum biocidal concentration (MBC) of AgNPs was found to be 32 μg ml^{-1} for *P. aeruginosa*, 48 μg ml^{-1} for *M. furfur*, 64 μg ml^{-1} for *E. coli* and *B. subtilis*, 256 μg ml^{-1} for *K. pneumoniae*, and 384 μg ml^{-1} for *S. aureus.* Furthermore, the synergistic effect of AgNPs along with antibacterial and antifungal antibiotics was determined by fractional inhibitory concentration (FIC) index. The tested bacteria were most sensitive to tetracycline, then kanamycin, and ampicillin, while *C. albicans* was susceptible to amphotericin B, and *M. furfur* was sensitive to fluconazole. It is worth mentioning that the biosynthesized AgNPs exhibited greater antibacterial activity than ampicillin against all Gram-negative bacteria and then followed by kanamycin against *E. coli* and *P. aeruginosa*. However, Gram-positive bacteria, i.e., *S. aureus* and *B. subtilis*, were more susceptible to tested antibiotics than to biosynthesized AgNPs. Similarly, AgNPs was more active against *E. coli* and *P. aeruginosa* than kanamycin. Also, biosynthesized AgNPs were found to be more effective against *C. albicans* than fluconazole and ketoconazole but not better than amphotericin B. Likewise, *M. furfur* was more susceptible to fluconazole than to AgNPs and was more resistant to amphotericin B and ketoconazole than to AgNPs. Moreover, 3-(4,5-dimethylthiazol-2-yl)-2,5 diphenyltetrazolium) (MTT) assay was performed to study cytotoxicity of both AgNPs alone and in combination with antibiotics/antifungals against mouse fibroblasts and HeLa cell line. The in vitro cytotoxicity of AgNPs against mouse fibroblasts and cancer HeLa cell lines revealed a dose-dependent potential. The IC50 value of AgNPs was attained at concentrations of 4 and

3.8 μg ml^{-1}, respectively. The remarkable synergistic effects of antibiotics with AgNPs recommended their valuable potential for nanomedicine and clinical applications as a combined therapy in the near future since the combination of both AgNPs and antibiotics decreased the needed concentrations of both while keeping their effect high. Rasool and Hemalatha (2017) conducted a study in which marine endophytic actinomycetes strain was isolated from seaweeds. Seaweeds were collected from Kovalam beach, Chennai, India. It was employed to mediate the synthesis of copper nanoparticles (CuNPs). The microbially synthesized CuNPs were characterized using UV/Vis spectroscopy. The synthesized CuNPs showed two SPR peaks at 370 nm and 690 nm, which might be attributed to presence of different sizes of CuNPs. FTIR revealed the presence of different bands which provided information about presence of various molecules like proteins, amides, and alcohols. EDX spectroscopy confirmed the presence of copper signals. The prepared CuNPs proved to exert antimicrobial action against five human pathogenic bacteria which opens the door for using it in medical applications.

In a study performed by Baygar et al. (2019), the cell-free extract of *Streptomyces* sp. AU2 was tested for its potential to mediate the synthesis of AgNPs. The microbially prepared AgNPs coated sutures as surgical sutures play an important role during wound healing in surgeries. It is known that surgical sutures are sensitive to microbial infections, and since AgNPs are promising agents against multiple-drug-resistant microorganisms (MDR), the prepared AgNPs were deposited on the sutures and coated them. Sutures coated with the *Streptomyces* sp. AU2 biosynthesized AgNPs were then characterized using SEM and EDX. The typical multi-filament structure of sutures was observed via the SEM images in both the control set (non-coated) and the bio-derived AgNPs coated sutures. Additionally, the deposition of AgNPs was strongly revealed onto the bio-AgNPs coated sutures and was clearly visible. EDS spectrum revealed the presence of the characteristic silver peaks. Inductively coupled plasma-mass spectroscopy (ICP-MS) was used to estimate the amount of released silver from the bio-AgNPs coated sutures which might have been occurred via a degradation process. Antimicrobial efficacy of the AgNPs coated sutures was observed against common pathogenic microorganisms involving *Escherichia coli*, *Candida albicans*, and *Staphylococcus aureus*. The highest ZOI was observed against *C. albicans*. *C. albicans* is pathogenic yeast which easily contaminates medical devices. According to Nobile and Johnson (2015), *C. albicans* is responsible for nearly 15% of sepsis in hospital-acquired cases. The MTT assay was carried out to further determine the biocompatibility/cytotoxicity effects of the bio-AgNPs coated sutures. Cytotoxicity test was applied on 3T3 murine fibroblasts, and the prepared AgNPs did not affect the cell viability at all.

Copper oxide nanoparticles (CuO-NPs) were biologically synthesized by two actinomycete strains isolated from leaves of the medicinal plant *Oxalis corniculata* L. (Hassan et al. 2019). Based on gene sequencing of 16S rRNA, the two isolates were identified as *Streptomyces pseudogriseolus* Acv-11 and *Streptomyces zaomyceticus* Oc-5. The SPR absorption peak was detected at 400 nm. XRD confirmed the crystalline nature of CuO-NPs. Spherical-shaped particles were demonstrated by TEM with an approximate size of 78 nm and 80.0 nm for strains Oc-5 and Acv-11, respectively. The antimicrobial potential of the CuO-NPs was evaluated against

Gram-positive bacteria, Gram-negative bacteria, and unicellular and multicellular fungi. Gram-positive bacteria were the most sensitive than Gram-negative bacteria. Additionally, *Aspergillus brasiliensis* was more resistant than *C. albicans*. The results were compared with copper sulfate ($CuSO_4.5H_2O$). It was noticed that copper sulfate exerted no antimicrobial activity within a range of 5–20 mM. The small size and the high surface area might be the main reason behind the inhibitory action of CuO-NPs as reported by Usmani et al. (2017). Besides, Yoon et al. (2007) assumed that the antimicrobial potential of copper oxide nanoparticles takes place because of hydroxyl radicals which cause distortion to the helical structure of DNA upon binding. Additionally, these hydroxyl radicals cause destruction to essential proteins via binding to carboxyl and sulfhydryl amino groups of amino acids which cause subsequent inactivation of essential enzymes. Santo et al. (2008) suggested that the inhibitory action of CuO-NPs is related to inactivation of surface proteins which are responsible for material transportation across cytoplasmic membrane, and hence, selective permeability becomes damaged. Fungi are among the most important microorganisms which colonize plants and cause plant diseases (Doehlemann et al. 2017). Interestingly, the biosynthesized CuO-NPs exhibited strong antimicrobial potential against some phytopathogenic fungal strains including *Pythium ultimum*, *Fusarium oxysporum*, *Alternaria alternata*, and *Aspergillus niger*. Wu and Cederbaum (2003) demonstrated that oxidation processes are responsible for energy production in biological systems, and in certain cases, reactive oxygen species (ROS) arise due to haphazard oxygen production which in turn causes damage to cellular components such as carbohydrates, proteins, lipids, and DNA. According to Sen et al. (2010), diseases like cardiovascular, cancer, renal failure, and inflammatory disease arise because of the generation of ROS. Thus, antioxidant substances are usually employed to decrease the harmful effects of oxidant reactions via limiting or inhibiting formation of free radicals or scavenging them through different strategies, and thereby the immune defense becomes much stronger. In this study, the antioxidant activity of the biosynthesized CuO-NPs was determined by three different assays represented in hydrogen peroxide scavenging activity, total antioxidant assay and reducing power assay. Comparison with control was carried out. It was found that by increasing CuO-NPs concentration (10.0–15.0 mM), the total antioxidant capacity of CuO-NPs was higher when compared with controls. But there was no significant difference when a maximum concentration of CuO-NPs reached 20 mM. Furthermore, testing the effect of the biosynthesized CuO-NPs against *M. domestica* houseflies and *Culex pipiens* was assessed after different reaction periods (i.e., after 24, 48, 72, and 96 h). The obtained results revealed that the biosynthesized CuO-NPs exerted a larvicidal efficacy against both *M. domestica* and *Culex pipiens*, and it was higher than $CuSO_4.5H_2O$ as control. This study is recommended as a great example for a safe, green, biological synthesis of CuO-NPs to be further used as a biological agent as well as in pharmaceutical industries.

As mentioned earlier nanoparticles synthesized by actinomycetes are employed in different applications. The different applications of NPs biologically synthesized by actinomycetes are summarized in Table 2.2.

Table 2.2 A representative list of actinomycetes mediated synthesis of NPs, types of NPs produced, size/shape, employed characterization techniques, and their different applications

Actinomycetes	Type of produced NPs	Isolation	Size (nm)	Shape	Characterization	Application	References
***Rhodococcus* NCIM 2891**	AgNPs	–	10	Spherical	UV/Vis spectrophotometry, SEM, EDX, DLS, TEM, SAED	–	Otari et al. (2012)
Halomonas Salina	AuNPs	Procured from NCIM	30–100	Triangular	UV/Vis spectrophotometry, SEM, TEM, XRD	–	Res et al. (2012)
***Streptomyces* sp. VITBT7**	AgNPs	Soil sample	20–70	Spherical	TEM	Antimicrobial activity	Subashini and Kannabiran (2013)
***Streptomyces* sp. JAR1**	AgNPs	Kodaikanal hill station, Tamil Nadu, India	68.13		UV/Vis spectrophotometry, FTIR, XRD, and AFM	Antibacterial activity	Chauhan et al. (2013)
***Streptomyces* sp. VITSTK7**	Ag_2O/AgNPs	Brine spring at Thoubal District, Manipur, India	20–60		UV/Vis spectrophotometry, XRD, AFM	Antifungal activity	Thenmozhi et al. (2013)
***Nocardiopsis* sp. MBRC-1**	AgNPs	The marine sediment from the Busan coast, South Korea	45 ± 0.15	Spherical	UV/Vis spectrophotometry TEM, field emission scanning electron microscopy (FESEM), EDX, FTIR, and XRD spectroscopy	Antimicrobial activity, cytotoxic activity	Manivasagan et al. (2013)

(continued)

Table 2.2 (continued)

Actinomycetes	Type of produced NPs	Isolation	Size (nm)	Shape	Characterization	Application	References
***Streptomyces* sp. VITPK1**	AgNPs	Brine spring at Thoubal District, Manipur, India	20–45		UV/Vis spectrophotometry, FTIR, XRD, EDX	Anticandidal activity	Sanjenbam et al. (2014)
***Streptomyces* sp.**	AgNPs	Soil	20–70	Spherical	UV/Vis spectrophotometry, XRD, EDX, FTIR	Anti-ESBL (extended-spectrum β-lactamase) activity against multi-drug-resistant (MDR) ESBL	Subashini et al. (2014)
Marine isolate *Streptomyces parvulus* SSNP11	AgNPs	–	2.1	Spherical	Zeta potential, FTIR, TEM, SAED analysis, particle size analysis	Antibacterial activity	Prakasham et al. (2014)
***Streptomyces* sp. SS2**	AgNPs	Soil sediment of Similipal Biosphere Reserve, India	67.95 ± 18.52	Rounded	UV/Vis spectrophotometry, FTIR, SEM, zeta potential, DLS	Antibacterial activity	Mohanta and Behera (2014)
***Streptomyces* MS 26**	AgNPs	Muttukadu estuary	50–76	Spherical	UV/Vis spectrophotometry, DLS, SEM, FTIR, AFM	Antibacterial activity	Zarina and Nanda (2014)
***Streptomyces* sp. LK3**	AgNPs	Soil sample, island of Nicobar	5	Spherical	UV/Vis spectrophotometry, TEM, XRD, FTIR, AFM	Antimicrobial activity	Karthik et al. (2014)

***Streptomyces minutiscleroticus* M10A62**	SeNPs	Soil sample collected from magnesite mine	10–250	Spherical	UV/Vis spectrophotometry, FTIR, XRD, EDX, HRTEM	Anti-biofilm, antioxidant, cytotoxic, anti-viral activities, and wound healing	Ramya et al. (2015)
***Saccharomyces cerevisiae* ATCC 24858**	MnO_2NPs	–	34.4	Hexagonal, spherical	UV/Vis spectrophotometry, FTIR, EDX, X-ray photoelectron spectroscopy (XPS)	–	Salunke et al. (2015)
***Streptomyces* sp. 09 PBT 005**	AgNPs	Soil sample from sugarcane rhizosphere, Vengodu, Tiruvannamalai district, Tamil Nadu, India	198–595	Spherical	UV/Vis spectrophotometry, SEM, FTIR	Antibacterial and cytotoxic activities	Saravana Kumar et al. (2015)
Acidophilic actinomycetes SL19 and SL24 strains	AgNPs	Pine forest soil ($pH < 4.0$)	12.7–15.9	Spherical	UV/Vis spectrophotometry, nanoparticle tracking analysis (NTA), Zeta potential, FTIR, TEM	Antibacterial activity	Golińska et al. (2016)
Nocardiopsis valliformis	AgNPs	Lonar crater, India	5–50	Spherical	UV/Vis spectrophotometry, FTIR, TEM	Antimicrobial and cytotoxic activities	Rathod et al. (2016)
***Gordonia amicalis* HS-11**	AgNPs, AuNPs	Oil contaminated soil, Pune, (India)	20	Spherical	UV/Vis spectrophotometry, XRD, TEM	Antioxidant scavenging activity	Sowani et al. (2016)

(continued)

Table 2.2 (continued)

Actinomycetes	Type of produced NPs	Isolation	Size (nm)	Shape	Characterization	Application	References
Marine *Streptomyces rochei* MHM13	AgNPs	Sediment of the Suez Gulf, Red Sea, Egypt	22–85	Spherical	UV/Vis spectrophotometry, XRD, FTIR, SEM	Antimicrobial activity, anticancer and antibiofouling activities	Abd-Elnaby et al. (2016)
***Streptomyces* sp. GUT 21 (KU500633)**	AgNPs	Soil sample, campus of Gulbarga University, India	23–48	Spherical	UV/Vis spectrophotometry, XRD, SEM, FTIR	Antibacterial activity	Desai et al. (2016)
Saccharomyces cerevisiae	AgNPs	Local market, Iran	2–20	Spherical	UV/Vis spectrophotometry, TEM, DLS,	–	Korbekandi et al. (2016)
***Streptomyces* sp.**	AgNPs and AuNPs	Acidic soil	AgNPs (45), AuNPs (10)	Spherical	UV/Vis spectrophotometry, TEM, FTIR, nanoparticle tracking analyzer, zeta potential, DLS, AFM	Antibacterial activity	Składanowski et al. (2017)
Pilimelia columellifera	AgNPs	Acidic pine forest soil	–		–	Antifungal activity	Wypij et al. (2017)
Streptomyces cyaneus	Tellurium dioxide NPs (TeO_2 NPs)	–	75	Spherical	UV/Vis spectrophotometry, XRD, FTIR, HRTEM, DLS EDX, and SEM mapping analysis	Antifungal potential	Mosallam et al. (2018)

***Streptacidiphilus durhamensis* FSCA67T**	AgNPs	–	8–84	Spherical	Elemental analyzer (EA), ICP-MS, DLS, zeta potential, FTIR, EDX, SEM, TEM, AFM	Antimicrobial activity	Buszewski et al. (2018)
***Streptomyces griseoplanus* SAI-25**	AgNPs	Rhizospheric soil of rice field at, Karnataka, India	19.5–20.9	Spherical	UV/Vis spectroscopy, SEM, HRTEM, EDX, FTIR, XRD, DLS, and Zeta potential	Antifungal activity	Vijayabharathi et al. (2018)
***Streptomyces capillispiralis* Ca-1, *Streptomyces zaomyceticus* Oc-5, and *Streptomyces pseudogriseolus* Acv-11**	AgNPs	Healthy leaves of *Convolvulus arvensis* L. and *Oxalis corniculata*	23.77–63.14	Spherical	UV/Vis spectrophotometry, XRD, FTIR, SEM-EDX, TEM and particle size analyzer	Antimicrobial, antioxidant and bio-insecticidal activities	Fouda et al. (2019)
C. albicans	ZnO NPs	–	20	Quasi spherical	UV/Vis spectrophotometry, XRD, SEM, TEM, PL, TGA, DTA,	Catalyst for synthesis of steroidal pyrazolines	Mashrai et al. (2017)

As mentioned above, several research articles have showed the ability of actinobacteria to mediate biosynthesis of NPs. Still, certain points need to be addressed and highlighted. First, still much efforts need to be done concerning the time required for NPs biosynthesis which usually takes several hours. Thus, it may be considered as slow process of synthesis when compared with the chemical and physical methodologies. Confinement of particle size, prevention of its agglomeration, and assuring its monodispersity are the important points during preparation of NPs which requires extensive exploration. Assuring the stability of the actinobacteria-derived NPs is a very essential aspect in order to be applicable in different disciplines. The extracellular synthesis would have a greater advantage over the intracellular one since additional operations such as chemical reaction and ultrasound are required to release the particles from cells during the intracellular synthesis of NPs. Among all, exploring and identifying the exact mechanism for biosynthesis of NPs is of a great concern. As by understanding the mechanism of synthesis, biosynthesis of NPs will have better applications in the near future.

In depth investigations are required to elucidate the actual antimicrobial potential of NPs. In a study performed by Fernández et al. (2016), authors suggested a possible mechanism for the antifungal activity of AgNPs against fruit pathogenic fungi. It was clearly illustrated that the antifungal activity was highly influenced by the size and shape of the prepared AgNPs. According to Prabhu and Poulose (2012) and Ortega et al. (2015), AgNPs anchor on the microbial cell wall, and then penetration takes place; hence structural changes occur that affect the cell membrane leading to eventual cell death. Fernández et al. (2016) suggested the close relation between the size of NPs and their surface area, as the decrease in size results in in increasing the surface area. Consequently, great number of atoms becomes exposed on the surface, permitting redox, photochemical, and physicochemical interactions to occur within cells.

References

Abd-Elnaby HM, Abo-Elala GM, Abdel-Raouf UM, Hamed MM (2016) Antibacterial and anticancer activity of extracellular synthesized silver nanoparticles from marine *Streptomyces rochei* MHM13. Egypt J Aquat Res 42:301–312

Afzal B, Yasin D, Husain S et al (2019) Biocatalysis and agricultural biotechnology screening of cyanobacterial strains for the selenium nanoparticles synthesis and their anti-oxidant activity. Biocatal Agric Biotechnol 21:101307–103114

Ahmad A, Mukherjee P, Senapati S et al (2003a) Extracellular biosynthesis of silver nanoparticles using the fungus *Fusarium oxysporum*. Colloids Surf B: Biointerfaces 28:313–318

Ahmad A, Senapati S, Khan MI et al (2003b) Intracellular synthesis of gold nanoparticles by a novel alkalotolerant actinomycete, *Rhodococcus* species. Nanotechnology 14:824–830

Ahmad A, Senapati S, Khan MI et al (2003c) Extracellular biosynthesis of monodisperse gold nanoparticles by a novel extremophilic actinomycete, *Thermomonospora* sp. Langmuir 19:3550–3553

Ahmed E, Kalathil S, Shi L et al (2018) Synthesis of ultra-small platinum, palladium and gold nanoparticles by *Shewanella loihica* PV-4 electrochemically active biofilms and their enhanced catalytic activities. J Saudi Chem Soc 22:919–929

Akpan EI, Shen X, Wetzel B, Friedrich K (2019) Design and synthesis of polymer nanocomposites. In: Polymer composites with functionalized nanoparticles, synthesis, properties, and applications, Micro and nano technologies. Elsevier Inc, Amsterdam, pp 47–83

Alani F, Moo-Young M, Anderson W (2012) Biosynthesis of silver nanoparticles by a new strain of *Streptomyces* sp. compared with *Aspergillus fumigatus*. World J Microbiol Biotechnol 28:1081–1086

Babauta J, Renslow R, Lewandowski Z, Beyenal H (2012) Electrochemically active biofilms: facts and fiction. A review. Biofouling 28:789–812

Bai HJ, Zhang ZM, Gong J (2006) Biological synthesis of semiconductor zinc sulfide nanoparticles by immobilized *Rhodobacter sphaeroides*. Biotechnol Lett 28:1135–1139

Bai HJ, Zhang ZM, Guo Y, Yang GE (2009) Biosynthesis of cadmium sulfide nanoparticles by photosynthetic bacteria *Rhodopseudomonas palustris*. Colloids Surf B: Biointerfaces 70:142–146

Bai H-J, Yang B-S, Chai C-J et al (2011) Green synthesis of silver nanoparticles using *Rhodobacter sphaeroides*. World J Microbiol Biotechnol 27:2723–2728

Baker S, Satish S (2015) Biosynthesis of gold nanoparticles by *Pseudomonas veronii* AS41G inhabiting *Annona squamosa* L. Spectrochim Acta A Mol Biomol Spectrosc 150:691–695

Baker S, Harini BP, Rakshith D, Satish S (2013) Marine microbes: invisible nanofactories. J Pharm Res 6:383–388

Bakhshi M, Hosseini MR (2016) Synthesis of CdS nanoparticles from cadmium sulfate solutions using the extracellular polymeric substances of *B. licheniformis* as stabilizing agent. Enzym Microb Technol 95:209–216

Barka EA, Vatsa P, Sanchez L et al (2016) Taxonomy, physiology, and natural products of Actinobacteria. Microbiol Mol Biol Rev 80:1–43

Barreto ML, Teixeira MG (2008) Dengue no Brasil: situação epidemiológica e contribuições para uma agenda de pesquisa. Estud avançados 22:53–72

Baygar T, Sarac N, Ugur A, Karaca IR (2019) Antimicrobial characteristics and biocompatibility of the surgical sutures coated with biosynthesized silver nanoparticles. Bioorg Chem 86:254–258

Becker N, Petric D, Zgomba M et al (2010) Mosquitoes and their control. Springer, Dordrecht/New York/Heidelberg

Beveridge TJ, Murray RG (1980) Sites of metal deposition in the cell wall of *Bacillus subtilis*. J Bacteriol 141:876–887

Bhattacharya S, Saha I, Mukhopadhyay A et al (2013) Role of nanotechnology in water treatment and purification: potential applications and implications. Int J Chem Sci Technol 3:59–64

Bintrim SB, Donohue TJ, Handelsman J et al (1997) Molecular phylogeny of Archaea from soil. Proc Natl Acad Sci 94:277–282

Bizuye A, Bii C, Erastus G, Maina N (2017) Antibacterial metabolite prospecting from Actinomycetes isolated from waste damped soils from Thika, central part of Kenya. Asian Pac J Trop Dis 7:757–764

Buszewski B, Railean-Plugaru V, Pomastowski P et al (2016) Antimicrobial activity of biosilver nanoparticles produced by novel *Streptacidiphilus durhamensis* strain. J Microbiol Immunol Infect 51:1–10

Buszewski B, Railean-Plugaru V, Pomastowski P et al (2018) Antimicrobial activity of biosilver nanoparticles produced by a novel *Streptacidiphilus durhamensis* strain. J Microbiol Immunol Infect 51:45–54

Calvano CD, Italiano F, Catucci L et al (2014) The lipidome of the photosynthetic bacterium *Rhodobacter sphaeroides* R26 is affected by cobalt and chromate ions stress. Biometals 27:65–73

Canfield DE (2005) The sulfur cycle. In: Canfield DE, Thamdrup B, Kristensen E (eds) Aquatic geomicrobiology, Advances in marine biology, vol 48. Academic, Amsterdam/London

Chandra G, Bhattacharjee I, Chatterjee S (2010) A review on *Anopheles subpictus* Grassi—a biological vector. Acta Trop 115:142–154

Chauhan R, Kumar A, Abraham J (2013) A biological approach to the synthesis of silver nanoparticles with *Streptomyces* sp JAR1 and its antimicrobial activity. Sci Pharm 81:607–621

Cheaburu-Yilmaz CN, Karasulu HY, Yilmaz O (2019) Nanoscaled dispersed systems used in drug-delivery applications. In: Polymeric nanomaterials in nanotherapeutics, Micro and nano technologies. Elsevier Inc, Amsterdam, pp 437–468
Clark DP, Pazdernik NJ, Clark DP, Pazdernik NJ (2016) Chapter 7: Nanobiotechnology. In: Biotechnology, United States of America, pp 219–248
Costo R, Bello V, Robic C et al (2011) Ultrasmall iron oxide nanoparticles for biomedical applications: improving the colloidal and magnetic properties. Langmuir 28:178–185
Das VL, Thomas R, Varghese RT et al (2014) Extracellular synthesis of silver nanoparticles by the Bacillus strain CS11 isolated from industrialized area. 3 Biotech 4:121–126
Dawson A, Kamat PV (2001) Semiconductor– metal nanocomposites. Photoinduced fusion and photocatalysis of gold-capped TiO_2 (TiO_2/gold) nanoparticles. J Phys Chem B 105:960–966
DeLong EF (1992) Archaea in coastal marine environments. Proc Natl Acad Sci 89:5685–5689
Desai PP, Prabhurajeshwar C, Chandrakanth KR (2016) Hydrothermal assisted biosynthesis of silver nanoparticles from *Streptomyces* sp. GUT 21 (KU500633) and its therapeutic antimicrobial activity. J Nanostruct Chem 6:235–246
Dhanda V, Kaul HN (1980) Mosquito vectors of Japanese encephalitis virus and their bionomics in India. Proc Indian Natl Sci Acad B 46B:759–768
Doehlemann G, Ökmen B, Zhu W, Sharon A (2017) Plant pathogenic fungi. Microbiol Spectr 5:1–23
Dong A, Prestrelski SJ, Allison SD, Carpenter JF (1995) Infrared spectroscopic studies of lyophilization and temperature-induced protein aggregation. J Pharm Sci 84:415–424
Drexler KE (2004) Nanotechnology: from Feynman to funding. Bull Sci Technol Soc 24:21–27
Durán N, Marcato PD, Alves OL et al (2005) Mechanistic aspects of biosynthesis of silver nanoparticles by several *Fusarium oxysporum* strains. J Nanobiotechnol 3:1–7
Edwards PP, Thomas JM (2007) Gold in a metallic divided Staten – from faraday to present-day nanoscience. Angew Chem Int Ed 46:5480–5486
El-Gendy NS, Omran BA (2019) Green synthesis of nanoparticles for water treatment. In: Nano and bio-based technologies for wastewater treatment. Beverly, MA Scrivener Publishing, LLC Hoboken, United States, pp 205–263
Faramarzi MA, Sadighi A (2013) Insights into biogenic and chemical production of inorganic nanomaterials and nanostructures. Adv Colloid Interf Sci 189:1–20
Fatemi M, Mollania N, Momeni-Moghaddam M, Sadeghifar F (2018) Extracellular biosynthesis of magnetic iron oxide nanoparticles by *Bacillus cereus* strain HMH1: characterization and in vitro cytotoxicity analysis on MCF-7 and 3T3 cell lines. J Biotechnol 270:1–11
Fernández JG, Fernández-Baldo MA, Berni E et al (2016) Production of silver nanoparticles using yeasts and evaluation of their antifungal activity against phytopathogenic fungi. Process Biochem 51:1306–1313
Fouda A, Hassan SE-D, Abdo AM, El-Gamal MS (2019) Antimicrobial, antioxidant and larvicidal activities of spherical silver nanoparticles synthesized by endophytic *Streptomyces* spp. Biol Trace Elem Res 195:1–18
Ganesh Babu MM, Gunasekaran P (2009) Production and structural characterization of crystalline silver nanoparticles from *Bacillus cereus* isolate. Colloids Surf B: Biointerfaces 74:191–195
Golińska P, Wypij M, Rathod D et al (2016) Synthesis of silver nanoparticles from two acidophilic strains of *Pilimelia columellifera* subsp. pallida and their antibacterial activities. J Basic Microbiol 56:541–556
Gowramma B, Keerthi U, Rafi M, Rao DM (2015) Biogenic silver nanoparticles production and characterization from native stain of *Corynebacterium* species and its antimicrobial activity. 3 Biotech 5:195–201
Gram C (1884) Ueber die isolirte Farbung der Schizomyceten in Schnitt-und Trockenpraparaten. Fortschritte der Med 2:185–189
Gurunathan S, Kalishwaralal K, Vaidyanathan R et al (2009) Biosynthesis, purification and characterization of silver nanoparticles using *Escherichia coli*. Colloids Surf B: Biointerfaces 74:328–335

Hassan SE-D, Salem SS, Fouda A et al (2018) New approach for antimicrobial activity and biocontrol of various pathogens by biosynthesized copper nanoparticles using endophytic actinomycetes. J Radiat Res Appl Sci 11:262–270

Hassan SED, Fouda A, Radwan AA et al (2019) Endophytic actinomycetes *Streptomyces* spp mediated biosynthesis of copper oxide nanoparticles as a promising tool for biotechnological applications. J Biol Inorg Chem 24:377–393

He S, Guo Z, Zhang Y et al (2007) Biosynthesis of gold nanoparticles using the bacteria *Rhodopseudomonas capsulata*. Mater Lett 61:3984–3987

He S, Zhang Y, Guo Z, Gu N (2008) Biological synthesis of gold nanowires using extract of *Rhodopseudomonas capsulata*. Biotechnol Prog 24:476–480

Headrick DH, Goeden RD (2001) Biological control as a tool for ecosystem management. Biol Control 21:249–257

Huang K, Ma H, Liu J et al (2012) Size-dependent localization and penetration of ultrasmall gold nanoparticles in cancer cells, multicellular spheroids, and tumors in vivo. ACS Nano 6:4483–4493

Hulkoti NI, Taranath TC (2014) Biosynthesis of nanoparticles using microbes-a review. Colloids Surf B: Biointerfaces 121:474–483

Huo S, Jin S, Ma X et al (2014) Ultrasmall gold nanoparticles as carriers for nucleus-based gene therapy due to size-dependent nuclear entry. ACS Nano 8:5852–5862

Husseiny MI, El-Aziz MA, Badr Y, Mahmoud MA (2007) Biosynthesis of gold nanoparticles using *Pseudomonas aeruginosa*. Spectrochim Acta A Mol Biomol Spectrosc 67:1003–1006

Ingle A, Gade A, Pierrat S et al (2008) Mycosynthesis of silver nanoparticles using the fungus *Fusarium acuminatum* and its activity against some human pathogenic bacteria. Curr Nanosci 4:141–144

Iranifam M, Fathinia M, Sadeghi Rad T et al (2013) A novel selenium nanoparticles-enhanced chemiluminescence system for determination of dinitrobutylphenol. Talanta 107:263–269

Iravani S (2011) Green synthesis of metal nanoparticles using plants. Green Chem 13:2638–2650

Iravani S (2014) Bacteria in nanoparticle synthesis: current status and future prospects. Int Sch Res Notices 2014:Article ID 359316, 18 pages. https://doi.org/10.1155/2014/359316

Italiano F, Rinalducci S, Agostiano A et al (2012) Changes in morphology, cell wall composition and soluble proteome in *Rhodobacter sphaeroides* cells exposed to chromate. Biometals 25:939–949

Italiano F, Agostiano A, Belviso BD et al (2018) Interaction between the photosynthetic anoxygenic microorganism *Rhodobacter sphaeroides* and soluble gold compounds. From toxicity to gold nanoparticle synthesis. Colloids Surf B: Biointerfaces 172:362–371

Itävaara M, Salavirta H, Marjamaa K, Ruskeeniemi T (2016) Geomicrobiology and metagenomics of terrestrial deep subsurface microbiomes. In: Advances in applied microbiology. Elsevier, Amsterdam, pp 1–77

James AA (1992) Mosquito molecular genetics: the hands that feed bite back. Science 80(257):37–39

Jeyaraj M, Murugan M, Anthony KJP, Gurunathan S (2015) An environmentally friendly and green approach for synthesis and applications of silver nanoparticles. In: Nanomaterials for environmental protection. Wiley, Hoboken, pp 313–327

Juibari MM, Yeganeh LP, Abbasalizadeh S (2015) Investigation of a hot-spring extremophilic *Ureibacillus thermosphaericus* strain thermo-BF for extracellular biosynthesis of functionalized gold nanoparticles. Bionanosciecne 5:233–241

Jurat-Fuentes JL, Jackson TA (2012) Bacterial entomopathogens. In: Insect pathology. Elsevier, San Diego, pp 265–349

Kalathil S, Lee J, Cho MH (2011) Electrochemically active biofilm-mediated synthesis of silver nanoparticles in water. Green Chem 13:1482–1485

Kalimuthu K, Suresh Babu R, Venkataraman D et al (2008) Biosynthesis of silver nanocrystals by *Bacillus licheniformis*. Colloids Surf B: Biointerfaces 65:150–153

Karner MB, DeLong EF, Karl DM (2001) Archaeal dominance in the mesopelagic zone of the Pacific Ocean. Nature 409:507–511

Khadivi Derakhshan F, Dehnad A, Salouti M (2012) Extracellular biosynthesis of gold nanoparticles by metal resistance bacteria: *Streptomyces griseus*. Synth React Inorg MetOrg NanoMetal Chem 42:868–871

Khan MM, Ansari SA, Lee J-H et al (2013) Mixed culture electrochemically active biofilms and their microscopic and spectroelectrochemical studies. ACS Sustain Chem Eng 2:423–432

Khan ME, Khan MM, Cho MH (2015a) Green synthesis, photocatalytic and photoelectrochemical performance of an Au–Graphene nanocomposite. RSC Adv 5:26897–26904

Khan ME, Khan MM, Cho MH (2015b) Biogenic synthesis of a Ag–graphene nanocomposite with efficient photocatalytic degradation, electrical conductivity and photoelectrochemical performance. New J Chem 39:8121–8129

Kibret M, Guerrero-Garzón JF, Urban E et al (2018) *Streptomyces* spp. from Ethiopia producing antimicrobial compounds: characterization via bioassays, genome analyses, and mass spectrometry. Front Microbiol 9:1270–1983

Killham K, Prosser JI (2015) The bacteria and archaea, 4th edn. Elsevier Inc., London

Kimber RL, Lewis EA, Parmeggiani F et al (2018) Biosynthesis and characterization of copper nanoparticles using *Shewanella oneidensis*: application for click chemistry. Small 14:1–8

Klaus-Joerger T, Joerger R, Olsson E, Granqvist C-G (2001) Bacteria as workers in the living factory: metal-accumulating bacteria and their potential for materials science. Trends Biotechnol 19:15–20

Klempner MS, Unnasch TR, Hu LT (2007) Taking a bite out of vector-transmitted infectious diseases. N Engl J Med 356:2567–2569

Korbekandi H, Mohseni S, Jouneghani RM et al (2016) Biosynthesis of silver nanoparticles using Saccharomyces cerevisiae. Artif Cells Nanomed Biotechnol 44:235–239

Kulkarni RR, Shaiwale NS, Deobagkar DN, Deobagkar DD (2015) Synthesis and extracellular accumulation of silver nanoparticles by employing radiation-resistant *Deinococcus radiodurans*, their characterization, and determination of bioactivity. Int J Nanomedicine 10:963–974

Kumaresan M, Vijai Anand K, Govindaraju K et al (2018) Seaweed Sargassum wightii mediated preparation of zirconia (ZrO_2) nanoparticles and their antibacterial activity against Gram-positive and Gram-negative bacteria. Microb Pathog 124:311–315

Li X, Chen S, Hu W et al (2009) In situ synthesis of CdS nanoparticles on bacterial cellulose nanofibers. Carbohydr Polym 76:509–512

Li X, Xu H, Chen Z-S, Chen G (2011) Biosynthesis of nanoparticles by microorganisms and their applications. J Nanomater 2011:Article ID 270974, 16 pages. https://doi.org/10.1155/2011/270974

Lloyd JR, Byrne JM, Coker VS (2011) Biotechnological synthesis of functional nanomaterials. Curr Opin Biotechnol 22:509–515

Mageswari A, Subramanian P, Ravindran V et al (2015) Synthesis and larvicidal activity of low-temperature stable silver nanoparticles from psychrotolerant *Pseudomonas mandelii*. Environ Sci Pollut Res 22:5383–5394

Makarov VV, Love AJ, Sinitsyna OV et al (2014) "Green" nanotechnologies: synthesis of metal nanoparticles using plants. Acta Nat 6:1–10

Malarkodi C, Rajeshkumar S, Paulkumar K et al (2013) Bactericidal activity of bio mediated silver nanoparticles synthesized by *Serratia nematodiphila*. Drug Invent Today 5:119–125

Malhotra A, Dolma K, Kaur N et al (2013) Biosynthesis of gold and silver nanoparticles using a novel marine strain of *Stenotrophomonas*. Bioresour Technol 142:727–731

Mallick K, Witcomb MJ, Scurrell MS (2004) Polymer stabilized silver nanoparticles: a photochemical synthesis route. J Mater Sci 39:4459–4463

Mandal D, Bolander ME, Mukhopadhyay D et al (2006) The use of microorganisms for the formation of metal nanoparticles and their application. Appl Microbiol Biotechnol 69:485–492

Manivasagan P, Venkatesan J, Senthilkumar K et al (2013) Biosynthesis, antimicrobial and cytotoxic effect of silver nanoparticles using a novel *Nocardiopsis* sp. MBRC-1. Biomed Res Int 2013:Article ID 287638, 9 pages. https://doi.org/10.1155/2013/287638

Manivasagan P, Venkatesan J, Sivakumar K, Kim S-K (2014) Pharmaceutically active secondary metabolites of marine actinobacteria. Microbiol Res 169:262–278

Manivasagan P, Venkatesan J, Sivakumar K, Kim SK (2016) Actinobacteria mediated synthesis of nanoparticles and their biological properties: a review. Crit Rev Microbiol 42:209–221

Marimuthu S, Rahuman AA, Kirthi AV et al (2013) Eco-friendly microbial route to synthesize cobalt nanoparticles using *Bacillus thuringiensis* against malaria and dengue vectors. Parasitol Res 112:4105–4112

Markus J, Mathiyalagan R, Kim YJ et al (2016) Intracellular synthesis of gold nanoparticles with antioxidant activity by probiotic *Lactobacillus kimchicus* DCY51T isolated from Korean kimchi. Enzym Microb Technol 95:85–93

Mashrai A, Khanam H, Aljawfi RN (2017) Biological synthesis of ZnO nanoparticles using *C. albicans* and studying their catalytic performance in the synthesis of steroidal pyrazolines. Arab J Chem 10:S1530–S1536

Mathivanan K, Selva R, Chandirika JU et al (2019) Biologically synthesized silver nanoparticles against pathogenic bacteria : synthesis, calcination and characterization. Biocatal Agric Biotechnol 22:101373–101380

Mohanta YK, Behera SK (2014) Biosynthesis, characterization and antimicrobial activity of silver nanoparticles by *Streptomyces sp. SS2*. Bioprocess Biosyst Eng 37:2263–2269–101380

Morozkina EV, Zvyagilskaya RA (2007) Nitrate reductases: structure, functions, and effect of stress factors. Biochemist 72:1151–1160

Mosallam FM, El-Sayyad GS, Fathy RM, El-Batal AI (2018) Biomolecules-mediated synthesis of selenium nanoparticles using *Aspergillus oryzae* fermented Lupin extract and gamma radiation for hindering the growth of some multidrug-resistant bacteria and pathogenic fungi. Microb Pathog 122:108–116

Mukherjee P, Ahmad A, Mandal D et al (2001) Bioreduction of $AuCl_4^-$ ions by the fungus, *Verticillium* sp. and surface trapping of the gold nanoparticles formed. Angew Chem Int Ed 40:3585–3588

Mukherjee P, Senapati S, Mandal D et al (2002) Extracellular synthesis of gold nanoparticles by the fungus *Fusarium oxysporum*. ChemBioChem 3:461–463

Mukhtar M, Herrel N, Amerasinghe FP et al (2003) Role of wastewater irrigation in mosquito breeding in South Punjab, Pakistan. Southeast Asian J Trop Med Public Health 34:72–80

Murphy CJ, Sundaresan A, Rao CNR, Murphy CJ (2008) Sustainability as an emerging design criterion in nanoparticle synthesis and applications. J Mater Chem 18:2173–2176

Naik MM, Prabhu MS, Samant SN et al (2017) Synergistic action of silver nanoparticles synthesized from silver resistant estuarine *Pseudomonas aeruginosa* strain SN5 with antibiotics against antibiotic resistant bacterial human pathogens. Thalassas 33:73–80

Najitha Banu A, Balasubramanian C, Moorthi PV (2014) Biosynthesis of silver nanoparticles using *Bacillus thuringiensis* against dengue vector, *Aedes aegypti* (Diptera: Culicidae). Parasitol Res 113:311–316

Nangia Y, Wangoo N, Goyal N et al (2009) A novel bacterial isolate *Stenotrophomonas maltophilia* as living factory for synthesis of gold nanoparticles. Microb Cell Factories 8:1–7

Narayanan KB, Sakthivel N (2010) Biological synthesis of metal nanoparticles by microbes. Adv Colloid Interf Sci 156:1–13

Nobile CJ, Johnson AD (2015) *Candida albicans* biofilms and human disease. Annu Rev Microbiol 69:71–92

Noman M, Shahid M, Ahmed T, Khan MB (2019) Use of biogenic copper nanoparticles synthesized from a native *Escherichia* sp. as photocatalysts for azo dye degradation and treatment of textile effluents. Environ Pollut 257:113514–113523

Omran BA, Nassar HN, Fatthallah NA et al (2018) Characterization and antimicrobial activity of silver nanoparticles mycosynthesized by *Aspergillus brasiliensis*. J Appl Microbiol 125:370–382

Omran BA, Nassar HN, Younis SA et al (2019) Physiochemical properties of *Trichoderma longibrachiatum* DSMZ 16517 synthesized silver nanoparticles for the mitigation of halotolerant sulphate reducing bacteria. J Appl Microbiol 126:138–154

Ortega FG, Fernández-Baldo MA, Fernández JG et al (2015) Study of antitumor activity in breast cell lines using silver nanoparticles produced by yeast. Int J Nanomedicine 10:2021–2031

Otari SV, Patil RM, Nadaf NH et al (2012) Green biosynthesis of silver nanoparticles from an actinobacteria *Rhodococcus* sp. Mater Lett 72:92–94

Oza G, Pandey S, Mewada A et al (2012a) Facile biosynthesis of gold nanoparticles exploiting optimum pH and temperature of fresh water algae *Chlorella pyrenoidusa*. Adv Appl Sci Res 3:1405–1412

Oza G, Pandey S, Shah R, Sharon M (2012b) Extracellular fabrication of silver nanoparticles using *Pseudomonas aeruginosa* and its antimicrobial assay. Pelagia Res Lib Adv Appl Sci Res 3:1776–1783

Parikh RY, Singh S, Prasad BLV et al (2008) Extracellular synthesis of crystalline silver nanoparticles and molecular evidence of silver resistance from *Morganella* sp.: towards understanding biochemical synthesis mechanism. ChemBioChem 9:1415–1422

Park TJ, Lee SY, Lee SJ et al (2006) Protein nanopatterns and biosensors using gold binding polypeptide as a fusion partner. Anal Chem 78:7197–7205

Peiris MK, Gunasekara CP, Jayaweera PM et al (2017) Biosynthesized silver nanoparticles: are they effective antimicrobials? Mem Inst Oswaldo Cruz 112:537–543

Pepper IL, Gentry TJ (2015) Earth environments. In: Environmental microbiology. Elsevier, China, pp 59–88

Prabhu S, Poulose EK (2012) Silver nanoparticles: mechanism of antimicrobial action, synthesis, medical applications, and toxicity effects. Int Nano Lett 2:32–42

Prakasham RS, Kumar BS, Kumar YS, Kumar KP (2014) Production and characterization of protein encapsulated silver nanoparticles by marine isolate *Streptomyces parvulus* SSNP11. Indian J Microbiol 54:329–336

Prasad K, Jha AK, Kulkarni AR (2007) *Lactobacillus* assisted synthesis of titanium nanoparticles. Nanoscale Res Lett 2:248–250

Priyadarshini S, Gopinath V, Meera Priyadharsshini N et al (2013) Synthesis of anisotropic silver nanoparticles using novel strain, *Bacillus flexus* and its biomedical application. Colloids Surf B: Biointerfaces 102:232–237

Puigdollers AR, Illas F, Pacchioni G (2016) Structure and properties of zirconia nanoparticles from density functional theory calculations. J Phys Chem C 120:4392–4402

Puja P, Kumar P (2019) A perspective on biogenic synthesis of platinum nanoparticles and their biomedical applications. Spectrochim Acta A Mol Biomol Spectrosc 211:94–99

Punjabi K, Mehta S, Chavan R et al (2018) Efficiency of biosynthesized silver and zinc nanoparticles against multi-drug resistant pathogens. Front Microbiol 9:1–11

Quinteros MA, Bonilla JO, Alborés SV et al (2019) Biogenic nanoparticles : synthesis, stability and biocompatibility mediated by proteins of *Pseudomonas aeruginosa*. Colloids Surf B: Biointerfaces 184:110517–110523

Rai A, Singh A, Ahmad A, Sastry M (2006) Role of halide ions and temperature on the morphology of biologically synthesized gold nanotriangles. Langmuir 22:736–741

Raja DARK, Balasubramani CKG, Arul PDD (2019) Biosynthesis of gold and silver nanoparticles from the symbiotic bacterium, *Photorhabdus luminescens* of entomopathogenic nematode : larvicidal properties against three mosquitoes and *Galleria mellonella* larvae. J Clust Sci 30:1051–1063

Rajeshkumar S, Naik P (2018) Synthesis and biomedical applications of cerium oxide nanoparticles – a review. Biotechnol Rep 17:1–5

Ramya S, Shanmugasundaram T, Balagurunathan R (2015) Biomedical potential of actinobacterially synthesized selenium nanoparticles with special reference to anti-biofilm, anti-oxidant, wound healing, cytotoxic and anti-viral activities. J Trace Elem Med Biol 32:30–39

Rasool U, Hemalatha S (2017) Marine endophytic actinomycetes assisted synthesis of copper nanoparticles (CuNPs): characterization and antibacterial efficacy against human pathogens. Mater Lett 194:176–180

Rathod D, Golinska P, Wypij M et al (2016) A new report of *Nocardiopsis valliformis* strain OT1 from alkaline Lonar crater of India and its use in synthesis of silver nanoparticles with special reference to evaluation of antibacterial activity and cytotoxicity. Med Microbiol Immunol 205:435–447

Res JMB, Shah R, Oza G et al (2012) Biogenic fabrication of gold nanoparticles using *Halomonas salina*. J Microbiol Biotechnol Res 2:485–492

Rotich MC (2018) Bio-prospecting for broad spectrum antibiotic producing actinomycetes isolated from virgin soils in Kericho county, Master of Science

Sadhasivam S, Shanmugam P, Yun KS (2010) Biosynthesis of silver nanoparticles by *Streptomyces hygroscopicus* and antimicrobial activity against medically important pathogenic microorganisms. Colloids Surf B: Biointerfaces 81:358–362

Sadhasivam S, Shanmugam P, Veerapandian M et al (2012) Biogenic synthesis of multidimensional gold nanoparticles assisted by *Streptomyces hygroscopicus* and its electrochemical and antibacterial properties. Biometals 25:351–360

Salunke BK, Sawant SS, Lee SI, Kim BS (2015) Comparative study of MnO_2 nanoparticle synthesis by marine bacterium Saccharophagus degradans and yeast *Saccharomyces cerevisiae*. Appl Microbiol Biotechnol 99:5419–5427

Sanjenbam P, Gopal JV, Kannabiran K (2014) Anticandidal activity of silver nanoparticles synthesized using *Streptomyces* sp.VITPK1. J Mycol Med 24:211–219

Santo CE, Taudte N, Nies DH, Grass G (2008) Contribution of copper ion resistance to survival of Escherichia coli on metallic copper surfaces. Appl Environ Microbiol 74:977–986

Sapkal MR, Deshmukh AM (2008) Biosynthesis of gold nanoparticles by *Streptomyces* species. Res J Biotechnol 3:36–39

Saratale GD, Saratale RG, Oh SE (2012) Production and characterization of multiple cellulolytic enzymes by isolated *Streptomyces* sp. MDS. Biomass Bioenergy 47:302–315

Saratale RG, Karuppusamy I, Saratale GD et al (2018) A comprehensive review on green nanomaterials using biological systems: recent perception and their future applications. Colloids Surf B: Biointerfaces 170:20–35

Saravana Kumar P, Balachandran C, Duraipandiyan V et al (2015) Extracellular biosynthesis of silver nanoparticle using *Streptomyces* sp. 09 PBT 005 and its antibacterial and cytotoxic properties. Appl Nanosci 5:169–180

Saravanan M, Kumar S, Mubarakali D (2018) Microbial pathogenesis synthesis of silver nanoparticles from *Bacillus brevis* (NCIM 2533) and their antibacterial activity against pathogenic bacteria. Microb Pthogenes 116:221–226

Sastry M, Ahmad A, Khan MI, Kumar R (2003) Biosynthesis of metal nanoparticles using fungi and actinomycete. Curr Sci 85:162–170

Sen S, Chakraborty R, Sridhar C et al (2010) Free radicals, antioxidants, diseases and phytomedicines: current status and future prospect. Int J Pharm Sci Rev Res 3:91–100

Shah M, Fawcett D, Sharma S et al (2015) Green synthesis of metallic nanoparticles via biological entities. Materials (Basel) 8:7278–7308

Shankar SS, Ahmad A, Sastry M (2003) Geranium leaf assisted biosynthesis of silver nanoparticles. Biotechnol Prog 19:1627–1631

Shankar SS, Rai A, Ankamwar B et al (2004) Biological synthesis of triangular gold nanoprisms. Nat Mater 3:482–488

Shanmugasundaram T, Radhakrishnan M, Gopikrishnan V et al (2013) A study of the bactericidal, anti-biofouling, cytotoxic and antioxidant properties of actinobacterially synthesised silver nanoparticles. Colloids Surf B: Biointerfaces 111:680–687

Shanthi S, David Jayaseelan B, Velusamy P et al (2016) Biosynthesis of silver nanoparticles using a probiotic *Bacillus licheniformis* Dahb1 and their antibiofilm activity and toxicity effects in *Ceriodaphnia cornuta*. Microb Pathog 93:70–77

Singh AK, Nakate UT (2014) Microwave synthesis, characterization, and photoluminescence properties of nanocrystalline zirconia. Sci World J 2014:Article ID 349457, 7 pages. https://doi.org/10.1155/2014/349457

Singh CP, Singh KN, Pandey MC (1996) Insect growth regulatory effect of neem derivative "Neemolin" on Spilosoma obligue Walker. Pestology 5:11–13

Sintubin L, De Windt W, Dick J et al (2009) Lactic acid bacteria as reducing and capping agent for the fast and efficient production of silver nanoparticles. Appl Microbiol Biotechnol 84:741–749

Sivalingam P, Joe J, Siva D et al (2012) Mangrove *Streptomyces* sp. BDUKAS10 as nanofactory for fabrication of bactericidal silver nanoparticles. Colloids Surf B: Biointerfaces 98:12–17
Składanowski M, Wypij M, Laskowski D et al (2017) Silver and gold nanoparticles synthesized from *Streptomyces* sp. isolated from acid forest soil with special reference to its antibacterial activity against pathogens. J Clust Sci 28:59–79
Sneha K, Sathishkumar M, Mao J et al (2010) Corynebacterium glutamicum-mediated crystallization of silver ions through sorption and reduction processes. Chem Eng J 162:989–996
Song JY, Jang H-K, Kim BS (2009) Biological synthesis of gold nanoparticles using *Magnolia kobus* and *Diospyros kaki* leaf extracts. Process Biochem 44:1133–1138
Sowani H, Mohite P, Munot H et al (2016) Green synthesis of gold and silver nanoparticles by an actinomycete *Gordonia amicalis* HS-11: mechanistic aspects and biological application. Process Biochem 51:374–383
Sreedharan SM, Gupta S, Saxena AK, Singh R (2019) *Macrophomina phaseolina*: microbased biorefinery for gold nanoparticle production. Ann Microbiol 4000:435–445
Srivastava SK, Constanti M (2012) Room temperature biogenic synthesis of multiple nanoparticles (Ag, Pd, Fe, Rh, Ni, Ru, Pt, Co, and Li) by *Pseudomonas aeruginosa* SM1. J Nanopart Res 14:831–841
Srivastava P, Braganca JM, Kowshik M (2014) In vivo synthesis of selenium nanoparticles by *Halococcus salifodinae* BK18 and their anti-proliferative properties against HeLa cell line. Biotechnol Prog 30:1480–1487
Subashini J, Kannabiran K (2013) Antimicrobial activity of *Streptomyces* sp. VITBT7 and its synthesized silver nanoparticles against medically important fungal and bacterial pathogens. Der Pharm Lett 5:192–200
Subashini J, Gopiesh Khanna V, Kannabiran K (2014) Anti-ESBL activity of silver nanoparticles biosynthesized using soil *Streptomyces* species. Bioprocess Biosyst Eng 37:999–1006
Suresh AK, Pelletier DA, Wang W et al (2011) Biofabrication of discrete spherical gold nanoparticles using the metal-reducing bacterium *Shewanella oneidensis*. Acta Biomater 7:2148–2152
Suriyaraj SP, Ramadoss G, Chandraraj K, Selvakumar R (2019) One pot facile green synthesis of crystalline bio-ZrO_2 nanoparticles using *Acinetobacter* sp. KCSI1 under room temperature. Mater Sci Eng C 105:110021–110031
Syed B, Nagendra NP, Dhananjaya BL et al (2016) Synthesis of silver nanoparticles by endosymbiont *Pseudomonas fluorescens* CA 417 and their bactericidal activity. Enzym Microb Technol 95:128–136
Tabrizi SG, Hamedi J, Mohammadipanah F (2013) Screening of soil actinomycetes against *Salmonella* serovar Typhi NCTC 5761 and characterization of the prominent active strains. Iran J Microbiol 5:356–365
Thakkar KN, Mhatre SS, Parikh RY (2010) Biological synthesis of metallic nanoparticles. Nanomed Nanotechnol Biol Med 6:257–262
Thenmozhi M, Kannabiran K, Kumar R, Gopiesh Khanna V (2013) Antifungal activity of *Streptomyces* sp. VITSTK7 and its synthesized Ag_2O/Ag nanoparticles against medically important *Aspergillus* pathogens. J Mycol Med 23:97–103
Usmani M, Khan I, Bhat A et al (2017) Current trend in the application of nanoparticles for waste water treatment and purification: a review. Curr Org Synth 14:206–226
Velmurugan P, Iydroose M (2014) Biosynthesis of silver nanoparticles using *Bacillus subtilis* EWP-46 cell-free extract and evaluation of its antibacterial activity. Bioprocess Biosyst Eng 37:1527–1534
Verma VC, Anand S, Ulrichs C, Singh SK (2013) Biogenic gold nanotriangles from *Saccharomonospora* sp., an endophytic actinomycetes of *Azadirachta indica* A. Juss. Int Nano Lett 3:1–7
Vijayabharathi R, Sathya A, Gopalakrishnan S (2018) Extracellular biosynthesis of silver nanoparticles using *Streptomyces griseoplanus* SAI-25 and its antifungal activity against *Macrophomina phaseolina*, the charcoal rot pathogen of sorghum. Biocatal Agric Biotechnol 14:166–171

Volpicella M, Costanza A, Palumbo O et al (2014) Rhodobacter sphaeroides adaptation to high concentrations of cobalt ions requires energetic metabolism changes. FEMS Microbiol Ecol 88:345–357

Wadhwani SA, Shedbalkar UU, Singh R, Chopade BA (2016) Biogenic selenium nanoparticles: current status and future prospects. Appl Microbiol Biotechnol 100:2555–2566

Waghmare SS, Deshmukh AM, Kulkarni SW, Oswaldo LA (2011) Biosynthesis and characterization of manganese and zinc nanoparticles. Univers J Environ Res Technol 1:64–96

Wang F, Li C, Sun L et al (2012) Porous single-crystalline palladium nanoparticles with high catalytic activities. Angew Chem Int Ed 51:4872–4876

Whitman WB, Coleman DC, Wiebe WJ (1998) Perspective prokaryotes: the unseen majority. Proc Natl Acad Sci U S A 95(12):6578–6583

Woese CR, Kandler O, Wheelis ML (1990) Towards a natural system of organisms: proposal for the domains Archaea, Bacteria, and Eukarya. Proc Natl Acad Sci 87:4576–4579

World Health Organization (2005) Sixth meeting of the technical advisory group on the global elimination of lymphatic Filariasis, Geneva, Switzerland, 20–23 September 2005. Wkly Epidemiol Rec Relev épidémiologique Hebd 80:401–408

Wu D, Cederbaum AI (2003) Alcohol, oxidative stress, and free radical damage. Alcohol Res Health 27:277–284

Wypij M, Czarnecka J, Dahm H et al (2017) Silver nanoparticles from *Pilimelia columellifera* subsp. pallida SL19 strain demonstrated antifungal activity against fungi causing superficial mycoses. J Basic Microbiol 57:793–800

Wypij M, Czarnecka J, Świecimska M et al (2018) Synthesis, characterization and evaluation of antimicrobial and cytotoxic activities of biogenic silver nanoparticles synthesized from *Streptomyces xinghaiensis* OF1 strain. World J Microbiol Biotechnol 34:23–36

Yin Y, Yang X, Hu L et al (2016) Superoxide-mediated extracellular biosynthesis of silver nanoparticles by the fungus *Fusarium oxysporum*. Environ Sci Technol Lett 3:160–165

Yoon K-Y, Byeon JH, Park J-H, Hwang J (2007) Susceptibility constants of *Escherichia coli* and *Bacillus subtilis* to silver and copper nanoparticles. Sci Total Environ 373:572–575

Zarina A, Nanda A (2014) Green approach for synthesis of silver nanoparticles from marine *Streptomyces* MS 26 and their antibiotic efficacy. J Pharm Sci Res 6:321–327

Zhang H, Hu X (2019) Biosynthesis of au nanoparticles by a marine bacterium and enhancing their catalytic activity through metal ions and metal oxides. Biotechnol Prog 35:1–10

Zhang H, Li Q, Lu Y et al (2005a) Biosorption and bioreduction of diamine silver complex by Corynebacterium. J Chem Technol Biotechnol 80:285–290

Zhang J, Wang H, Yan X, Zhang L (2005b) Comparison of short-term toxicity between Nano-Se and selenite in mice. Life Sci 76:1099–1109

Zonooz NF, Salouti M, Shapouri R, Nasseryan J (2012) Biosynthesis of gold nanoparticles by *Streptomyces* sp. ERI-3 supernatant and process optimization for enhanced production. J Clust Sci 23:375–382

Zotchev SB (2012) Marine actinomycetes as an emerging resource for the drug development pipelines. J Biotechnol 158:168–175

Chapter 3
Inspired Biological Synthesis of Nanomaterials Using Eukaryotic Microbial Nano-Machinery

3.1 Introduction

Fungi are usually introduced as a preferred and a suitable biological nano-factory for synthesis of nanomaterials (Rauwel et al. 2015). The use of a fungal system in synthesizing nanomaterials is referred to as mycosynthesis. Mycosynthesis completely fulfills the required demands of green chemistry principles, as the synthesis procedure is completed at the surrounding ambient pressure, temperature, and neutral pH (Chhipa 2019). Several fungal strains have been investigated to synthesize metal and metal oxide nanoparticles. Different extracellular and intracellular enzymes are produced by fungi (Omran et al. 2018). These enzymes play a major role in nanomaterial bioreduction, capping, and stabilization for long periods. Interestingly, fungi growing in the mycelial form excrete high quantities of extracellular enzymes and proteins, which make them suitable candidates to produce large amounts of nanomaterials. Besides, fungi express high tolerance and bioaccumulation capabilities of metals (Omran et al. 2019a). Fungi are more advantages than the other microbial strains regarding the synthesis of metal and metal oxide nanoparticles. Indeed, fungal mycelia are able to resist agitation, flow pressure, and several harsh conditions in bioreactors when compared with bacteria. They are easy to handle and easy to grow. Fungal mycelia produce more extracellular metabolic secretions of reductive proteins than bacteria. Furthermore, they can be easily handled in downstream processing. Out of microbes, fungi are the most effective candidates for metal NPs synthesis in a large-scale production due to the large amounts of NPs produced. This is due to the fact that fungi secrete more amounts of proteins which are directly translated to higher productive enzymes leading to high NP formation. They have very high wall binding capacity (Zomorodian et al. 2016). Mycosynthesis of NPs is a straightforward route for attaining easy and stable biologically fabricated NPs (Omran et al. 2019b).

B. A. Omran, *Nanobiotechnology: A Multidisciplinary Field of Science*, Nanotechnology in the Life Sciences, https://doi.org/10.1007/978-3-030-46071-6_3

3.2 Evolution of Eukaryotes

Glancing through history, oxygen atmosphere is thought to have occurred since 2.4 billion years long ago. Looking into history, the initial eukaryotic existing life seemed to occur since approximately1.85 billion years ago. Different eukaryotic life forms required an oxygen atmosphere in order to exist. Algae are considered as the first multicellular life form and have evolved since 1.2 billion years ago (Cole 2016). Eukaryotes include animal and plant cells, however the presence of chloroplasts is characteristic to plant cells. Fungal cells belong to eukaryotes, and they resemble animal cells, but they differ from each other as fungal cells contain a cell wall which is comprised of chitin.

Fungi constitute a large, eukaryotic diverse group of heterotrophic organisms. They represent a group of saprophytes or parasitic spore producing eukaryotic organisms which lack chlorophyll (Tan and Hsu 2018). Fungal saprophytes usually exist in soil and on decomposed organic matter. Their cells belong to eukaryotic cells as they possess membrane-bound organelles, internal membrane systems, and a well-defined cell wall which is largely made up of polysaccharides (glucan and mannan) in addition to chitin. Fungi reproduce sexually or asexually. Fungi exist everywhere; they float around us in the air; on hair, feathers, horn, bones, dung, leaves; and in digestive tracts of humans, animals, insects, fish, and other organisms. They can even be found digesting other fungi. Fungi possess the capability of tolerating a wide range of temperatures (Samson 2016). However, active growth is generally associated with limited ranges of temperature. Fungi which grow between 15 and 35 °C are usually referred to as mesophilic, and those which grow higher than this range are called thermophilic. On the other hand, psychrophilic fungi grow at low temperatures (< 5 °C). Interestingly, different fungi remain alive for long periods even at temperatures which are unsuitable for their growth. Temperature influences the time of lag phase, growth rate, and fungal yield in quite different ways for each fungus. Fungal spores help them to survive under extreme environmental conditions, particularly extreme temperature. This capacity is termed as "thermostability," and it is very common among fungi (Samson 2016).

Fungi are basically categorized into three main categories:

- Molds: molds refer to multicellular filamentous fungi, which consist of branched filaments named "hyphae" that grow to form extensions referred to as "mycelia." Regular shaped cross walls called "septa" may be found in hyphae of higher fungi. While, lower fungi usually lack these septa (Tan and Hsu 2018). Spores are the means by which molds reproduce via an asexual reproduction process (involving mitosis) or via sexual reproduction (involving meiosis) which is usually preceded by fusing nuclei of two haploid cells. Mushrooms are classified as higher fungi, and their spores reproduce sexually producing these macroscopic forms. While, under laboratory conditions, molds reproduce chiefly via asexual spores. Molds such as *Aspergillus* sp., *Alternaria* sp., *Penicillium* sp., etc.
- Yeasts*:* yeasts are principally unicellular and are spherical or oval in shape. They mostly propagate via asexual route via a process referred to as" budding." In this

process, a bud or a protuberance is developed on the surface of a yeast cell and keeps enlarging till being separated from the parent cell (Warnock 2012). A very small number of yeasts reproduce through cell fusion. Certain yeast strains generate elongated cells in form of chains referred to as pseudo hyphae which are similar to the hyphae of molds. *Candida* spp. and *Cryptococcus* spp. are examples of yeasts.

- Dimorphic fungi: dimorphic fungi refer to the types of fungi that may exist in both morphologies (molds and yeasts). This mainly relies on the temperature and the surrounding environmental conditions. Among, these fungi are *Histoplasma* spp., *Coccidioides* spp., and *Sporothrix* spp. which have the ability to alter their mode of growth to either a mycelial or yeast phase.

3.3 Myconanotechnology

Myconanotechnology is a term which refers to the biological synthesis of nanomaterials using fungi (Nayantara and Kaur 2018). Fungi have a number of advantages for nanoparticle synthesis compared with other microorganisms, as they are relatively easy to isolate, fungal cultures secrete large amounts of extracellular enzymes, have a wide range and diversity and the downstream processing of the synthesized NPs is much simple. When compared with bacteria, fungi produce large quantities of NPs as a consequence of their high secretion of enzymes which are directly involved in the bioreduction and stabilization of NPs. Fungal biomass cell surface carries a negative charge and contains sticky components which facilitate the attachment of metal ions upon the cell surface along with the electrostatic interaction (Khandel and Shahi 2018). Figures 3.1 and 3.2 show the extracellular bioreduction of silver ions into AgNPs as reported by Omran et al. (2018). Table 3.1 shows the potential of different fungal strains to synthesize NPs. Location of the synthesized NPs, size, shape, employed characterization techniques, and their possible applications are illustrated as well.

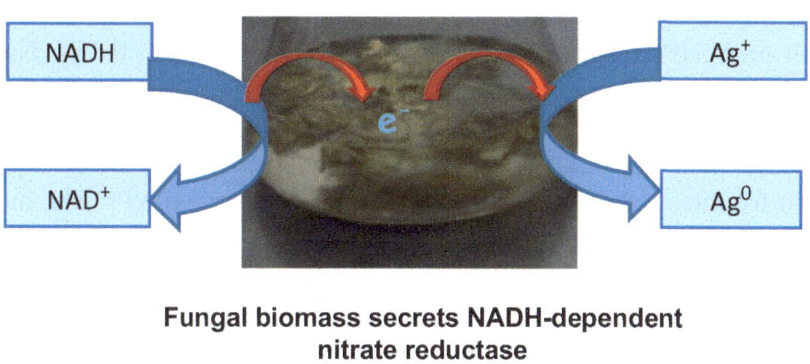

Fig. 3.1 Extracellular enzymatic reduction of silver ions by fungal biomass

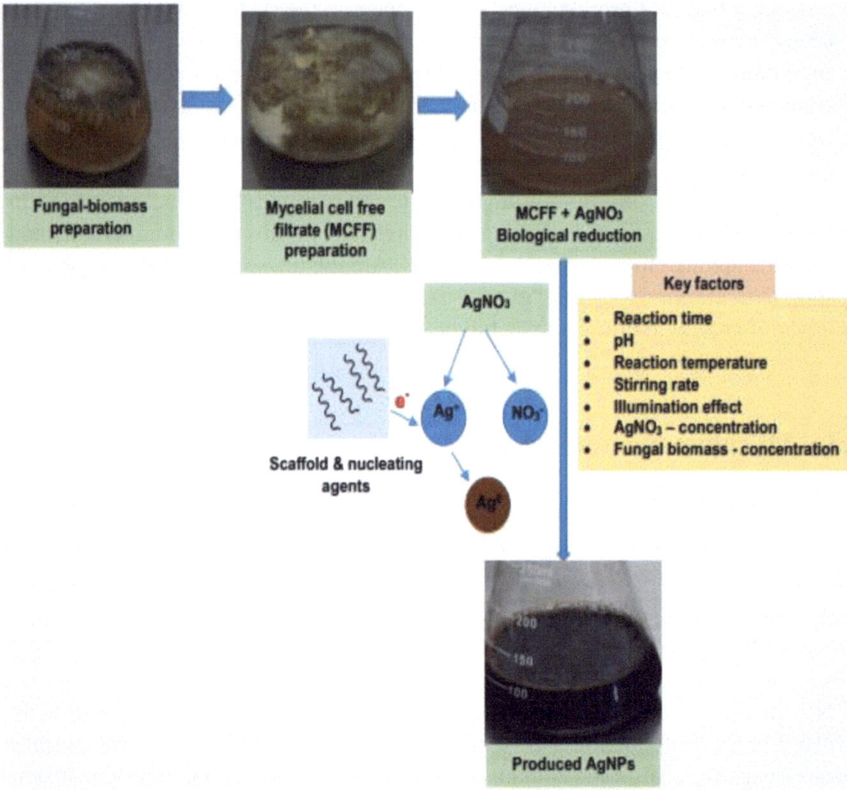

Fig. 3.2 Bio-production of AgNPs via (mycelial cell free filtrate) MCFF of *A. brasiliensis*

3.4 Mycosynthesis of Nanomaterials Using Fungi

In 2009, extracellular synthesis of AgNPs by a marine fungus was performed for the first time as described by Kathiresan et al. (2009). This study evaluated the potential of rhizospheric fungus *Penicillium fellutanum* which was isolated from mangrove root soil of *Rhizophora annamalayana*, India, to mycosynthesize AgNPs. Optimum production of AgNPs was attained at 24 h of incubation time, 0.35% NaCl, and pH 6, at a temperature of 5 °C and 1 mM $AgNO_3$. Authors used sodium dodecyl sulfate-polyacrylamide gel electrophoresis (SDS-PAGE)) to determine the molecular mass of protein bands existed within the fungal filtrate to provide a predicted mechanism for the synthesis and stabilization of the prepared NPs. The molecular mass of the noticeable protein bands within the filtrate of *P. fellutanum* was found to be at 70 kDa. Two fungal strains were isolated and identified as *Aspergillus tubingensis* (AY585546) and *Bionectria ochroleuca* (GU56625231). They exhibited an excellent extracellular capability to mycosynthesize AgNPs as reported by Rodrigues et al. (2013). The prepared AgNPs were characterized via TEM, XRD, and photon correlation spectroscopy for particle size and zeta potential. SDS-PAGE

Table 3.1 Overview of fungal-mediated genesis of AuNPs, AgNPs, SeNPs, SiNPs, GeNPs, TiNPs, ZnONPs, AgCl, CuONPs, and cobalt oxide NPs

Fungi	Type of produced NPs	Type of synthesis	Size/shape	Characterization techniques	Application	Reference
Aspergillus niger NCIM 616	AuNPs	Extracellular	Spherical, elliptical, 12.79 ± 5.61 nm	UV/Vis spectrophotometry, TEM, SEM, XRD	–	Bhambure et al. (2009)
Aspergillus fumigatus, *A. flavus*	AuNPs	*A. fumigatus* (extracellular), *A. flavus* (intracellular)	Triangles, spherical, hexagonal 17.76 nm	XRD, TEM	–	Gupta and Bector (2013)
Tricholoma crassum	AuNPs	Intracellular	Hexagonal, equilateral triangular, rhomboidal, polyhedral, isosceles triangular shapes, 5-25 nm	UV/Vis spectroscopy, XRD, SEM, TEM, AFM, DLS	Antimicrobial activity against bacteria, fungi, and multi-drug-resistant pathogenic bacteria, apoptotic properties on eukaryotic cells	Basu et al. (2018)
Macrophomina phaseolina	AuNPs	Extracellular	Spherical, 14–16 nm	UV/Vis spectroscopy, EDX, SEM, TEM, FTIR, DLS	–	Sreedharan et al. (2019)
Alternaria alternata	AgNPs	Extracellular	Spherical, 20–60 nm	UV/Vis spectrophotometry, FTIR, SEM, EDX	Antifungal activity against *Phoma glomerata*, *Phoma herbarum*, *Fusarium semitectum*, *Trichoderma* sp., and *Candida albicans*	Gajbhiye et al. (2009)
Inonotus obliquus (Chaga mushroom) extract	AgNPs	Extracellular	Spherical, 14.7–35.2 nm	UV/Vis spectrophotometry, FTIR, EDS, TEM, SEM	Antibacterial activity, anti-proliferative activity against A549 human lung cancer (CCL-185) and MCF-7 human breast cancer (HTB-22) cell lines	Nagajyothi et al. (2014)

(continued)

Table 3.1 (continued)

Fungi	Type of produced NPs	Type of synthesis	Size/shape	Characterization techniques	Application	Reference
Guignardia mangiferae	AgNPs	Extracellular	Spherical, 5–30 nm	HRTEM, SAED, XRD, EDX	Antifungal activity, cytotoxic activity against normal African monkey kidney (Vero), HeLa (cervical) and MCF-7 (breast) cells	Balakumaran et al. (2015)
Fusarium tricinctum (MTCC-1978), *F. culmorum* (MTCC-1893), *F. moniliforme* (MTCC- 7375), and *Pterolepis glomerata* (MTCC- 2710)	AgNPs	Extracellular	*F. tricinctum* (56 nm), *F. culmorum* (47 nm), *F. moniliforme* (46 nm), and *P. glomerata* (71 nm) Spherical-shaped particles	UV/Vis spectrophotometry, FTIR, XRD, TEM	Antibacterial activity	Gupta et al. (2015)
Monascus purpureus	AgNPs	Extracellular	Spherical, cuboid, 2-8 nm	UV/Vis spectrophotometry, FTIR, DLS, TEM, and XRD	Anti-candida activity	El-Baz et al. (2016)
Pleurotus cornucopiae var. citrinopileatus	AgNPs	Extracellular	Spherical, less than 100 nm	UV/Vis spectrophotometry, FTIR, FESEM, HRTEM, EDX	Anti-candida activity (*C. albicans*, *C. kruse*, *C. glabrata*, and *C. pseudotropicalis*)	Owaid et al. (2015)
Pleurotus djamor var. roseus	AgNPs	Extracellular	Spherical, 5–50 nm	UV/Vis spectrophotometry, TEM, XRD, FTIR	Cytotoxicity effect on PC3 Cells	Raman et al. (2015)
Pleurotus cornucopiae var. citrinopileatus	AgNPs	Extracellular	Spherical, less than 100 nm	UV/Vis spectrophotometry, HRTEM, FESEM, EDX, FTIR	Anti-candidal effect (*C. albicans*, *C. glabrata*; *C. krusei*, and *C. pseudotropicalis*)	Owaid et al. (2015)

Endophytic *Colletotrichum sp.* ALF2–6 inhabiting *Andrographis paniculata*	AgNPs	Extracellular	(Spherical, near to spherical, triangular and hexagonal) 20–50 nm	UV/Vis spectrophotometry, FTIR, XRD, TEM	Antibacterial activity against *E. coli* (MTCC 7410), *S. typhi* (MTCC 733), *B. subtilis* (MTCC 121), and *S. aureus* (MTCC 7443)	Azmath et al. (2016)
Aspergillus flavus (KF934407)	AgNPs	Extracellular	Spherical, 50 nm	UV/Vis spectrophotometry, DLS, TEM	Antimicrobial activity (*Staphylococcus aureus*, *Bacillus subtilis*, *Escherichia coli* and *Trichoderma* spp.)	Fatima et al. (2016)
Endophytic Colletotrichum sp. ALF2-6 inhabiting *Andrographis paniculata*	AgNPs	Extracellular	Myriad shapes (spherical, near to spherical, triangular, and hexagonal), 20–50 nm	UV/Vis spectrophotometry, FTIR, XRD, TEM	Antibacterial activity against *Escherichia coli* (MTCC 7410), *Salmonella typhi* (MTCC 733), *Bacillus subtilis* (MTCC 121), and *Staphylococcus aureus* (MTCC 7443)	Azmath et al. (2016)
Marine endophytic fungus *Penicillium polonicum* isolated from algal species (*Chaetomorpha antennina*)	AgNPs	Extracellular	Spherical or near spherical 10–15 nm	UV/Vis spectrophotometry, Raman spectroscopy, FTIR, HRTEM, EDS	Antibacterial efficacy and anti-biofilm against multidrug-resistant *Acinetobacter baumannii*	Neethu et al. (2018)
Aspergillus brasiliensis ATTC 16404	AgNPs	Extracellular	Spherical, 6–21 nm	UV/Vis spectrophotometry, EDX, DLS, FTIR, XRD, TEM, SEM	Antimicrobial activity against *Bacillus subtilis* ATCC 6633 and *Staphylococcus aureus* ATCC 35556, *Pseudomonas aeruginosa* ATCC 10145 and *Escherichia coli* ATCC 23282, *Candida albicans* IMRU3669	Omran et al. (2018)

(continued)

Table 3.1 (continued)

Fungi	Type of produced NPs	Type of synthesis	Size/shape	Characterization techniques	Application	Reference
Aspergillus niger, *Fusarium semitectum*	AgNPs	Extracellular	Spherical (10–25 nm)	UV/Vis spectrophotometry, SEM, XRD,	Antimicrobial activity against *E. coli* (ATCC- 8739), *P. aeruginosa* (ATCC-15442), and *S. aureus* (ATCC-6538)	Madakka et al. (2018)
Trichoderma longibrachiatum DSMZ 16517	AgNPs	Extracellular	Spherical, triangular, cuboid, 5–11 nm	UV/Vis spectrophotometry, EDX, DLS, FTIR, XRD, TEM, SEM	Biocidal activity against SRB mixed culture	Omran et al. (2019a)
Aspergillus brasiliensis ATCC 16404	Co_3O_4-NPs	Extracellular	Quasi-spherical, 20–25	UV/Vis spectrophotometry, EDX, DLS, FTIR, XRD, TEM, SEM	Antimicrobial activity	Omran et al. (2019a, b)
Marine endophytic fungus – *Cladosporium cladosporioides* isolated from brown algae, *Sargassum wightii*	AgNPs	Extracellular	Spherical, 30–60 nm	UV/Vis spectrophotometry, FTIR, AFM, FESEM, XRD, DLS	Antioxidant and antimicrobial activity against *E. coli* (MTCC 118), *S. aureus* (MTCC 7443), *B. subtilis* (MTCC 441), *S. epidermis* (MTCC 435), and a yeast pathogen, *C. albicans* (MTCC 183)	Hulikere and Joshi (2019)
Agaricus bisporus	AuNPs, AgNPs, SeNPs, SiNPs, germanium nanoparticles (GeNPs)	Extracellular	AuNPs nanospheres (2–5 nm), AgNPs nanospheres (1–10 nm), SeNPs spherical (100–250 nm), SiNPs (50–200 nm), GeNPs (50–250 nm)	TEM, EDX, X-ray fluorescence analysis	–	Loshchinina et al. (2018)

Agaricus arvensis	AuNPs, AgNPs, SeNPs, SiNPs, GeNPs	Extracellular	AuNPs nanospheres (2–10 nm), AgNPs nanospheres (1–10 nm), SeNPs spherical (150–550 nm), SiNPs (50–250 nm), GeNPs (50–250 nm)	TEM, EDX, X-ray fluorescence analysis	–	Loshchinina et al. (2018)
Aspergillus terreus	AgNPs, AuNPs	Extracellular	AgNPs (spherical, 8–20 nm), AuNPs (spherical shape and anisotropic morphology, 10–50 nm)	UV/Visspectrophotometry HRTEM, FTIR, XRD, EDX, SAED, ICP-AES, Zetasizer analyses	Antimicrobial activity against *E. coli* ATCC 8739, *Pseudomonas aeruginosa* ATCC 27853, *Klebsiella pneumoniae* ATCC 2719, *Enterococcus faecalis* ATCC 29212, *Bacillus subtilis* ATCC 6633, methicillin-resistant *S. aureus* (MRSA), *S. aureus* ATCC 29736, *C. albicans* ATCC 2091	Balakumaran et al. (2016)
Penicillium sp. isolated from *Calophyllum apetalum*	AgCl NPs	Extracellular	Cubic, 8.63–30.91 nm	UV/Vis spectroscopy, XRD, SEM	–	Chandrappa et al. (2016)
Aspergillus parasiticus	Silica NPs	Intracellular	Cubic shape (85 nm), spherical shape (24 ± 8 nm)	SEM, scanning transmission electron microscope (STEM), EDX, and FTIR	–	Zielonka et al. (2018)
Metarhizium anisopliae (ATCC 60336)	TiNPs	Extracellular	Near spherical, 46.82 nm	UV/Vis spectrophotometry, EDX, TEM	Larvicidal effect against *Galleria mellonella* larvae	Yosri et al. (2018)

(continued)

Table 3.1 (continued)

Fungi	Type of produced NPs	Type of synthesis	Size/shape	Characterization techniques	Application	Reference
Penicillium chrysogenum	CuO NPs	Extracellular	Spherical, 9.70 nm	UV/Vis spectrophotometry, FTIR, XRD, DLS, TEM, SEM, EDX, and mapping images	Antifungal activity against *Fusarium oxysporum*, *Alternaria solani*, and *Aspergillus niger* antibacterial activity against *Ralstonia solanacearum* and *Erwinia amylovora*	El-sayyad (2019)
Aspergillus niger YESM1	Magnetic Fe and magnetic Fe_3O_4 NPs	Extracellular	Spherical, 18–50 nm	XRD, SEM	–	Abdeen et al. (2016)
Cochliobolus geniculatus	ZnO NPs	Extracellular	Quasi-spherical, 2–6 nm	UV/Vis spectrophotometry, EDX, XRD, FTIR, SDS-PAGE, zeta potential	–	Kadam et al. (2019)
Stereum hirsutum	Cu and CuO NPs	Extracellular	Spherical, 5–20 nm	UV/Vis spectrophotometry, TEM, XRD, FTIR	–	Cuevas et al. (2015)

identified the proteins existed in the fungal filtrate. *A. tubingensis* filtrate and colloidal suspension of AgNPs showed the presence of electrophoretic protein bands at 75, 122, 191, and 328 (kDa). In the case of microbial filtrate of *B. ochroleuca*, three visible bands were detected at 70, 100, and 174 kDa, while four bands were revealed to surround AgNPs involving 25, 30, 44, and 49 kDa. AgNPs exerted a potent antifungal activity against *Candida* sp. and *Pseudomonas aeruginosa*.

Metarhizium robertsii is an entomo-pathogen which can be exploited as an alternative to the chemical insecticides and also it acts as an efficient xenobiotic degrader. Both Różalska et al. (2013) and Różalska et al. (2015) documented that *M. robertsii* owns a degrading potential regarding endocrine disrupting compound (EDC), e.g., nonylphenol. Although biodegradation of toxic compounds is highly efficient and environmentally friendly methodology, the disposal of the remaining waste biomass after degradation of pollutants still represents a problem which needs to be solved (Ahluwalia and Goyal 2007; Tamboli et al. 2010). Interestingly, fungal waste biomasses are comprised of numerous organic and inorganic substances whose recovery can be extremely beneficial. Różalska and coauthors reported for the first time an innovative approach to synthesize AgNPs by using the waste biomass of the filamentous fungus *M. robertsii* after being exploited in nonylphenol degradation (Różalska et al. 2016). UV/Vis, FTIR, EDX, TGA, XPS spectroscopy, STEM and DLS techniques were employed to characterize AgNPs. AgNPs derived from waste biomass of *M. robertsii* exhibited a high antibacterial activity against Gram-positive and Gram-negative bacteria at relatively low concentrations. Authors assumed that the mechanism behind this antimicrobial potential was significantly dependent upon cell membrane damage as was documented via laser scanning confocal microscopy. Besides, AgNPs proved to act as an efficient biocatalyst in degrading toxic compounds like nitrophenol and methylene blue with a high catalytic potential.

Gold nanoparticles (AuNPs) were successfully biosynthesized by the marine endophytic fungus *Cladosporium cladosporioides* which was isolated from the seaweed *Sophora wightii* (Joshi et al. 2017). The extracellular cell free filtrate (CFF) was mixed with 1 mM $HAuCl_4$. A change in color occurred from yellow to reddish violet color. A narrow and sharp peak characteristic to AuNPs appeared at 540 nm. The mycosynthesized AuNPs were stable for 6 months. Field emission scanning electron microscopy (FESEM) illustrated the average size as 60 nm. Four different XRD peaks were displayed at 2θ [111], [200], [220], and [311] which proved the crystalline *fcc* of the prepared AuNPs. AFM demonstrated the heterogeneous lateral and vertical dimensions of *C. cladosporioides*-derived AuNPs. Surface roughness was around 36 nm as observed via the three-dimensional images of the AuNPs, and the average size was approximately 40 nm. EDX profile revealed strong Au signals at 1.6 and 2.2 keV. Two extra peaks were identified corresponding to oxygen and carbon which according to previous studies might be attributed to biomolecules within the CFF (Priyadarshini et al. 2014). Based on an experiment to determine the role of NADH- dependent enzymes, authors were assured that AuNPs were not biosynthesized via the reaction between NADH and dialyzed cell filtrate along with $HAuCl_4$. Instead, the FTIR study revealed the possible role of polyphenols as well as proteins in the bioreduction process. These results were in accordance with the interpretation reported by Sanna et al. (2014). *C. cladosporioides* derived AuNPs were tested for their antimicrobial potential against a panel of human pathogenic bacteria, namely, *E. coli* MTCC 118, *S. aureus* MTCC 7443, *B. subtilis* MTCC 441, *P. aeruginosa* MTCC 424, and a fungal pathogen, *A. niger* MTCC 281. The most sensitive pathogen to AuNPs was *S. aureus* (MTCC 7443), and the most resistant was *B. subtilis* (MTCC 441). SEM was employed in a trial to track the mode of action of AuNPs. Images showed the accumulation of AuNPs upon the cell membrane. Rupture in cell wall was observed as well. According to Rai et al. (2010), an interaction between AuNPs and DNA inhibits growth of microbial cells in addition to the leakage of cellular constituents due to the effect of AuNPs. In other studies reported by Hussain et al. (2005) and Fatima et al. (2015), AuNPs were suggested to generate free radicals and inactivate enzymes owing to the interaction of AuNPs with enzymes' thiol group. This might clarify the mechanism behind the mode of action of AuNPs against pathogenic microorganisms. The high antimicrobial potential of *C. cladosporioides* AuNPs makes them an ideal candidate for developing a broad-spectrum antimicrobial agent. In order to assess the radical scavenging ability of a specific material, 2, 2-diphenyl-1-picrylhydrazyl (DPPH) is usually performed because of its simplicity (Ahmad et al. 2003). In this study, the mycosynthesized AuNPs exhibited an outstanding antioxidant activity which was almost comparable to ascorbic acid as a standard and it was found to be dose-dependent.

An effective extracellular biosynthesis procedure for rapid production of AgNPs was presented by Elamawi et al. (2018) via the fungus *Trichoderma longibrachiatum*. Bioreduction was attained via the CFF of the tested fungus. Optimization of different physical parameters like fungal biomass concentration (1, 5, 10, 15, and 20 g), temperature (25, 28, and 33 °C), incubation time (0–120 h), and agitation (shaken or not shaken) was performed. The maximum production of AgNPs was obtained at 10 g of fungal biomass, 28 °C for reaction temperature, 72 h for incubation time, and at static conditions. Visual appearance of brown color primarily indicated the mycosynthesis of AgNPs. UV/Vis spectroscopy displayed a maximum absorption peak at 385 nm. TEM demonstrated the presence of monodispersed, spherical-shaped particles with a mean diameter of 10 nm. Bands at1634.92 and 3269.31 cm^{-1} were detected via FTIR which corresponded to vibrational bending of amide I and amide II protein bands as previously demonstrated by Gole et al. (2001). DLS supported that the average size distribution of the particles was 24.43 and the polydispersed index (PdI) value was 0.420. Zeta potential value was 19.7 mV. The mycosynthesized AgNPs was relatively stable up to 2 months. The prepared AgNPs was investigated as an antifungal against pathogenic fungi including *A. alternate*, *F. oxysporum*, *F. verticillioides*, *F. moniliforme*, *Aspergillus flavus*, *A. heteromorphus*, *P. glabrum*, *P. brevicompactum*, *P. grisea*, and *H. oryzae*. AgNPs were the most sensitive against *P. grisea*, *verticillioides*, *H. oryzae*, *F. moniliforme*, and *A. alternate* with high inhibition percentages. Contrary, *F. oxysporum* was the most resistant towards AgNPs as it had 68.2% inhibition efficiency. Cobalt oxide nanoparticles have been successfully synthesized using the endophytic fungus *Aspergillus nidulans* as revealed by Vijayanandan and Balakrishnan (2018). Four different fungal isolates were isolated from plant surfaces involving stems, roots, leaves, and flowers. The obtained fungal isolates were subjected to a tolerance assay in which they were exposed to 500 ppm and 1000 ppm concentrations of cobalt acetylacetonate salt on potato dextrose agar (PDA) plates. The selected strain was chosen based upon the highest tolerance index. Tolerance study showed that *A. nidulans* was more tolerable than the other isolated strains. The first observation was the change in color from pink to orange. The UV/Vis spectrophotometer revealed the presence of cobalt oxide NPs absorption peak at 315 nm. The peak sharpness indicated the small size of the prepared particles. The band gap of cobalt oxide crystallite was 3.63 eV. FTIR spectrum pointed out the existence of wavenumbers corresponding to amide and carboxylic bands of proteins that might have been attached to nanoparticles via amine and cysteine groups. XRD pattern displayed peaks centered at 2Θ value of 36.47, 44.10, and 65 which corresponds to cobalt oxide NPs in spinel phase. The mean size of the prepared particles was calculated via Debye-Scherrer formula and was found to be 41.97 nm. DLS estimated the average size distribution and was found to be 123.8 nm. The zeta potential value was 10.8 mV. The mycosynthesized particles were spherical in shape as demonstrated by TEM. Two significant weight loss behaviors were demonstrated via the thermogram of thermogravimetric analysis (TGA) during the increase in temperature up to 530 °C. The first decomposition was observed at approximately 30 °C and continued till reaching 135 °C with a weight loss of 7.12%. It was characterized by the loss of absorbed water and dehydration of residual moisture. The second weight loss was observed at 280 °C (4526 mg) and ended at 530 °C (1430 mg). The second weight loss might be attributed to desorption and subsequent evaporation of bio-organic compounds within sample as explained by Mani et al. (2013). Beyond 530 °C, the weight of the sample remained constant which confirmed the completion of decomposition of the residues as demonstrated by Farhadi et al. (2013).

Seetharaman et al. (2018) conducted a study in which eco-friendly AgNPs was synthesized from the entophytic fungus *Phomopsis liquidambaris* extracellular filtrate. Healthy leaves of *Salacia chinensis* were used as a source for isolating the entophytic fungus *P. liquidambaris*. 18Sr RNA was employed to molecularly identify the isolated fungus. UV/visible spectrophotometer revealed the presence of a SPR peak of AgNPs at 430 nm. Different parameters were optimized to guarantee maximum production of AgNPs. Optimum conditions were 10 ml *P. liquidambaris* supernatant, 1 mM $AgNO_3$, 37 °C, 24 h, and pH 8. FTIR revealed the presence of flavonoids, proteins, and triterpenoids which might have been responsible for reducing and capping the biosynthesized AgNPs. The zeta potential value of the mycosynthesized AgNPs was −25.7 mV. The zeta potential value demonstrated the long-term stability and high dispersity with no agglomeration. Spherical-shaped particles were revealed via TEM with an average size of 18.7 nm. EDX confirmed the presence of elemental silver signals at 3 keV. XRD confirmed the crystallinity of the mycosynthesized AgNPs. The antimicrobial potential of *P. liquidambaris*-derived AgNPs was assessed against *Vibrio cholera* (MTCC 3906), *Salmonella typhi* (MTCC 531), *Micrococcus luteus* (MTCC 1809), *Proteus mirabilis* (MTCC 425), *Escherichia coli* (MTCC 1687), *Klebsiella pneumonia* (MTCC 4031), *Shigella flexneri* (MTCC 9543), and *Pseudomonas putida* (MTCC 1194). The prepared AgNPs exhibited the highest ZOI against *E. coli* (27 mm) followed by *P. mirabilis* (18 mm), *M. luteus* (17 mm), *S. flexneri* (16 mm), *K. pneumoniae* (15 mm), *V. cholera* (15 mm), and *P. putida* (14 mm), and the least activity was noticed in case of *S. typhi* (10 mm.). Additionally, a hemocompatibility assay was carried out to identify biocompatibility and damage of the synthesized AgNPs upon red blood cells (RBCs). In this study, AgNPs exhibited low release of red hemoglobin. Based on previous research, surface passivation of nanomaterials improves their biocompatibility. A larvicidal assay was carried out against instar larvae of *Culex quinquefasciatus* and *Aedes aegypti.* The obtained results revealed that the mycosynthesized AgNPs exhibited 100% mortality at a concentration of 10 ppm against the two tested mosquito species. In a previous research performed by Benelli et al. (2016), Benelli and coauthors suggested that AgNPs breach the exoskeleton and bind to sulfur groups originated from proteins or phosphorous derived from DNA resulting in prevention of translation/ transcription, failure of respiratory chain reaction, further reduction in membrane permeability, disruption in proton motive forces causing loss of cellular functions and eventually results in their death. Two fungal strains, namely, *Fusarium chlamydosporum* NG30 and *Penicillium chrysogenum* NG85 were investigated for their potential to synthesize AgNPs extracellularly by using their CFF (Khalil et al. 2019). Different authors support the extracellular synthesis by means of CFF as less steps are required to isolate and purify the targeted NPs (Kalpana and Lee 2013). The two fungal strains were isolated from an agricultural soil in Giza, Egypt, via the soil dilution plate technique. The identification of isolates was performed using nuclear ribosomal DNA internal transcribed spacer (ITS) sequencing. They were deposited in the GenBank of the National Center for Biotechnological Information (NCBI) database and were given the accession numbers of MH141316 and MH119131 for *F. chlamydosporum* NG30 and *P. chrysogenum* NG85, respectively. The synthesized AgNPs were characterized by means of TEM, DLS, and FTIR. The size of the AgNPs was in the range of 6–26 nm for *F. chlamydosporum* AgNPs (FAgNPs) and from 9 to 17.5 nm for *P. chrysogenum* AgNPs (PAgNPs). DLS proved that the average size distribution of FAgNPs was 16.1 nm, while that of PAgNPs was 11.1 nm. FTIR spectroscopy showed the presence of strong vibrations of hydroxyl functional groups which might be assigned to proteins as well as presence of amide linkages. Authors assumed that such functional groups have aided in stabilizing the prepared AgNPs. The antifungal potential of the prepared AgNPs was assessed against the mycotoxigenic fungi *Aspergillus flavus* and *Aspergillus ochraceus.* Results confirmed the inhibition of fungal spore germination. The minimum inhibitory concentration (MIC) was found to be 48, 45, and 50 mg/mL in cases of FAgNPs, PAgNPs and the antifungal compound itraconazole, respectively, against *A. flavus*, while, in the case of *A. ochraceus*, the MIC values were recorded to be 51, 47, and 49 mg/mL in cases of FAgNPs, PAgNPs, and itraconazole, respectively. A cytotoxicity assay of the prepared AgNPs was performed on human normal melanocytes (HFB4). Results revealed the survival of 80% and 75% of cells at a concentration of 6 mg/mL in cases of FAgNPs and PAgNPs, respectively. Bocate and co-authors performed a study in which *Fusarium oxysporum* was tested for its ability to mycosynthesize AgNPs (Bocate et al. 2019). Characteristic absorption peak at 440 nm corresponding to AgNPs was detected by the UV/Vis spectrophotometer. XRD pattern revealed typical Bragg reflections at 2Θ = 39.18, 44.81, 65.40, and 77.59, which were assigned to 111, 200, 220, 311, and 222 XRD patterns of AgNPs. The mycosynthesized AgNPs were spherical in shape with an average size of 93 ± 3 nm as demonstrated by TEM. The predicted value of zeta potential was −37.1 nm. The antimicrobial potential of *F. oxysporum*-derived AgNPs was tested against some toxicogenic species of filamentous fungi including *Aspergillus flavus* MCT 00335, *Aspergillus parasiticus* MCT 00336, *Aspergillus nomius* MCT 00328 which are known to produce aflatoxins. In addition, it was tested against *Aspergillus melleus* MCT 00144 and *Aspergillus ochraceus* MCT 00435 which are known to be potential producers of ochratoxins. These pathogenic fungi were provided from the Mycological Collection of Trichocomaceae (MCT) in Manguinhos, Rio de Janeiro, Brazil. Moreover, it was compared with simvastatin (SIM) and the action of both together was tested. It was observed that SIM alone exerted an inhibitory effect against the pathogenic fungal strains. Moreover, the combination of both SIM and AgNPs exhibited an antifungal synergistic effect. Hence, it is strongly recommended to use both of AgNPs and SIM as alternatives for agrochemicals and as food preservatives. Ovais et al. (2018) reported that different fungal strains release extracellular enzymes such as acetyl xylan esterase, glucosidase, cellobiohydrolase D, and β-glucosidase which might be involved in biosynthesis of metal NPs. Studies performed by Kumar et al. (2007a, b) revealed the involvement of fungal nitrate reductase in the extracellular mycosynthesis of AgNPs. Other studies conducted by Ingle et al. (2008) and Durán et al. (2005) demonstrated the significant role of NADH-reliant reductase enzymes in the bioreduction of Ag^+ ions to their nanoparticle state Ag^0. Senapati et al. (2005) demonstrated the capability of *Fusarium oxysporum* to secrete reducing agents (i.e., reductase) for the mycosynthesis of AuNPs in addition to AgNPs. Beside the nitrate-reliant reductases, quinone shuttle aids in the extracellular synthesis of NPs. Durán et al. (2005) reported the failure of *Fusarium moniliforme* to mycosynthesize AgNPs, even during the release of reductases which indicated the reduction of Ag^+ via the involvement of conjugated oxidation-reduction reactions. Furthermore, in studies carried out by Kumar et al. (2007a) and Karbasian et al. (2008), *F. oxysporum* was employed to mycosynthesize AgNPs in the presence of a cofactor (NADPH), a stabilizing protein (phytochelatin) along with an electron carrier (4-hydroxyquinoline). *Cladosporium cladosporioides* and *Coriolus versicolor* proteins, organic acids, and polysaccharides helped effectively in the extracellular synthesis of AgNPs as revealed by Balaji et al. (2009). AgNPs were synthesized using the extract of reishi mushroom (*Ganoderma lucidum*) as demonstrated by Aygün et al. (2019). UV/Vis spectrum displayed a broad absorption peak between 400 and 460 nm which indicated the formation of AgNPs. TEM images showed the synthesis of spheroid-shaped particles with a diameter range of 15–22 nm. XRD showed that the AgNPs were *fcc* in structure. The antioxidant activity of the prepared AgNPs towards DPPH was studied.

The highest DPPH scavenging activity percentage was found to be 76.45% at 250 mg/L. Additionally, authors demonstrated that the obtained results of DNA cleavage activity indicated that the green synthesized AgNPs caused single strain DNA cleavage activity within 30 and 60 min at concentrations of 50 and 100 mg/L, respectively. Furthermore, the prepared AgNPs exhibited a strong antibacterial activity against Gram-positive (*S. aureus*, *E. hirae*, *B. cereus*) and Gram-negative (*E. coli*, *P. aeruginosa*, *L. pneumophila subsp. pneumophila*) bacteria as well as a high antifungal activity against *C. albicans*.

Gupta and Chundawat (2019) proposed a study in which the mycelial cell-free filtrate (MCFF) of *Fusarium oxysporum* mediated the biogenic synthesis of PtNPs. UV/Vis spectroscopy revealed the disappearance of H_2PtCl_6 characteristic peak at 261 nm, thus indicated the formation of PtNPs which proved to be stable for 3 months. Functional groups related to several biomolecules such as amides, long-chain fatty acids, and polysaccharides were revealed from the FTIR spectra and were assumed to be behind the capping and stabilizing of the biologically fabricated PtNPs. SEM illustrated the synthesis of spherical, cubic, and truncated triangular shaped nanosized PtNPs with an average size of 28.96 nm. The biosynthesized PtNPs were further examined for photocatalytic, biological, and antioxidant activities. PtNPs exhibited maximum degradation of methyl orange which could be very helpful in treating wastewater resulting from textile industry. The antimicrobial action of the prepared PtNPs is evaluated against *Staphylococcus aureus* (MTCC96), *Pseudomonas aeruginosa* (MTCC441), *Klebsiella pneumonia* (MTCC109), *Escherichia coli* (MTCC442), and one fungal strain *Aspergillus niger* (MTCC282). The antimicrobial potential against the tested bacterial and fungal strains was assessed via the agar well diffusion assay by measuring the ZOI. *Staphylococcus aureus* was found to be the least resistant (24 mm), followed by *Pseudomonas aeruginosa* (18 mm) and then *Klebsiella pneumoniae* (16 mm), and *Escherichia coli* (14 mm) was the most resistant. The activity of PtNPs recorded to be very low towards the tested fungus *Aspergillus niger* when compared with the other bacterial strains. Besides, DPPH method was employed to determine the antioxidant activity. The *Fusarium oxysporum*-derived PtNPs exhibited an outstanding antioxidant activity which reached 79% scavenging activity.

3.4.1 Role of Biomolecules in Mycosynthesis Mechanism of Nanoparticles

According to Gudikandula et al. (2017), several biomolecules play a significant role in the biological synthesis of nanoparticles, for instance, amino acids, enzymes/proteins, vitamins, and polysaccharides. Additionally, secondary metabolites are involved such as alkaloids, amides, amines, flavones, carbonyl groups, phenolics, and terpenoids (Klaus et al. 1999). In 2000, Newman and Kolter (2000) spotted the light upon the significant role of quinones which acts as an electron shuttle. In a study carried out by Sastry et al. (2003), authors speculated that AgNPs mycosynthesis via *Verticillium* sp. took place by means of the electrostatic trapping of Ag^+ ions by the negatively charged carboxylate groups of enzymes existed on the cell wall. Subsequently, this aided in bioreduction of silver ions. Zirconia NPs were mycosynthesized via *F. oxysporum* (Bansal et al. 2004). Authors proved that 24–28 kDa cationic proteins were responsible for the hydrolysis of ZrF_6^{2-} anions. Durán et al. (2005) suggested the mechanism of mycosynthesis of AgNPs via different strains of *F. oxysporum* and assumed that bioreduction of Ag^+ took place through the nitrate reductase enzyme. Bansal et al. (2005) investigated the mycosynthesis of silica (SiO_2 NPs) and titania (TiO_2 NPs) via the extracellular proteins of *F. oxysporum*. Authors revealed that 21 and 24 kDa molecular weight proteins had the capability to hydrolyze SiF_6^{2-} into silica NPs and TIF_6^{2-} into titania NPs. AgNPs were mycosynthesized by the white-rot fungus *Phanerochaete chrysosporium* as investigated by Vigneshwaran et al. (2006). The study spotted the light on the lead role of reducing sugars as reducing agents. Mukherjee et al. (2008) clarified that sulfur-containing amino acids are principally responsible for reducing silver ions via *Trichoderma asperellum*. The study suggested that the S-H functional group in cysteine played a lead role in bioreduction of Ag^+ by the effect of hydrogen radical and one electron from β-carbon of cysteine in the presence of NADPH reductase enzyme. Jain et al. (2011) also demonstrated the extracellular synthesis and stabilization of AgNPs by the help of 32 and 35 kDa protein molecules present in extracellular extract of the fungus *A. flavus*. The interaction between nanoparticles and protein molecules was explained due to the presence of the free amine groups or cysteine residues in proteins and negatively charged carboxylic group of the enzyme. In another study conducted by Govender et al. (2009), platinum nanoparticles (PtNPs) were successfully mycosynthesized via F. *oxysporum*. Authors clarified the role of hydrogenase enzyme in the bioreduction process. Liang et al. (2011) demonstrated the role of peptides in confining the morphological structure of ZnO NPs. It was reported that the morphology of ZnO crystal majorly depended upon the interactions between peptides and surface planes of the prepared ZnO. In studies performed by Faramarzi and Forootanfar (2011) and Durán et al. (2014), the important role of purified laccase enzyme isolated from white-rot fungi *Paraconiothyrium variabile* and *Trametes versicolor* was clearly identified in synthesizing AuNPs and AgNPs, respectively. Likewise, Sanghi et al. (2011) demonstrated the role played by laccase enzyme for the extracellular synthesis and ligninase enzyme in the synthesis of AuNPs via *P. chrysosporium*. Barwal et al. (2011) identified that carbonic anhydrase, ATPase, sedoheptulose-1,7- bisphosphatase, ferredoxin $NADP^+$ reductase, oxygen-evolving enhancer protein ribulose bisphosphate, nuclear histone proteins, and superoxide dismutase played a substantial role in mycosynthesis of AgNPs via *Chlamydomonas reinhardtii*. On the other side, Raliya et al. (2013) revealed the capability of fungi to secrete hydrolytic enzymes, for instance, cellulases, amylases, and proteases, which facilitate the synthesis of NPs. Quester et al. (2016) also assumed the role of proteins in mycosynthesis of AgNPs by *Neurospora crassa*. Similarly, Gudikandula et al. (2017) explained the lead role of α-NADPH-dependent nitrate reductase enzyme in mycosynthesis of AgNPs from white-rot fungi through the conversion of nitrate to nitrite; electrons were shuttled to the incoming Ag^+. Zielonka and Klimek-Ochab (2017) reviewed that exposure of fungal mycelia to metal solution results in releasing metabolic compounds as well as extracellular enzymes to aid in their survival. This results in subsequent reduction of toxic metal solutions into non-toxic NPs by the effect of the secreted metabolites. Basu et al. (2018) demonstrated that AuNPs were capped with 40 kDa protein. The prepared AuNPs via *Tricholoma crassum* were washed and allowed to run on SDS-PAGE along with *T. crassum* cell-free extracts. The surface bound proteins were detached from the AuNPs via boiling in SDS. A single strong band occurred at 40 kDa which was similar to a protein band present the cell filtrate. Besides, However, a faint protein band bound to AuNPs was observed at 116 kDa level.

Ahmad et al. (2003) and Basavaraja et al. (2008) speculated the promising role of NADPH-dependent reductase in bioreduction of Ag^+ in cases of *Fusarium oxysporum* and *Fusarium semitectum*. Basavaraja et al. (2008) revealed the presence of a band at 260 nm which was attributed to electronic excitations of tryptophan and tyrosine residues in the proteins as demonstrated by Eftink and Ghiron (1981). Consequently, this suggested the release of proteins within the CFF of *F. semitectum*, indicating a possible coupling of the electron shuttle along with the involvement of NADH-dependent reductase. Likewise, in 2006, authors suggested the synthesis of AgNPs by *Aspergillus fumigatus* via secretion of protein components into the medium by the fungal biomass (Bhainsa and D'souza 2006). Durán et al. (2005) confirmed the role of nitrate reductase in silver ion bioreduction and pointed out the significant role of the electron shuttle as well as quinones, 2-acetyl-3,8-dihydroxy-6-methoxy anthraquinone, or its isomers such as 2-acetyl-2,8- dihydroxy-6-methoxy anthraquinone as was detected via thin layer chromatography (TLC). This was in agreement with the findings of Newman and Kolter (2000).

A mechanism was proposed by Omran et al. (2019b) in which the mycosynthesis of Co_3O_4-NPs was elucidated (Fig. 3.2). It was verified that the MCFF had an essential role in the biotransformation of $CoSO_4$ to Co_3O_4-NPs. The increment of pH proved the role of hydroxyl groups (OH^-) in the biotransformation process. Thus, under basic conditions, the yield of Co_3O_4-NPs was highly elevated owing to the existence of extra amount of OH^- groups, via the reaction of $CoSO_4$ with the NaOH in the MCFF producing cobalt hydroxide which would be oxidized by the sodium nitrate in the MCFF by the reaction of Na^+ and NO_3^-. Accordingly, upon dehydration reaction of cobalt hydroxide and cobalt oxyhydroxide, cobalt oxide NPs were formed and then capped and stabilized by the functional groups of the MCFF, via the presence of van der Waals forces which might have held the positively charged Co_3O_4-NPs and the negatively charged groups in the molecular structure of the MCFF. Figures 3.3 and 3.4 represent an illustration of Co_3O_4-NPs and AgNPs mycosynthesis by *Aspergillus brasiliensis* ATCC 16404 (Omran et al. 2019b). Figures 3.5 and 3.6 show a schematic representation of Co_3O_4-NPs and AgNPs mycosynthesis via the MCFF of *A. brasiliensis* ATCC 16404 and *Trichoderma longibrachiatum* ATCC DSMZ 16517.

$$CoSO_4 + 2NaOH \longrightarrow Co(OH)_2 + Na_2SO_4$$

$$3Co(OH)_2 + NO_3^- \longrightarrow Co(OH)_2 + 2\,CoOOH + H_2O + NO_2$$

$$Co(OH)_2 + 2\,CoOOH \longrightarrow Co_3O_4 + 2H_2O$$

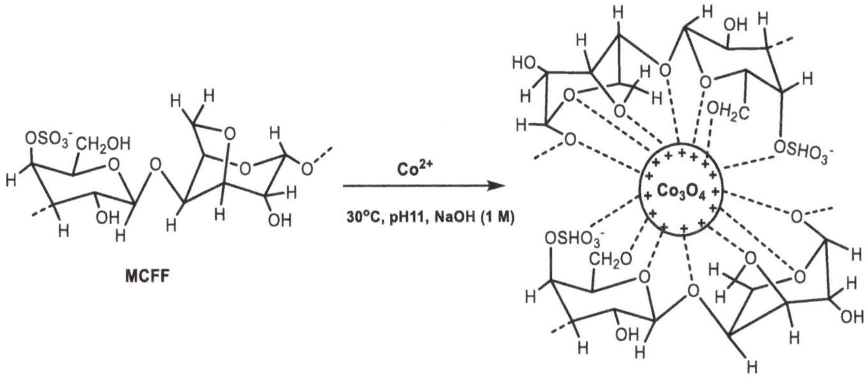

Fig. 3.3 A presumptive elucidated mechanism for the mycosynthesis of Co_3O_4-NPs by MCFF of *A. brasiliensis*

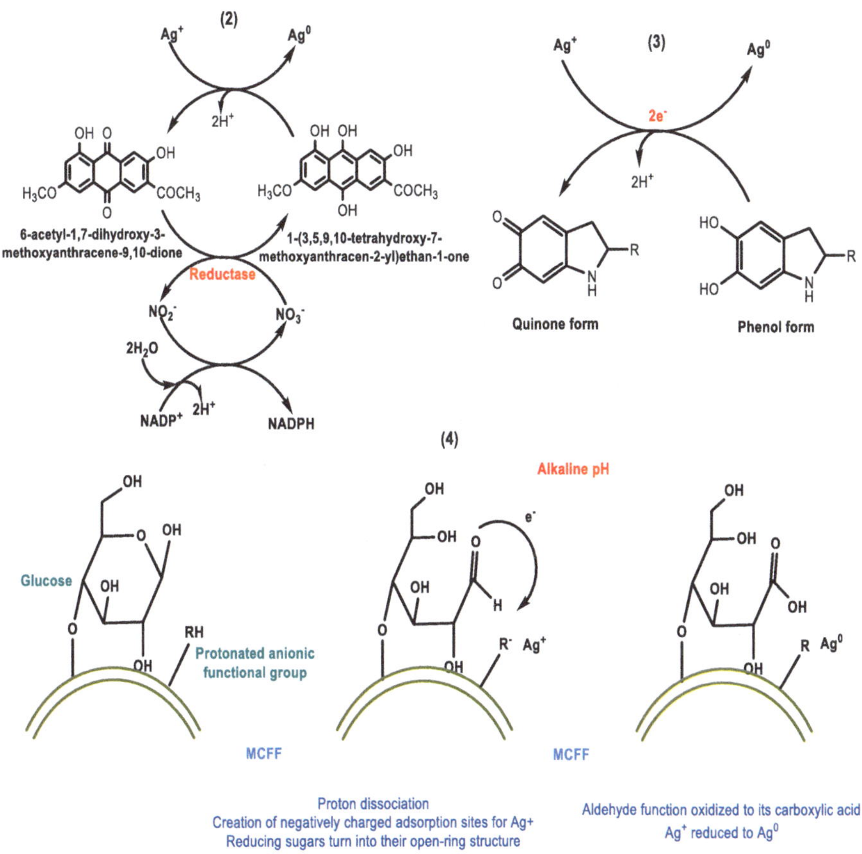

Fig. 3.4 Proposed explanation for the suggested mechanism of MCFF-mediated synthesis of AgNPs

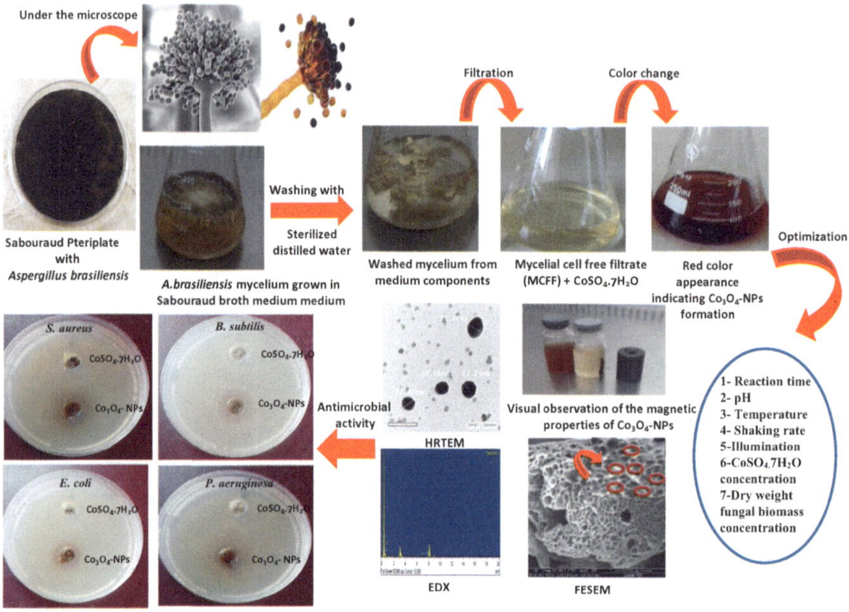

Fig. 3.5 A schematic representation showing the mycosynthesis, optimization, characterization and the different applications of Co_3O_4 NPs mycosynthesized via *A. brasiliensis* ATCC 16404 MCFF

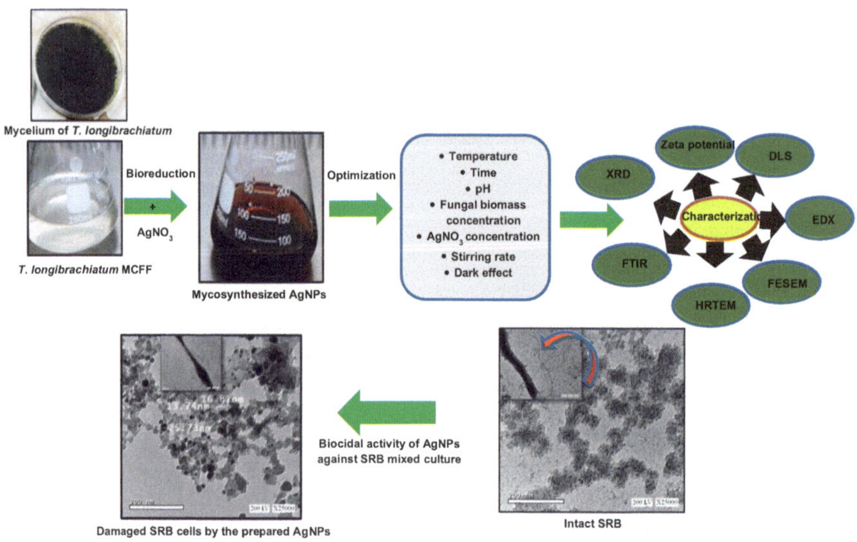

Fig. 3.6 A schematic representation showing the mycosynthesis, optimization, characterization and the different applications of AgNPs mycosynthesized via *Trichoderma longibrachiatum* ATCC DSMZ 16517

A proposed mechanism for the fungal-mediated synthesis of Co_3O_4-NPs is illustrated in Fig. 3.3 as follows: uptake of silver ions and activation of silver reduction machinery; (2) electron shuttle system involving various cofactors and enzymes; (3) electrostatic interaction between silver ions and MCFF components; and (4) reduction through extracellular enzymes and other organic molecules released in the characterization and the different applications of Co_3O_4-NPs mycosynthesized via *A. brasiliensis* ATCC 16404 MCFF.

3.5 Mycogenesis of Nanomaterials Using Yeasts

Marine microorganisms including marine originated yeasts possess the capability to live in extreme conditions; as they have the potential to synthesize several functional biomolecules (Connell et al. 2008). Their bio-products can be used in pharmaceutical industries, bioethanol production, and for industrial enzyme production (Zaky et al. 2014). The secreted biomolecules can be employed in potential applications like detoxification of metals and decreasing greenhouse gases (Zaky et al. 2014). Marine yeasts are the yeasts which can be isolated from marine environments, and seawater is preferred during the preparation of their culture medium (Chi et al. 2010). In 1894, Bernhard Fischer managed to isolate the first marine yeasts from the Atlantic Ocean. The first isolated yeasts were identified as *Torula* sp. and *Mycoderma* sp. (Kutty and Philip 2008). Following this first isolation process, different marine yeasts were isolated from different parts worldwide and were obtained from different sources like seaweeds, seawater, mammals, marine fish, as well as mammals. Some of these (Kutty and Philip 2008).

In a study conducted by Fernández and coworkers, two yeast strains referred to as *Rhodotorula glutinis* and *Cryptococcus laurentii* are were tested for their capability to mycosynthesize AgNPs (Fernández et al. 2016). These two tested yeast strains have been reported to act as biological control agents (Sansone et al. 2005, Calvo et al. 2010). Both of *C. laurentii* (BNM 0525) and *R. glutinis* (BNM 0524) were isolated from apple peel and were deposited in the National Bank of Microorganisms, Argentina. Both yeast strains are commonly known for displaying nitrate reductase activity. In order to confirm their nitrate activity, the enzyme was assayed in the yeast culture supernatants based on the methodology provided by Harley (1993) and Saifuddin et al. (2009). High values of nitrate reductase activity were attained by the two yeast strains and were 266.56 ± 13.11 nmol h^{-1} mL^{-1} and 216.85 ± 11 nmol h^{-1} mL^{-1} for *C. laurentii* and *R. glutinis*, respectively. Based on the above-mentioned result, the two yeasts were selected for synthesizing NPs. Incubation of $AgNO_3$ with the yeast supernatant (cell-free filtrate) resulted in color change from pale yellow to dark brown color. The production of AgNPs was performed extracellularly via using the culture supernatants of the two yeast strains. Verification of AgNPs biosynthesis was confirmed by the appearance of the distinctive silver SPR peak.

DLS was employed to determine the particle size distribution. In the case of AgNPs derived from *C. laurentii*, two different sizes were detected 400 nm with a percentage of 13% and 35 nm with a percentage of 74%. On the other side, in the case of *R. glutinis*, the size distribution was 15 nm, 160 nm, and 220 nm with percentages of 65%, 17%, and 18%, respectively. Fluorescence microscopy was employed to determine the fluorescence emission spectra of the AgNPs biosynthesized by *R. glutinis* and *C. laurentii*. An emission band centered between 345 and 355 nm was noticed. This result indicated the presence of proteins within the supernatant. Similar results were attained by Durán et al. (2005). The crystalline nature was further confirmed via XRD. FTIR spectrum of the biosynthesized AgNPs from *C. laurentii* displayed a pattern of peaks which were somehow similar to that of *R. glutinis* with few differences. The biosynthesized AgNPs were tested for their antifungal activity against pathogenic fungi which cause rotting to pears, apples, and grapes thus shortening their lifetime in storage and limit their marketing. Among these fungi are *Botrytis cinerea*, *Penicillium expansum*, *Aspergillus niger*, *Alternaria* sp., and *Rhizopus* sp. It was observed that AgNPs derived from *R. glutinis* exhibited a high antifungal activity rather than *C. laurentii*, and both were incredibly more effective when compared with the chemically synthesized AgNPs. Besides, authors demonstrated that at a concentration of 3 ppm, the AgNPs derived from *R. glutinis* was very much similar in efficiency to iprodione (a common fungicide used to combat postharvest diseases caused by phytopathogenic fungi and is usually employed at a concentration of 500 ppm) against all the tested fungi with the exception of *Rhizopus* sp. Authors recommended that further studies are in need to investigate the applicability to incorporate AgNPs in the packaging materials of such fruits for long preservation time.

Jalal et al. (2018) reported one pot, extracellular biosynthesis of AgNPs via the supernatant of *Candida glabrata* which was isolated from oropharyngeal mucosa of human immunodeficiency virus (HIV) patients. After the addition of 1 mM of $AgNO_3$ to *C. glabrata* supernatant, a brown color was observed. UV/Vis spectrophotometer showed the presence of a sharp peak at 460 nm. Hydroxyl groups derived from carbohydrates and amides arising from proteins and enzymes were observed via FTIR. Similar observations were noticed by Bhat et al. (2015). TEM micrographs showed the biosynthesis of uniform-, oval-, and spherical-shaped AgNPs with a size ranging between 5 and 12 nm. Antimicrobial potential of *C. glabrata* was tested against some bacterial strains like *S. aureus*, *E. coli*, *S. typhimurium*, *S. flexneri*, *K. pneumoniae*, *P. aeruginosa*, and some *Candida* spp. such as *C. albicans*, *C. dubliniensis*, *C. parapsilosis*, *C. tropicalis*, *C. krusei*, and *C. glabrata* using agar well diffusion assay. The prepared AgNPs exhibited higher ZOI against the bacterial species rather than the yeast strains. Minimum inhibitory concentration (MIC), Minimum bactericidal concentration (MBC), and minimal fungicidal concentration (MFC) were determined. The MIC and MBC values against the bacterial strains ranged from 31.25–125 μg/ml to 62.5–250 μg/ml, respectively. The MIC and MFC values were in the range of 62.5–250 μg/ml and 125–500 μg/ml, respectively. Jalal and coauthors attributed the difference in the antimicrobial potential against bacteria and yeast strains of the prepared AgNPs to the difference in cell structure of both of them. According to Panáček et al. (2009), bacterial cell possesses a simpler structure compared with the eukaryotic yeast cells whose structures are more organized and possess a superior detoxification system. Moreover, authors tracked the alteration in the morphological and ultrastructure exerted by AgNPs upon *C. albicans* as an example. This was elucidated via TEM. TEM analysis clearly demonstrated that the untreated *C. albicans* was normal with a well-conserved cell wall which was significantly made up of an outer layer followed by an intermediate space then a thin innermost layer of the cell membrane. Contrary, after being treated with AgNPs, apparent changes in the morphological structure took place. Severe pits or pores were formed causing damage to cell wall and cell membrane. General deformation took place to *Candida* cells upon exposure to AgNPs, and cells were extensively shrunken. Similar observations were noticed by several authors (Kim et al. 2009; Nasrollahi et al. 2011; Hwang et al. 2012; Vazquez-Muñoz et al. 2014; Ishida et al. 2014).

Since many years ago, selenium sulfide was used in treating skin diseases. Additionally, sulfur is an active moiety which exerts a fungicidal effect along with its characteristic therapeutic features. When selenium combines with sulfur, they both act as an effective activator with high therapeutic activity. Different diseases are treated with selenium sulfide including folliculitis, dermatitis, and pityriasis which are caused by fungal infections (Mitchell et al. 1993). Asghari-Paskiabi et al. (2019) introduced a study in which selenium sulfide (SeS) was biologically generated by *Saccharomyces cerevisiae. S. cerevisiae* Persian Type Culture Collection (PTCC) 5052 was provided from the Iranian Research Organization for Science and Technology. It was noticed that after the addition of selenium salt (at a concentration of 1 mM) to a 24 old *S. cerevisiae* culture, a change in color from yellow to brownish red was observed indicating the biosynthesis of SeS NPs. Micrographs of TEM showed the clear presence of the prepared particles within the yeast cells. Authors clarified the role of sulfite and sulfate in facilitating the movement of the metal ion precursors towards the sulfur cycle. Additionally, they were reduced by the activity of sulfite reductase. Besides, glutathione pathway was activated by the interaction of the yeast cells with selenium ions as a defense mechanism to resist toxicity. Crystallinity was improved by applying heat during purification as the crystalline phase is greatly influenced by thermal conditions. Antifungal activity of SeS NPs was tested against *Aspergillus flavus* (PTCC 5004), *Aspergillus fumigatus* (Af239), *Alternaria alternata* (PFCC 116), *Fusarium oxysporum* (PFCC 38e115), *Trichophyton rubrum* (PFCC 1762), *Microsporum canis* (PTCC 5069), *Candida albicans* (ATCC 10231), *Candida glabrata* (ATCC 90030), and *Candida krusei* (PTCC 5295). It was found that the prepared NPs inhibited the growth of *A. flavus*, *A.* alternata, A. *fumigatus*, *M. canis*, *T. rubrum*, and *C. krusei.*

Polluted and contaminated water is of a great concern all over the world (Thomas et al. 2016). Approximately 10% of dyes and synthetic colorants used in textile industries are directly released into the environment and surrounding water bodies. Dyes are toxic and contaminate water where they are released into; in addition they are carcinogenic and mutagenic to humans and marine organisms (Jadhav et al. 2010). Moreover, they cause intestinal cancer and influences seed germination of plants. Blue 71 dye is an azo dye which is widely used in textile industry as a coloring agent and displays a bright blue color (Kirankumar and Sumathi 2017). Palladium nanoparticles were successfully synthesized via *Saccharomyces cerevisiae* as reported by Sriramulu and Sumathi (2018). Dried *S. cerevisiae* was purchased from Vellore, India. Five grams of dry yeast were added to 100 ml of water and stirred for 30 min at room temperature. Following this, 10 ml of the aqueous solution was taken and added to 90 ml of 1 mM of palladium acetate solution ($C_4H_6O_4Pd$), and the whole mixture was then kept at room temperature overnight. A color change was noticed from colorless to brown color. The biosynthesized PdNPs were characterized via UV/Vis spectroscopy. Biosynthesis of PdNPs was confirmed by the disappearance of the 405 nm peak which is characteristic to palladium acetate solution. This was in agreement with Sheny et al. (2012) and Veisi et al. (2017). FTIR spectrum revealed the occurrence of different bands which corresponded to functional groups that aided in the bioreduction of Pd (II) to PdNPs. Bands assigned to phenolic and hydroxyl groups were detected. Particles were found to exhibit hexagonal shape with small size as revealed by SEM. The photocatalytic degradation of PdNPs was determined against blue 71 dye under UV light. Blue 71 dye shows a characteristic peak at 591 nm. Degradation of the dye was confirmed by the decrease in the peak intensity with 98% within 60 min.

References

Abdeen M, Sabry S, Ghozlan H, et al (2016) Microbial-physical synthesis of fe and Fe_3O_4 magnetic nanoparticles using *Aspergillus Nige*r YESM1 and supercritical condition of ethanol. J Nanomater 2016: 9174891, 7 p. https://doi.org/10.1155/2016/9174891

Ahluwalia SS, Goyal D (2007) Microbial and plant derived biomass for removal of heavy metals from wastewater. Bioresour Technol 98:2243–2257

Ahmad A, Mukherjee P, Senapati S et al (2003) Extracellular biosynthesis of silver nanoparticles using the fungus *Fusarium oxysporum*. Colloids Surf B Biointerfaces 28:313–318

Asghari-Paskiabi F, Imani M, Rafii-Tabar H, Razzaghi-Abyaneh M (2019) Physicochemical properties, antifungal activity and cytotoxicity of selenium sulfide nanoparticles green synthesized by *Saccharomyces cerevisiae*. Biochem Biophys Res Commun 516:1078–1084

Aygün A, Özdemir S, Gülcan M et al (2019) Synthesis and characterization of Reishi mushroom-mediated green synthesis of silver nanoparticles for the biochemical applications. J Pharm Biomed Anal 178:1129170–1129177

Azmath P, Baker S, Rakshith D, Satish S (2016) Mycosynthesis of silver nanoparticles bearing antibacterial activity. Saudi Pharm J 24:140–146

Balaji DS, Basavaraja S, Deshpande R et al (2009) Extracellular biosynthesis of functionalized silver nanoparticles by strains of *Cladosporium cladosporioides* fungus. Colloids Surf B Biointerfaces 68:88–92

Balakumaran MD, Ramachandran R, Kalaichelvan PT (2015) Exploitation of endophytic fungus, *Guignardia mangiferae* for extracellular synthesis of silver nanoparticles and their in vitro biological activities. Microbiol Res 178:9–17

Balakumaran MD, Ramachandran R, Balashanmugam P et al (2016) Mycosynthesis of silver and gold nanoparticles: optimization, characterization and antimicrobial activity against human pathogens. Microbiol Res 182:8–20

Bansal V, Rautaray D, Ahmad A, Sastry M (2004) Biosynthesis of zirconia nanoparticles using the fungus *Fusarium oxysporum*. J Mater Chem 14:3303–3305

Bansal V, Rautaray D, Bharde A et al (2005) Fungus-mediated biosynthesis of silica and titania particles. J Mater Chem 15:2583–2589

Barwal I, Ranjan P, Kateriya S, Yadav SC (2011) Cellular oxido-reductive proteins of *Chlamydomonas reinhardtii* control the biosynthesis of silver nanoparticles. J Nanobiotechnol 9:56

Basavaraja S, Balaji SD, Lagashetty A et al (2008) Extracellular biosynthesis of silver nanoparticles using the fungus *Fusarium semitectum*. Mater Res Bull 43:1164–1170

Basu A, Ray S, Chowdhury S et al (2018) Evaluating the antimicrobial, apoptotic, and cancer cell gene delivery properties of protein-capped gold nanoparticles synthesized from the edible mycorrhizal fungus *Tricholoma crassum*. Nanoscale Res Lett 13:154–170

Benelli G, Caselli A, Canale A (2016) Nanoparticles for mosquito control: challenges and constraints. J King Saud Univ Sci 29:424–435

Bhainsa KC, D'souza SF (2006) Extracellular biosynthesis of silver nanoparticles using the fungus Aspergillus fumigatus. Colloids Surf B Biointerfaces 47:160–164

Bhambure R, Bule M, Shaligram N et al (2009) Extracellular biosynthesis of gold nanoparticles using *Aspergillus niger* – its characterization and stability. Chem Eng Technol 32:1036–1041

Bhat MA, Nayak BK, Nanda A (2015) Evaluation of bactericidal activity of biologically synthesised silver nanoparticles from *Candida albicans* in combination with ciprofloxacin. Mater Today Proc 2:4395–4401

Bocate KP, Reis GF, de Souza PC et al (2019) Antifungal activity of silver nanoparticles and simvastatin against toxigenic species of *Aspergillus*. Int J Food Microbiol 291:79–86

Calvo J, Calvente V, Orellano ME et al (2010) Control of *Penicillium expansum* and *Botrytis cinerea* on apple fruit by mixtures of bacteria and yeast. Food Bioprocess Technol 3:644–650

Chandrappa CP, Govindappa M, Chandrasekar N et al (2016) Endophytic synthesis of silver chloride nanoparticles from *Penicillium* sp. of *Calophyllum apetalum*. Adv Nat Sci Nanosci Nanotechnol 7:025106–025111

Chhipa H (2019) Mycosynthesis of nanoparticles for smart agricultural practice: a green and eco-friendly approach. Elsevier

Chi Z-M, Liu G, Zhao S et al (2010) Marine yeasts as biocontrol agents and producers of bio-products. Appl Microbiol Biotechnol 86:1227–1241

Cole LA (2016) Biology of life: biochemistry, physiology and philosophy. Biol Life 93–99

Connell L, Redman R, Craig S et al (2008) Diversity of soil yeasts isolated from South Victoria land, Antarctica. Microb Ecol 56:448–459

Cuevas R, Durán N, Diez MC et al (2015) Extracellular biosynthesis of copper and copper oxide nanoparticles by *Stereum hirsutum*, a native white-rot fungus from chilean forests. J Nanomater

Durán N, Marcato PD, Alves OL et al (2005) Mechanistic aspects of biosynthesis of silver nanoparticles by several *Fusarium oxysporum* strains. J Nanobiotechnol 3:8–15. https://doi.org/10.1186/1477-3155-3-8

Durán N, Cuevas R, Cordi L et al (2014) Biogenic silver nanoparticles associated with silver chloride nanoparticles (Ag@AgCl) produced by laccase from Trametes versicolor. Springerplus 3:645–652

Eftink MR, Ghiron CA (1981) Fluorescence quenching studies with proteins. Anal Biochem 114:199–227

Elamawi RM, Al-Harbi RE, Hendi AA (2018) Biosynthesis and characterization of silver nanoparticles using *Trichoderma longibrachiatum* and their effect on phytopathogenic fungi. Egypt J Biol Pest Control 28:1–11

El-Baz AF, El-Batal AI, Abomosalam FM et al (2016) Extracellular biosynthesis of anti-*Candida* silver nanoparticles using *Monascus purpureus*. J Basic Microbiol 56:531–540

El-Sayyad AIEGS (2019) *Penicillium chrysogenum* -mediated mycogenic synthesis of copper oxide nanoparticles using gamma rays for in vitro antimicrobial activity against some plant pathogens. J Clust Sci 3:1–12

Faramarzi MA, Forootanfar H (2011) Biosynthesis and characterization of gold nanoparticles produced by laccase from *Paraconiothyrium variabile*. Colloids Surf B Biointerfaces 87:23–27

Farhadi S, Safabakhsh J, Zaringhadam P (2013) Synthesis, characterization, and investigation of optical and magnetic properties of cobalt oxide (Co_3O_4) nanoparticles. J Nanostruct Chem 3:69–78

Fatima F, Bajpai P, Pathak N et al (2015) Antimicrobial and immunomodulatory efficacy of extracellularly synthesized silver and gold nanoparticles by a novel phosphate solubilizing fungus Bipolaris tetramera. BMC Microbiol 15:52–62

Fatima F, Verma SR, Pathak N, Bajpai P (2016) Extracellular mycosynthesis of silver nanoparticles and their microbicidal activity. J Glob Antimicrob Resist 7:88–92

Fernández JG, Fernández-Baldo MA, Berni E et al (2016) Production of silver nanoparticles using yeasts and evaluation of their antifungal activity against phytopathogenic fungi. Process Biochem 51:1306–1313

Gajbhiye M, Kesharwani J, Ingle A et al (2009) Fungus-mediated synthesis of silver nanoparticles and their activity against pathogenic fungi in combination with fluconazole. Nanomed Nanotechnol Biol Med 5:382–386

Gole A, Dash C, Ramakrishnan V et al (2001) Pepsin– gold colloid conjugates: preparation, characterization, and enzymatic activity. Langmuir 17:1674–1679

Govender Y, Riddin T, Gericke M, Whiteley CG (2009) Bioreduction of platinum salts into nanoparticles: a mechanistic perspective. Biotechnol Lett 31:95–100

Gudikandula K, Vadapally P, Singara Charya MA (2017) Biogenic synthesis of silver nanoparticles from white rot fungi: their characterization and antibacterial studies. OpenNano 2:64–78

Gupta S, Bector S (2013) Biosynthesis of extracellular and intracellular gold nanoparticles by *Aspergillus fumigatus* and *A. flavus*. Antonie Van Leeuwenhoek 103:1113–1123

Gupta K, Chundawat TS (2019) Bio-inspired synthesis of platinum nanoparticles from fungus *Fusarium oxysporum*: its characteristics, potential antimicrobial, antioxidant and photocatalytic activities. Mater Res Express 6:1050d6–1050d18

Gupta IR, Anderson AJ, Rai M (2015) Toxicity of fungal-generated silver nanoparticles to soil-inhabiting *Pseudomonas putida* KT2440, a rhizospheric bacterium responsible for plant protection and bioremediation. J Hazard Mater 286:48–54

Harley SM (1993) Use of a simple, colorimetric assay to demonstrate conditions for induction of nitrate reductase in plants. Am Biol Teach 55:162–164

Hulikere MM, Joshi CG (2019) Characterization, antioxidant and antimicrobial activity of silver nanoparticles synthesized using marine endophytic fungus-Cladosporium cladosporioides. Process Biochem 82:199–204

Hussain SM, Hess KL, Gearhart JM et al (2005) In vitro toxicity of nanoparticles in BRL 3A rat liver cells. Toxicol In Vitro 19:975–983

Hwang IS, Lee J, Hwang JH et al (2012) Silver nanoparticles induce apoptotic cell death in *Candida albicans* through the increase of hydroxyl radicals. FEBS J 279:1327–1338

Ingle A, Gade A, Pierrat S et al (2008) Mycosynthesis of silver nanoparticles using the fungus *Fusarium acuminatum* and its activity against some human pathogenic bacteria. Curr Nanosci 4:141–144

Ishida K, Cipriano TF, Rocha GM et al (2014) Silver nanoparticle production by the fungus *Fusarium oxysporum*: nanoparticle characterisation and analysis of antifungal activity against pathogenic yeasts. Mem Inst Oswaldo Cruz 109:220–228

Jadhav JP, Phugare SS, Dhanve RS, Jadhav SB (2010) Rapid biodegradation and decolorization of direct Orange 39 (Orange TGLL) by an isolated bacterium *Pseudomonas aeruginosa* strain BCH. Biodegradation 21:453–463

Jain N, Bhargava A, Majumdar S et al (2011) Extracellular biosynthesis and characterization of silver nanoparticles using *Aspergillus flavus* NJP08: a mechanism perspective. Nanoscale 3:635–641

Jalal M, Ansari MA, Alzohairy MA et al (2018) Biosynthesis of silver nanoparticles from oropharyngeal *Candida glabrata* isolates and their antimicrobial activity against clinical strains of bacteria and fungi. Nanomaterials 8:586–598

Joshi CG, Danagoudar A, Poyya J et al (2017) Biogenic synthesis of gold nanoparticles by marine endophytic fungus-*Cladosporium cladosporioides* isolated from seaweed and evaluation of their antioxidant and antimicrobial properties. Process Biochem 63:137–144

Kadam VV, Ettiyappan JP, Mohan Balakrishnan R (2019) Mechanistic insight into the endophytic fungus mediated synthesis of protein capped ZnO nanoparticles. Mater Sci Eng B Solid State Mater Adv Technol 243:214–221

Kalpana D, Lee YS (2013) Synthesis and characterization of bactericidal silver nanoparticles using cultural filtrate of simulated microgravity grown *Klebsiella pneumoniae*. Enzym Microb Technol 52:151–156

Karbasian M, Atyabi SM, Siadat SD et al (2008) Optimizing nano-silver formation by *Fusarium oxysporum* PTCC 5115 employing response surface methodology. Am J Agric Biol Sci 3:44–437

Kathiresan K, Manivannan S, Nabeel MA, Dhivya B (2009) Studies on silver nanoparticles synthesized by a marine fungus, *Penicillium fellutanum* isolated from coastal mangrove sediment. Colloids Surf B Biointerfaces 71:133–137

Khalil NM, Abd El-Ghany MN, Rodríguez-Couto S (2019) Antifungal and anti-mycotoxin efficacy of biogenic silver nanoparticles produced by *Fusarium chlamydosporum* and *Penicillium chrysogenum* at non-cytotoxic doses. Chemosphere 218:477–486

Khandel P, Shahi SK (2018) Mycogenic nanoparticles and their bio-prospective applications: current status and future challenges. J Nanostructure Chem 8:369–391

Kim KJ, Sung WS, Suh BK et al (2009) Antifungal activity and mode of action of silver nanoparticles on Candida albicans. Biometals 22:235–242

Kirankumar VS, Sumathi S (2017) Structural, optical, magnetic and photocatalytic properties of bismuth doped copper aluminate nanoparticles. Mater Chem Phys 197:17–26

Klaus T, Granqvist CG, Joerger R et al (1999) Silver-based crystalline nanoparticles, microbially fabricated. Proc Natl Acad Sci 96:13611–13614

Kumar SA, Abyaneh MK, Gosavi SW et al (2007a) Nitrate reductase-mediated synthesis of silver nanoparticles from $AgNO_3$. Biotechnology letters, 29:439–445

Kumar SA, Ansary AA, Ahmad A et al (2007b) Extracellular biosynthesis of CdSe quantum dots by the fungus, *Fusarium oxysporum*. J Biomed Nanotechnol 3:190–194

Kutty SN, Philip R (2008) Marine yeasts-a review. Yeast 25:465–483

Liang MK, Deschaume O, Patwardhan SV et al (2011) Direct evidence of ZnO morphology modification via the selective adsorption of ZnO-binding peptides. J Mater Chem 21: 80–89

Loshchinina EA, Vetchinkina EP, Kupryashina MA et al (2018) Nanoparticles synthesis by Agaricus soil basidiomycetes. J Biosci Bioeng 126:44–52

Madakka M, Jayaraju N, Rajesh N (2018) Mycosynthesis of silver nanoparticles and their characterization. MethodsX 5:20–29

Mani U, Dhanasingh S, Arunachalam R et al (2013) A simple and green method for the synthesis of silver nanoparticles using Ricinus communis leaf extract. Prog Nanotechnol Nanomater 2:21–25

Mitchell SC, Nickson RM, Waring RH (1993) The biological activity of selenium sulfide. Sulfur Rep 13:279–289

Mukherjee P, Roy M, Mandal BP et al (2008) Green synthesis of highly stabilized nanocrystalline silver particles by a non-pathogenic and agriculturally important fungus *T. asperellum*. Nanotechnology 19:075103–075110

Nagajyothi PC, Sreekanth TVM, Lee JIL, Lee KD (2014) Mycosynthesis: antibacterial, antioxidant and antiproliferative activities of silver nanoparticles synthesized from *Inonotus obliquus* (Chaga mushroom) extract. J Photochem Photobiol B Biol 130:299–304

Nasrollahi A, Pourshamsian KH, Mansourkiaee P (2011) Antifungal activity of silver nanoparticles on some of fungi. Int J Nano Dimens 1:233–239

Nayantara, Kaur P (2018) Biosynthesis of nanoparticles using eco-friendly factories and their role in plant pathogenicity: a review. Biotechnol Res Innov 2:63–73

Neethu S, Midhun SJ, Radhakrishnan EK, Jyothis M (2018) Green synthesized silver nanoparticles by marine endophytic fungus *Penicillium polonicum* and its antibacterial efficacy against biofilm forming, multidrug-resistant *Acinetobacter baumanii*. Microb Pathog 116:263–272
Newman DK, Kolter R (2000) A role for excreted quinones in extracellular electron transfer. Nature 405:94–97
Omran BA, Nassar HN, Fatthallah NA et al (2018) Characterization and antimicrobial activity of silver nanoparticles mycosynthesized by *Aspergillus brasiliensis*. J Appl Microbiol 125:370–382
Omran BA, Nassar HN, Younis SA et al (2019a) Physiochemical properties of *Trichoderma longibrachiatum* DSMZ 16517-synthesized silver nanoparticles for the mitigation of halotolerant sulphate-reducing bacteria. J Appl Microbiol 126:138–154
Omran BA, Nassar HN, Younis SA et al (2019b) Novel mycosynthesis of cobalt oxide nanoparticles using *Aspergillus brasiliensis* ATCC 16404: optimization, characterization and antimicrobial activity. J Appl Microbiol 128:438–457
Ovais M, Khalil AT, Islam NU et al (2018) Role of plant phytochemicals and microbial enzymes in biosynthesis of metallic nanoparticles. Appl Microbiol Biotechnol 102:6799–6814
Owaid MN, Raman J, Lakshmanan H et al (2015) Mycosynthesis of silver nanoparticles by *Pleurotus cornucopiae* var. citrinopileatus and its inhibitory effects against *Candida* sp. Mater Lett 153:186–190
Panáček A, Kolář M, Večeřová R et al (2009) Antifungal activity of silver nanoparticles against *Candida* spp. Biomaterials 30:6333–6340
Priyadarshini E, Pradhan N, Sukla LB, Panda PK (2014) Controlled synthesis of gold nanoparticles using *Aspergillus terreus* IF0 and its antibacterial potential against gram negative pathogenic bacteria. J Nanotechnol 2014: 653198, 9 p. https://doi.org/10.1155/2014/653198
Quester K, Avalos Borja M, Castro Longoria E (2016) Controllable biosynthesis of small silver nanoparticles using fungal extract J Biomater Nanobiotechnol 7:118–125
Rai A, Prabhune A, Perry CC (2010) Antibiotic mediated synthesis of gold nanoparticles with potent antimicrobial activity and their application in antimicrobial coatings. J Mater Chem 20:6789–6798
Raliya R, Rathore I, Tarafdar JC (2013) Development of microbial nanofactory for zinc, magnesium, and titanium nanoparticles production using soil fungi. J Bionanosci 7:590–596
Raman J, Reddy GR, Lakshmanan H et al (2015) Mycosynthesis and characterization of silver nanoparticles from *Pleurotus djamor* var. roseus and their in vitro cytotoxicity effect on PC3 cells. Process Biochem 50:140–147
Rauwel P, Küünal S, Ferdov S, Rauwel E (2015) A review on the green synthesis of silver nanoparticles and their morphologies studied via TEM. Adv Mater Sci Eng 2015: 682749, 9 p. https://doi.org/10.1155/2015/682749
Rodrigues AG, Ping LY, Marcato PD et al (2013) Biogenic antimicrobial silver nanoparticles produced by fungi. Appl Microbiol Biotechnol 97:775–782
Różalska S, Pawłowska J, Wrzosek M et al (2013) Utilization of 4-n-nonylphenol by *Metarhizium* sp. isolates. Acta Biochim Pol 60:677–682
Różalska S, Soboń A, Pawłowska J et al (2015) Biodegradation of nonylphenol by a novel entomopathogenic *Metarhizium robertsii* strain. Bioresour Technol 191:166–172
Różalska S, Soliwoda K, Długoński J (2016) Synthesis of silver nanoparticles from *Metarhizium robertsii* waste biomass extract after nonylphenol degradation, and their antimicrobial and catalytic potential. RSC Adv 6:21475–21485
Saifuddin N, Wong CW, Yasumira AA (2009) Rapid biosynthesis of silver nanoparticles using culture supernatant of bacteria with microwave irradiation. J Chem 6:61–70
Samson RA (2016) Cellular constitution, water and nutritional needs, and secondary metabolites. In: Environmental mycology in public health. Elsevier, pp 5–15
Sanghi R, Verma P, Puri S (2011) Enzymatic formation of gold nanoparticles using *Phanerochaete chrysosporium*. Adv Chem Eng Sci 1:154–162

Sanna V, Pala N, Dessì G et al (2014) Single-step green synthesis and characterization of gold- conjugated polyphenol nanoparticles with antioxidant and biological activities. Int J Nanomedicine 9:4935–4951

Sansone G, Rezza I, Calvente V et al (2005) Control of *Botrytis cinerea* strains resistant to iprodione in apple with rhodotorulic acid and yeasts. Postharvest Biol Technol 35:245–251

Sastry M, Ahmad A, Khan MI, Kumar R (2003) Biosynthesis of metal nanoparticles using fungi and actinomycete. Curr Sci 85:162–170

Seetharaman PK, Chandrasekaran R, Gnanasekar S et al (2018) Antimicrobial and larvicidal activity of eco-friendly silver nanoparticles synthesized from endophytic fungi *Phomopsis liquidambaris*. Biocatal Agric Biotechnol 16:22–30

Senapati S, Ahmad A, Khan MI et al (2005) Extracellular biosynthesis of bimetallic Au–Ag alloy nanoparticles. Small 1:517–520

Sheny DS, Philip D, Mathew J (2012) Rapid green synthesis of palladium nanoparticles using the dried leaf of *Anacardium occidentale*. Spectrochim Acta Part A Mol Biomol Spectrosc 91:35–38

Sreedharan SM, Gupta S, Saxena AK, Singh R (2019) *Macrophomina phaseolina* : microbased biorefinery for gold nanoparticle production. Ann Microbiol 69:435–445

Sriramulu M, Sumathi S (2018) Biosynthesis of palladium nanoparticles using *Saccharomyces cerevisiae* extract and its photocatalytic degradation behaviour. Adv Nat Sci Nanosci Nanotechnol 9:025018–025024

Tamboli DP, Kagalkar AN, Jadhav MU et al (2010) Production of polyhydroxyhexadecanoic acid by using waste biomass of *Sphingobacterium* sp. ATM generated after degradation of textile dye direct red 5B. Bioresour Technol 101:2421–2427

Tan GSE, Hsu LY (2018) Overview of fungal infections. Drugs 1:1–12

Thomas M, Naikoo GA, Sheikh MUD et al (2016) Effective photocatalytic degradation of Congo red dye using alginate/carboxymethyl cellulose/TiO_2 nanocomposite hydrogel under direct sunlight irradiation. J Photochem Photobiol A Chem 327:33–43

Vazquez-Muñoz R, Avalos-Borja M, Castro-Longoria E (2014) Ultrastructural analysis of Candida albicans when exposed to silver nanoparticles. PLoS One 9:e108876–e108886

Veisi H, Hemmati S, Qomi M (2017) Aerobic oxidation of benzyl alcohols through biosynthesized palladium nanoparticles mediated by oak fruit bark extract as an efficient heterogeneous nanocatalyst. Tetrahedron Lett 58:4191–4196

Vigneshwaran N, Kathe AA, Varadarajan PV et al (2006) Biomimetics of silver nanoparticles by white rot fungus, *Phaenerochaete chrysosporium*. Colloids Surf B Biointerfaces 53:55–59

Vijayanandan AS, Balakrishnan RM (2018) Biosynthesis of cobalt oxide nanoparticles using endophytic fungus *Aspergillus nidulans*. J Environ Manag 218:442–450

Warnock DW (2012) Fungi: superficial, subcutaneous and systemic mycoses. In: Medical microbiology. Elsevier, pp 616–641

Yosri M, Abdel-Aziz MM, Sayed RM (2018) Larvicidal potential of irradiated myco-insecticide from *Metarhizium anisopliae* and larvicidal synergistic effect with its mycosynthesized titanium nanoparticles (TiNPs). J Radiat Res Appl Sci 11:328–334

Zaky AS, Tucker GA, Daw ZY, Du C (2014) Marine yeast isolation and industrial application. FEMS Yeast Res 14:813–825

Zielonka A, Klimek-Ochab M (2017) Fungal synthesis of size-defined nanoparticles. Adv Nat Sci Nanosci Nanotechnol 8:043001–043010

Zielonka A, Żymańczyk-Duda E, Brzezińska-Rodak M et al (2018) Nanosilica synthesis mediated by *Aspergillus parasiticus* strain. Fungal Biol 122:333–344

Zomorodian K, Pourshahid S, Sadatsharifi A, et al (2016) Biosynthesis and characterization of silver nanoparticles by *Aspergillus* species. BioMed Res Int 2016: 5435397, 6 p. https://doi.org/10.1155/2016/5435397

Chapter 4
Current Trends in Algae-Mediated Synthesis of Metal and Metal Oxide Nanoparticles (Phyconanotechnology)

4.1 Introduction

Algae have become substantial organisms in diverse biological applications involving waste water purification as they are capable of accumulating plant nutrients, radioactive materials, heavy metals, pesticides, and organic and inorganic highly toxic chemicals (e.g., selenium, arsenic, and zinc) in their cells (Alp et al. 2012; Labib et al. 2015). Furthermore, algal biomass can be used for compost manufacturing, methane production, and animal feed in addition to production of liquid fuels and fine chemicals (Agarwal et al. 2019). Many algae can tolerate, take up, accumulate metals and bioreduce metal ions into nanoparticles. Synergy between science and industry has opened the gates for innovative frontiers in the ever-growing domain of nanobiotechnology which aims at genesis, implementation, and use of nanomaterials (NMs) which are biologically originated (Khanna et al. 2019). The main aspect of nanobiotechnology is the biogenesis of nanoscale particles by virtue of biological moieties (El-Gendy and Omran 2019). Several methodologies have been adopted to biosynthesize metallic nanoparticles using algae from their corresponding aqueous salt solutions (LewisOscar et al. 2016). Agarwal et al. (2019) elucidated four routes to biosynthesize nanoparticles from microalgae: (i) direct exploitation of extracted biomolecules from disrupted cells, (ii) exploitation of cell-free supernatants made of microalgae culture media, (iii) biosynthesis of nanoparticles of different natures from whole cells of microalgae, and (iv) using living cells of microalgae. This chapter summarizes previous and recent studies of algae-mediated synthesis of metal and metal oxide nanoparticles (micro- and macroalgae). Additionally, it provides a highlight upon the mechanism of synthesis and the possible applications.

B. A. Omran, *Nanobiotechnology: A Multidisciplinary Field of Science*,
Nanotechnology in the Life Sciences, https://doi.org/10.1007/978-3-030-46071-6_4

4.2 Diversity of Algae

Algae are divided into either micro- or macroalgae. Phytoplankton which usually survive suspended in the water column are referred to as microalgae (Fawcett et al. 2017). Microalgae are unicellular, colony-forming, and filamentous photosynthetic microorganisms (Dahoumane et al. 2016). All microalgae, apart from prokaryotic cyanobacteria, possess a nucleus which is surrounded by a membrane, and thus they belong to eukaryotes. Microalgae constitute an important substantial part planet's biodiversity. According to certain estimations, there are hundreds of thousands to millions of microalgal species, most of which have not been isolated nor characterized (Metting 1996; Sheehan et al. 1998). Microalgae inhabit either aquatic or semi-aquatic environments like lakes, seas, swamps, and rivers. Microalgae are emerging as substantial and cost-effective cell factories to generate commercial products like carotenoids and fatty acids (Borowitzka 2013). Algae also have a great ability to convert lipids into biofuels (Borowitzka 2013). Macroalgae sometimes referred to as seaweeds. They are a miscellaneous group of marine, multicellular, photosynthetic, chlorophyll-containing, eukaryotic organisms (Levine 2016). They lack the presence of true roots, stems, and leaves, and they multiply via simple reproductive structures. They usually exist in intertidal zone up to 30 m deep. Nearly, 10,000 marine macroalgal species have been divided based upon their content of photosynthetic pigments, cell wall components, construction and/or orientation of flagella (Levine 2016). They are plant-like organisms which range in size from few centimeters up to several meters in length (Fawcett et al. 2017).

Algae are basically classified into three groups based on either the algal body or thallus pigmentation. They are brown algae (phaeophytes), green algae (chlorophytes), and red algae (rhodophytes) (Kim 2011). Usually, brown and red algae are marine based. Certain species of red algae are found in water depths where light levels are very low. Contrary, green algae exist in both marine and freshwater environments.

4.3 Biogenesis of Nanomaterials Using Algae (Phyconanotechnology)

One of the shining and escalating scientific disciplines in biological synthesis of NPs is "phyconanotechnology" (Narayanan and Sakthivel 2011). Phyconanotechnology refers to the science in which (micro- and macro-) algae are exploited as bio-factories for synthesis of NPs (Shankar et al. 2016). Among the different reported biological entities, algae possess unique advantages owing to their highly elevated capacity of metal uptake, low cost, and macroscopic structures (Davis et al. 2003). Fawcett et al. (2017) demonstrated the great capability of algae to accumulate heavy metal ions in addition to their outstanding ability to modify them into more malleable formulae. Consequently, algae have been looking for as excellent candidates for synthesizing biologically derived nanomaterials (Khanna et al. 2019). Numerous reports proved the presence of high amounts of biologically active compounds and secondary metabolites within algae (Chanda et al. 2010; Mohamed et al. 2012). These algal biologically active components were proved to possess medicinal properties which can be employed in conventional medical treatments as well as alternative therapies (Miyashita 2009; Mohamed et al. 2012; Wada et al. 2011). Moreover, it was found that they have inhibitory (Nishino et al. 1999), anti-inflammatory (Khan et al. 2008) potential as well as the capability to suppress some kinds of cancer (Namvar et al. 2012).

Two main synthesis methodologies are usually employed to implement synthesis of metallic NPs using microalgae and the corresponding metal salt solution. So far, researchers depend on either the intracellular "takes place inside the cell" or the extracellular "takes place in the cell free filtrate with no involvement of cells but with direct interaction with the extracted biomolecules" (Dahoumane et al. 2016). The biosynthesis of nanomaterials using microalgae is categorized into four main routes; the first route resides in exploiting biomolecules extracted from microalgal disrupted cells; the second route involves the use of supernatant which is made of culture media in which the cells were removed but without being exposed to treatments other than filtration or centrifugation; the third route relies on harvesting whole microalgal cells and then resuspending them in distilled water; and the last route depends on the use of living microalgal cells with maintenance of their ordinary culturing conditions (Dahoumane et al. 2016). In 2007, AuNPs and AgNPs were successfully biosynthesized by a Singaporean team who managed to extract biomolecules from *Chlorella vulgaris*, a unicellular green microalga (Xie et al. 2007a). First, in the case of AuNPs, *C. vulgaris* biomass was subjected to lyophilization and afterwards was subjected to reverse-phase high-performance liquid chromatography (RP-HPLC), till isolating the protein responsible for shaping the desired type of NPs. Isolated protein which was referred to as gold shaping protein (GSP) was then reacted with 1 mM $HAuCl_4$, and this interaction yielded gold nanoplates. The produced gold nanoplates exhibited different shapes, i.e., triangles, truncated triangles and hexagons, and spherical shapes. In case of AgNPs, a dialysis membrane was employed to extract low and high molecular weight proteins that existed within *C. vulgaris* biomass. It was observed that fabrication of AgNPs was possibly attained by the effect of proteins with high molecular weight only. Different shapes were obtained like nanospheres, nanodisks, and triangular nanoplates. Afterwards, several microalgal species were investigated to biosynthesize nanomaterials via using their biomolecules. AgNPs were biosynthesized via the cell free supernatant as demonstrated by Patel and colleagues using several species of cyanobacteria and Chlorophyta (Patel et al. 2015). Though, authors did not introduce data regarding the characteristic features of the prepared AgNPs nor the nature of the involved biomolecules, the production of AgNPs took place only when the reaction mixture was exposed to light. Consequently, it can be concluded that the biochemical pathways in this case were light-driven. Table 4.1 describes a representative list of algae used in biosynthesis of NPs.

Table 4.1 A representative list of algae-mediated synthesis of metal and metal oxide nanoparticles

Algae	Algal source	Type of produced NPs	Size (nm)	Shape	Characterization techniques	Application	Reference
Cyanobacteria (prokaryotic blue-green algae)							
Calothrix pulvinata, strain ALCP 745A, *Anabaena flos-aquae*, strain ALCP B24, *Leptolyngbya foveolarum*, strain ALCP 671B	Black soiling developing on a building near Genève (Switzerland), MNHN Culture Collection, Black soiling developing on a wall in Strasbourg (France)	Au	5.4 ± 0.6, 12 ± 0.7, 6 ± 0.6	Spherical	UV/visible spectrophotometry, TEM, SEM, EDX, XRD, XPS	–	Brayner et al. (2007)
Plectonema boryanum UTEX 485	Culture collection at the University of Texas at Austin, Texas	Ag	1–15	Spherical	TEM, EDX, SAED	–	Lengke et al. (2007)
Spirulina platensis	Freshwater lake of Vellore in South India	Ag, Au, Au core-Ag shell NPs	7–16, 6–10, 17–25	Spherical	HRTEM, XRD, FTIR	–	Govindaraju et al. (2008)
Lyngbya majuscula, *Spirulina subsalsa*	National Facility for Marine Cyanobacteria, Tiruchirapalli, Tamil Nadu, India	Au	<20 nm	Spherical	TEM	–	Chakraborty et al. (2009)
Oscillatoria willei NTDM01	Kurusadai Island, Gulf of Mannar, Tamil Nadu, India	Ag	100–200	Spherical	FTIR, SEM, EDX		Mubarak Ali et al. (2011)

Lyngbya majuscula, *Spirulina subsalsa*	National Facility for Marine Cyanobacteria, Tiruchirapalli, Tamilnadu, India.	Au	*L. majuscula* (5–30), *S. subsalsa* (25)	*L. majuscula* (spherical, nanorods), *S. subsalsa* (spherical, hexagonal)	UV/Visible spectrophotometry, DLS, TEM, XRD, FTIR	–	Parial and Pal (2011)
Phormidium sp.	Germplasm of Department of Microbiology, Bharathidasan University, India	Au	25	Triangular	UV/Visible spectrophotometry, FTIR, TEM, SEM, EDX	Antioxidant activity, biolabeling, medical applications	MubarakAli et al. (2013)
Cylindrospermum stagnale NCCU-104	Germplasm Collection Centre of India, New Delhi, India	Ag	48	Cubic	UV/Visible spectrophotometry, SEM	–	(Husain et al. (2015)
Spirulina platensis	Jal Mahal, Jaipur, Rajasthan, India	Ag	30–50	Spherical	UV/Visible spectrophotometry, SEM, TEM, FTIR	Antibacterial activity	Sharma et al. (2015b)
Leptolyngbya tenuis, *Coleofasciculus chthonoplastes*, and *Nostoc ellipsosporum*	*L. tenuis* and *C. chthonoplastes* (National Facility for Marine Cyanobacteria (NFMC)), Tiruchirappalli, Tamil Nadu, India, *N. ellipsosporum* (National Institute of Cyanobacteria, IARI, PUSA, New Delhi, India)	Au	10–40, 10–30, 8–42	Spherical, spherical, spherical and oval	HRTEM, EDX, FTIR-ATR, FESEM, Zeta potential	Antibacterial activity	Parial et al. (2016)

(continued)

Table 4.1 (continued)

Algae	Algal source	Type of produced NPs	Size (nm)	Shape	Characterization techniques	Application	Reference
Anabaena cylindrica (SAG 1403.2)	–	Au	10	Spherical	XRD, TEM, laser-induced breakdown spectroscopy (LIBS)	–	Rösken et al. (2016)
Calothrix sp.	Universidad de las Fuerzas Armadas ESPE, Sangolqui, Ecuador	Au	30–120	Truncated shape	UV/Visible spectrophotometry, DLS, TEM, XRD	Catalytic activity	Kumar et al. (2016)
Nostoc sp. *strain* HKAR-2	Hot spring of Rajgir, India	Ag	51–100	Spherical-shaped flower-like structure	UV/Visible spectrophotometry, FTIR, XRD, DLS, zeta potential SAED, TEM, SEM	Antitumor, antibacterial, antifungal	(Sonker et al. (2017)
Microchaete NCCU-342	Centre for Conservation and Utilization of Blue Green Algae, Indian Agricultural Research Institute, New Delhi	Ag	40–200	Spherical	UV/Vis spectrophotometry, TEM, DLS	Catalytic activity	Husain et al. (2019)
Leptolyngbya strain JSC-1	–	Ag	50–100	Spherical	Scanning transmission electron microscopy (STEM), SEM, EDX, XPS	–	Khan et al. (2019)
Oscillatoria limnetica	Algal culture collection of phycology lab, faculty of science, Mansoura University	Ag	3.30–17.97	Quasi-spherical	UV/Vis spectrophotometry, FTIR, SEM, TEM	Antibacterial, anticancer and hemolytic activities	Hamouda et al. (2019)

Anabaena cylindrica	Biotechnology Institute	CuO	40–60	Rod	Zetasizer, XRD, FTIR, FESEM, TEM, XPS, BET by Quantachrome surface area analyzer	Antimicrobial activity, cellular toxicity, micronuclei assay on fish blood	Bhattacharya et al. (2019)
Microalgae and macroalgae (AgNPs)							
Sargassum wightii Greville	Mandapam Camp, Southeast Coast, Tamil Nadu, India	Ag	8–27	Spherical	UV/Vis spectrophotometry, FTIR, XRD, HRTEM	Antibacterial activity	Govindaraju et al. (2009)
Porphyra vietnamensis	–	Ag	13 ± 3	Spherical	UV/Visible spectrophotometry, FTIR, XRD, HRTEM, Zeta potential and particle size analyzer	Antibacterial activity	Venkatpurwar and Pokharkar (2011)
Chaetoceros calcitran, Chlorella salina, Isochrysis galbana, Tetraselmis gracilis	Central Marine Fisheries Research Institute, Tuticorin	Ag	53.1–71.9	Spherical	UV/Visible spectrophotometry, SEM	Antimicrobial activity	Merin et al. (2010)
Chlamydomonas reinhardtii	Chlamydomonas Center, Duke University, USA	Ag	5–35	Rounded, rectangular	UV/Visible spectrophotometry, TEM, SEM, ICP-MS, SDS-PAGE	–	Barwal et al. (2011)
Padina tetrastromatica	Mandapam coastal region, Gulf of Mannar, Tamil Nadu, South India	Ag	20	Spherical	UV/Visible spectrophotometry, XRD, SEM, TEM, EDX, FTIR	–	Jegadeeswaran et al. (2012)
Cystophora moniliformis	Glenelg, Adelaide, South Australia	Ag	50–100	–	UV/Visible spectrophotometry, SEM, EDX, XRD, DLS, Zeta potential	–	Prasad et al. (2013)

(continued)

Table 4.1 (continued)

Algae	Algal source	Type of produced NPs	Size (nm)	Shape	Characterization techniques	Application	Reference
Chaetomorpha linum	Intertidal region of Kanyakumari coast, southeast coast of India	Ag	3–44	Cubic	UV/Vis spectrophotometry, FTIR, SEM	–	Kannan et al. (2013a)
Codium capitatum	Intertidal region, Rocky Bay, east coast of South Africa	Ag	~30	Spherical	UV/Vis spectrophotometry, EDX, FTIR	–	Chlorophyceae et al. (2013)
Scenedesmus sp.	Culture Collection Centre, CSIR-Institute of Minerals and Materials Technology, India	Ag	15–20	Spherical	UV/Vis spectrophotometry, TEM, XRD, FTIR, TGA, DLS, atomic absorption spectroscopy (AAS), SAED	Antimicrobial activity	Jena et al. (2014)
Euglena gracilis, *Euglena intermedia*	Institute of hydrobiology, Chinese Academy of Sciences	Ag	*E. gracilis* (6–24), *E. intermedia* 15–60	Spherical	UV/Vis spectrophotometry, TEM, ICP-AES, FTIR, TEM	–	Li et al. (2014)
Caulerpa racemosa	Gulf of Mannar, southeast coast of India	Ag	5–25	Triangular	UV/Vis spectrophotometry, FTIR, XRD, TEM	Antibacterial activity	Kathiraven et al. (2015)
Ulva lactuca	Coastal areas of Rameshwaram, Southern India	Ag	20–35	Cubic	UV/Vis spectrophotometry, FTIR, EDX, XRD, SEM	Larvicidal activity	Murugan et al. (2015)

Pithophora oedogonia	Freshwater pond of town Hooghly, West Bengal, India	Ag	34	Cubic, hexagonal	UV/Vis spectrophotometry, SEM, XRD, EDX, DLS	Antibacterial activity	Sinha et al. (2015)
Spirogyra varians	Sweet water areas of Kerman, Iran	Ag	17.6	Quasi-spheres	UV/Vis spectrophotometry, XRD, SEM, FTIR	Antibacterial activity	Salari et al. (2016)
Caulerpa racemosa	Mandapam (Tamil Nadu, India) coastal area	Ag	25	Spherical	UV/Vis spectrophotometry, FTIR, XRD, EDX, HRTEM	Catalytic degradation activity	Nesakumar et al. (2016)
Acutodesmus dimorphus	–	Ag	2–20	Spherical	UV/Vis spectrophotometry, FTIR, SEM, TEM, AFM, EDX	Antioxidant potential	Chokshi et al. (2016)
Laminaria japonica	Seaweed industry, Korea	Ag	31	Spherical, oval	UV/Vis spectrophotometry, Zeta sizer, HRTEM, DLS, XRD	Effect on seed germination	Kim et al. (2018)
Neochloris oleoabundans	Culture Collection of Algae, University of Texas, Austin.	Ag	16.63	Spherical	UV/Vis spectrophotometry, TEM	–	Bao and Lan (2018)
Gelidium amansii	Coastal belt of South Korea	Ag	27–54	Spherical	UV/Vis spectrophotometry, FTIR, SEM	Antimicrobial activity	Pugazhendhi et al. (2018)

(continued)

Table 4.1 (continued)

Algae	Algal source	Type of produced NPs	Size (nm)	Shape	Characterization techniques	Application	Reference
Chlamydomonas reinhardtii	*Chlamydomonas* Resource Center, University of Minnesota, St. Paul, MN, USA	Ag	19.2 ± 5.0 (whole cultures), 3.5 ± 0.6 nm and 17.4 ± 2.6 nm (EPS)	Spherical	TEM, SAED, FTIR	–	Rahman et al. (2019)
Microalgae and macroalgae (AuNPs)							
Sargassum wightii	Mandapam Camp Southeast coast of Tamil Nadu, India	Au	8–12	Spherical	UV/Vis spectrophotometry, XRD, TEM	–	Singaravelu et al. (2007)
Tetraselmis suecica	–	Au	51–120	Spherical	UV/Vis spectrophotometry, XRD, TEM, FTIR		Shakibaie et al. (2010)
Stoechospermum marginatum	Tuticorin coast, Tamil Nadu, India	Au	18.7–93.7	Spherical, hexagonal, triangular	Photoluminescence spectroscopy, UV/Vis spectrophotometry, FTIR, SEM, TEM, XRD, EDX, particle size analyzer	Antibacterial activity	Rajathi et al. (2012)
Tetraselmis kochinensis	–	Au	5–35		UV/Vis spectrophotometry, XRD, TEM, DLS	–	Senapati et al. (2012)

Turbinaria conoides	Gulf of Mannar, South India	Au	6–10	Spherical, triangular, pseudo-spherical-	UV/Vis spectrophotometry, TEM, XRD, FTIR, TEM, SAED	–	Rajeshkumar et al. (2013a)
Chondrus crispus	Algamar plant, Redondela (Pontevedra, Spain)	Au, Ag	~30	AuNPs (triangular, hexagonal, spherical) AgNPs (spheres)	UV/Vis spectrophotometry, FTIR, TEM, SEM	–	Castro et al. (2013)
Ecklonia cava	Busan, South Korea	Au	30 ± 0.25	Triangular, spherical	UV/Vis spectrophotometry, FTIR, XRD, Zeta potential	Antimicrobial activity, biocompatibility action	Venkatesan et al. (2014)
Rhizoclonium fontinale	Eastern part of India (Sundarbans, West Bengal)	Au	13–22	Spherical	UV/Vis spectrophotometry, DLS, HRTEM, EDX, ESEM		Parial and Pal (2015)
Chlorella vulgaris	Algal Culture Collection, University of Madras, Chennai, India	Au	2–10	Self-assembled spheres	UV/Vis spectrophotometry, FTIR, XRD, TEM	Anti-pathogenic activity	Annamalai and Nallamuthu (2015)
Turbinaria ornata	Beaches of Mandapam, region of Tamil Nadu, India	Au	7–11	Spherical	UV/Vis spectrophotometry, FTIR, XED, EDX, SEM	–	Ashokkumar and Vijayaraghavan (2016)
Sargassum tenerrimum, Turbinaria conoides	South coast of Tamil Nadu, India	Au	27–35	Spherical	UV/Vis spectrophotometry, DLS, HRTEM, Zeta potential, FTIR, DLS	Catalytic activity	Ramakrishna et al. (2016b)

(continued)

Table 4.1 (continued)

Algae	Algal source	Type of produced NPs	Size (nm)	Shape	Characterization techniques	Application	Reference
Galaxaura elongata	Northwest coast of Al-Haraa, Umluj City, Red Seashore, Kingdom of Saudi Arabia	Au	3.85–77.13	Spherical, rod, triangular, truncated triangular and hexagonal	FTIR, TEM, Zeta potential	Antibacterial activity	Abdel-raouf (2017)
Dunaliella salina		Au	22.4	Spherical	TEM, SAED, XRD	Anticancer activity	Singh et al. (2019)
Microalgae and macroalgae (other metal and metal oxide nanoparticles)							
Sargassum muticum	Coastal areas Of Persian Gulf Waters	Fe_3O_4	18 ± 4	Cubic	XRD, FESEM, FTIR, energy dispersive X-ray fluorescence spectroscopy (EDXRF), vibrating sample magnetometry (VSM)		Mahdavi et al. (2013)
Sargassum plagiophyllum	Coast of Mandapam Rameswaram, Tamil Nadu, India	AgCl	18–42	Spherical	UV/Vis spectrophotometry, FTIR, XRD, HRTEM, FESEM	Antibacterial activity	Dhas et al. (2014)
Kappaphycus alvarezii	Sabah, Malaysia	Cu@ Cu_2O core shell	53	Quasi-spheres	UV/Vis spectrophotometry, XRD, FTIR, TEM, AFM	–	Khanehzaei et al. (2014)
Bifurcaria bifurcata	Morocco's Atlantic Coast in Rose Marie near Rabat	CuO	5–45	Spherical, elongated	UV/Vis spectrophotometry, FTIR, XRD	Antibacterial activity	Abboud et al. (2014)

Sargassum muticum	–	ZnO	30–57	Hexagonal	UV/Vis spectrophotometry, FTIR, FESEM	–	Azizi et al. (2015)
Sargassum bovinum	Bushehr coast of the Persian Gulf (Southwestern Iran)	Pd	5–10	Octahedral	UV/Vis spectrophotometry, EDX, FTIR, TEM, XRD	Electrochemical reduction of hydrogen peroxide (H2O2)	Momeni and Nabipour (2015)
Chlorella vulgaris	Culture collection of Algae, University of Texas (UTEX), Austin, USA	AgCl	9.8 ± 5.7	Spherical	UV/Vis spectrophotometry, EDX, XRD, TEM	Antimicrobial activity	da Silva Ferreira et al. (2017)
Botryococcus braunii	Udaisagar Lake, Udaipur (Rajasthan, India)	Cu, Ag	Ag (40–100) Cu (10–70)	Ag (cubic, spherical, and truncated triangular), Cu (cubic, spherical)	UV/Vis spectrophotometry, FTIR, XED, SEM	Antimicrobial potency	Arya et al. (2018)
Colpomenia sinuosa, *Pterocladia capillacea*	Coastal area of Abu Qir Gulf	Fe_3O_4	*C. sinuosa* (11.24–33.71), *P. capillacea* (16.85–22.47)	–	UV/Vis spectrophotometry, EDX, FTIR, SEM	Antibacterial potency	Salem et al. (2019)

4.3.1 Biosynthesis of Nanoparticles Using Cyanobacteria

Cyanobacteria are prokaryotic blue-green algae. They possess unique characters such as high yield of biomass; inhabit diverse environments including polluted and contaminated ones, as well as their ability to carry out oxygenic photosynthesis. They release beneficial byproducts and aid in generating biofuel, enhancing soil fertility, and reducing greenhouse gas emissions (Pasula and Lim 2017). Cyanobacterial surface is made up of complex structures assembled into two distinct layers; each layer contains unique molecular functional groups and specific metal-binding properties. For more details, the external cyanobacterial cell wall in cyanobacteria is enclosed with S-layers which consist of glycoproteins, beneath which the outer cell membrane exists formed of the peptidoglycan layer, and then followed by the cytoplasmic membrane (Hoiczyk and Hansel 2000). The presence of negatively charged extracellular polymeric substances (EPS) in addition to the negatively charged groups on cell membrane facilitates the rapid binding of metals onto the cell surface of the cyanobacterial cell. Interestingly, EPS regulate the needed balance of metal requirements and the probability of toxic overexposure through the uptake and trafficking (Hudek et al. 2012). Yee et al. (2004) reported that *Calothrix*, a filamentous cyanobacterium, distributes metal-binding sites all over its surface. Yee and co-authors illustrated that proton-active surface phosphoryl, carboxyl, hydroxyl, and amine functional groups present on the cell wall in addition to the exopolymer sheath interact with the surrounding metal ions (Yee et al. 2004). Cyanobacteria synthesize NPs both intracellularly and extracellularly. Yet, few reports regarding the biological synthesis of noble metal NPs by cyanobacteria exist (El-Naggar et al. 2017). Cyanobacterial strains, namely, *Leptolyngbya*, *Anabaena* and *Calothrix* were reported by Brayner et al. (2007) to mediate the biological synthesis of well-confined size of Au, Ag, Pd, and Pt NPs. These NPs were synthesized intracellularly. The presence of algal bioactive molecules contributed to their stabilization. It was found that the size of the recovered particles and the reaction yield were dependent upon the used cyanobacterial genus. Studies revealed the involvement of the intracellular enzyme "nitrogenase" within the process of metal bioreduction and subsequent colloidal growth of the prepared NPs. Chakraborty et al. (2009) elucidated the capability of *Spirulina* to biosynthesize AuNPs, AgNPs, and Au/Ag bimetallic NPs. *Nostoc linckia* was used to synthesize selenium NPs from selenite, both extracellularly and intracellularly with a size range from 10 to 80 nm (Cepoi et al. 2014). Bioreduction of selenite to elemental selenium was attained via the influence of reactive thiol groups of proteins/peptides. Such reaction is referred to as "painter-type" reaction (Zinicovscaia et al. 2016).

In a study performed by Lengke et al. (2006a), the cyanobacterium *Plectonema boryanum* UTEX 485 was obtained from the culture collection at the University of Texas, Austin, USA. The interaction between the tested cyanobacterium strain and $AuCl_3$ resulted in the precipitation of AuNPs. Octahedral platelets were deposited on the cell wall of *P. boryanum*. X-ray absorption spectroscopy (XAS) illustrated the synthesis of Au (I) species. In another study performed by Lengke et al. (2006b) investigated the interaction between *Plectonema boryanum* UTEX 485 with aqueous $PtCl_4$. The interaction prompted the subsequent precipitation of amorphous and spherical-shaped Pt (II) NPs. Screening the biological potential of three cyanobacteria strains to mediate synthesis of NPs was introduced by Parial et al. (2012a). The tested cyanobacterial strains were *Phormidium valderianum*, *P. tenue*, and *Microcoleus chthonoplastes*. All strains were exposed to $HAuCl_4$ solution and were screened for their capability to biosynthesize AuNPs. All three cyanobacterial genera were capable of synthesizing AuNPs intracellularly. This was indicated via the purple coloration within 72 h of treatment at 20 °C. The obtained AuNPs were examined by UV/Vis spectroscopy, TEM, and XRD. *P. valderianum*-derived AuNPs exhibited various shapes including spherical, triangular, and hexagonal shapes at basic and neutral pH (pH 9 and pH 7, respectively). TEM revealed the synthesis of spherical and irregular shapes of AuNPs by *P. tenue*.

AuNPs exhibiting a rod shape possess distinctive optical features which can be extremely useful during the manufacture of bio-imaging tools as well as in biosensors. Moreover, several publications illustrated their use as hyperthermia and tumor destructive agents (Huang et al. 2006; Hauck et al. 2008). Gold nanorods possess a strong SPR, and hence they produce scattered heat and light; thus they are extremely useful as therapeutic agents for imaging and other diagnostic targets (Pissuwan et al. 2008). According to Hauck et al. (2008), the efficiency of gold nanorods to kill tumor cells exceeded the therapeutic agent cisplatin and combined therapy of both cisplatin and gold nanorods exhibited a killing efficiency of 78% of cancer cells. In another study performed by Parial and colleagues, the cyanobacterium *Nostoc ellipsosporum* was found to possess the power to biosynthesize gold nanorods intracellularly (Parial et al. 2012b). The Au nanorods were biosynthesized within the cell after being exposed to 15 mg L^{-1} gold (III) solution at pH 4.5 for 48 h at 20 °C. The AuNPs were extracted via sodium citrate solution and were analyzed using UV/Vis spectroscopy. Characteristic surface-multiple plasmon peaks at 560, 610, and 670 nm were detected. The nature and size of the particles were determined via TEM, DLS, XRD, and zeta potential studies. The nanorod size ranged from 137 to 209 nm in length and 33 to 69 nm in diameter. The average hydrodynamic size was 435 nm as revealed by DLS. XRD showed the reduction of Au^{3+} to Au^0. Suganya et al. (2015) demonstrated the biosynthesis of AuNPs via the protein extract of *platensis*. *S. platensis* was isolated from a culture pond at Thandalam, near Sriperumbudur, Tamil Nadu, India. Synthesis of AuNPs was evident by the appearance of ruby red color after 48 h of reaction time. Three distinctive peaks of AuNPs were noticed at 685 nm, 524 nm, and 385 nm using *S. platensis* protein, and maximum excitation was observed at 620 nm. The size and shape of the prepared AuNPs were determined via HRTEM. The prepared particles were monodispersed and exhibited a spherical shape with a size ranging from 2 to 8 nm. EDX spectrum revealed strong signals of gold along with other elements like carbon, oxygen, potassium, and sodium. FTIR confirmed the protein involvement in the bioreduction, capping, and stabilization of the biosynthesized particles. The *S. platensis*-derived AuNPs were investigated for their antibacterial potential against Gram-positive bacteria *B. subtilis* and *S. aureus* using colony-forming unit (CFU). Additionally, TEM was employed to study the morphological changes exerted by the prepared AuNPs upon the bacterial cells. It was evident from the TEM micrographs that AuNPs caused damage to cells via disrupting the bacterial cell wall causing pits at the exterior of cell in addition to cell membrane deformation.

Bakir et al. (2018) investigated for the first time the potentiality of a cyanobacterial strain isolated from the Arabian Gulf to mediate production of NPs. *Lyngbya majuscula* was isolated from the AlOqair area, Al-Ahsa Government, Eastern Province, Kingdom of Saudi Arabia. The cyanobacterium was initially incubated with 1500 mg/mL of $HAuCl_4$ for 2 days. The blue-green strain changed into purple, which designated the intracellular biosynthesis of AuNPs. Additionally, the prolonged incubation for more than 2 months induced the extracellular production of AuNPs. UV/visible spectroscopy revealed an absorption spectrum peak at ~535 nm, whereas SEM revealed the size of the prepared particles with an average diameter of 41.7 ± 0.2 nm. Authors further investigated the antioxidant and anti-myocardial infarction activities of the cyanobacterial extract, colloidal solution of AuNPs, and a combination of both in animal models. First, isoproterenol (100 mg/kg, SC (subcutaneous)) was injected into experimental rats subcutaneously for 3 days to prompt the state of myocardial infarction. Afterwards, animals were given cyanobacterial extract (200 mg/kg/day, IP (intraperitoneal)), AuNPs (200 mg/kg/day, IP), and then an oral mixture of both for 14 days. Different indicators of myocardial infraction were determined including; blood pressure, cardiac biomarkers, electrocardiogram (ECG) and antioxidant enzymes. It was found that isoproterenol elevated ST and QT segments as well as a notable increase was observed in heart rate and serum activities of creatine kinase-myocardial bound (CP-MB), creatine phosphokinase (CPK), and cardiac troponin T (cTnT). Additionally, it reduced the content of glutathione peroxidase (GRx), superoxide dismutase (SOD), diastolic arterial pressure (DAP), arterial pressure (DAP), and mean arterial pressure (MAP) in heart tissues. Interestingly, AuNPs alone or in combination with cyanobacterial extract of *L. majuscula* exerted an inhibitory effect on isoproterenol-induced changes. Afzal et al. (2019) screened the ability of 20 cyanobacterial strains procured from CFTRI, Mysore, New Delhi, and Tiruchirapalli. Cyanobacteria-mediated synthesis of SeNPs was observed via color changes from sky blue to different degrees of orange. All the tested cyanobacterial strains were capable of synthesizing SeNPs. The prepared SeNPs were scanned via the UV/Vis spectroscopy. The λ_{max} ranged from 259 to 274 nm. Additionally, SEM revealed the formation of spherical shaped particles. The smallest size was 11.8 nm, and the biggest size was 60 nm produced by *Arthrospira indica* SOSA-4 and *Arthrospira indica* SAE-84, respectively. Selenium signals were detected using EDX at 1.4 keV in case of all the tested cyanobacterial strains.

4.3.2 Biosynthesis of Nanoparticles Using Diatoms

Diatoms constitute a major group of algae particularly microalgae. They are microscopic algae whose length ranges between 10 and 100 μm, unicellular, and they are usually distributed in marine, brackish, freshwaters, oceans, waterways, and soils

(Bock and Norris 2016). They are photosynthesizing microorganisms which belong to brown algae (division Chromophyta, class Bacillariophyceae). Diatoms possess two overlapping external cell walls, namely, valves or tests, and they are made up of silica. These valves are referred to as epitheca and hypotheca (Zachleder 1998). They form together a structure referred to as "frustule". Frustule surface is a fine structure filled with extensions, perforations, and thickenings. This silica shield has the potential to withstand severe environmental conditions including acid and heat exposure (Houck et al. 2018). Approximately 8000 species have been identified till now as reported by Bock and Norris (2016). They are responsible for nearly 20% of worldwide carbon cycle fixation (Kröger and Poulsen 2008). Diatoms are sometimes employed as abrasives, filters, and carrier materials in industries (Pytlik et al. 2017). Schröfel et al. (2011) described an innovative synthesis of AuNPs, EPS-gold, and silica-gold bio-nanocomposites via two diatom strains (*Navicula atomus* and *Diadesmis gallica*). Diatom cultures were obtained from the Culture Collection of Algology Centre, Institute of Hydrobiology, Czech Republic. The fabricated AuNPs were analyzed via UV/Vis spectroscopy, light microscopy, TEM, and SEM. The change in color was observed after 2 h of incubation time to dark reddish color. It was found that *D. gallica*-derived AuNPs exhibited a large size (22 nm) which was bigger than that derived from *N. atomus* (9 nm). Pascual García et al. (2014) used the marine diatom *Thalassiosira pseudonana* to biosynthesize AgNPs. This marine diatom is characterized by the presence of cylindrical shell made up of two valves joined together by girdle bands. The strain was purchased from Provasoli-Guillard National Center for Culture of Marine Phytoplankton, West Boothbay Harbor, Maine, USA. This study depended on using TEM along with electron and ion microscopic techniques to detect the exact location of the synthesized AgNPs. The study revealed for the first time the presence of AgNPs inside the cell membrane. Aqueous extract of diatom *Amphora* sp. was employed to biosynthesize AgNPs (Jena et al. 2015). It was found that the reaction became faster when it was exposed only to light. UV/Vis spectroscopy revealed the presence of AgNPs sharp peak at 415 nm. TEM illustrated the formation of spherical-shaped particles with an average size of 20–25 nm. Compositional elemental analysis by EDX displayed strong characteristic silver signals. The XRD affirmed the polycrystalline nature of the prepared particles. A photosynthetic pigment, namely, fucoxanthin, was the lead biomolecule behind the bioreduction of silver ions to the nanoparticle state. Additionally, *Amphora* sp.-derived AgNPs exhibited significant antimicrobial potential against Gram-positive and Gram-negative bacteria. Pytlik et al. (2017) demonstrated the biological fabrication of AuNPs using the seawater diatom *Stephanopyxis turris*. *S. turris* with a cell diameter ranging between 10 and 25 μm and a cell length ranging between 50 and 20 μm was isolated from the North Sea. The biosynthesized AuNPs were characterized via SEM, EDX, TEM, and SAED. AuNPs were biosynthesized intracellularly as revealed by 3D Raman measurement. These nanoparticles were successfully used to measure surface enhanced Raman spectroscopy (SERS) spectra.

Ammonia is a chemical agent which is broadly used in various industries such as pharmaceutical, fertilizer, explosive, plastics' manufacturing, refrigeration, synthetic fibers, and paper (Pandey et al. 2013; Cannilla et al. 2014; Edison et al. 2016; Das et al. 2017). But, unfortunately, ammonia causes irritation to skin and eyes of humans, and high concentrations might cause permanent blindness and lung disorders. Moreover, it is hazardous to aquatic species and crustaceans even at very low concentrations, and more importantly it cannot be changed into less toxic compounds. Moreover, presence of ammonia in the human body is an indicator of disorders or diseases like liver and kidney collapse, stomach infection of bacteria, cancer, ulcer, diabetes, etc. (Edison et al. 2016). So, sensing and monitoring of ammonia is an important issue. Lots of techniques have been carried out to monitor and sense ammonia presence. In a study reported by Chetia et al. (2017), AgNPs were green synthesized using the diatom species *Navicula*. *Navicula* sp. which was collected from local freshwater bodies in Assam, India. UV/Vis spectroscopy revealed the presence of a broad absorption peak at 338 nm. Additionally, an absorption peak occurred between 200 and 260 nm was related to diatom frustules. The size of the prepared AgNPs was found to be 70 nm according to SEM images. EDX showed the presence of strong silver signals around 3Kev. Average size of AgNPs was calculated via Debye Scherrer equation and was found to be 30.35 nm. FTIR analysis showed that hydroxyl groups (–OH) of fucoxanthin and carboxyl groups (–COOH) of proteins reduced Ag^+ ions to metallic AgNPs and aided in their stabilization without any agglomeration. The fabricated AgNPs were employed to sense ammonia. Sensing of ammonia was first observed by visualizing the color change via the addition of 0.5 ml of AgNPs to ammonia solution with different concentrations ranging from 0 ppm to 100 ppm. The color changed gradually from purple to light yellow.

4.3.3 Biosynthesis of Nanoparticles Using Micro- and Macroalgae

Many published studies revealed the ability of green algae to biosynthesize AgNPs (Barwal et al. 2011; Jena et al. 2013; Annamalai and Nallamuthu 2016). In a study performed by Xie et al. (2007b), crystalline Au nanoplates were biosynthesized via the algal extract of green microalgae *Chlorella vulgaris*. Proteins were suggested to play a major dual function in reducing gold ions and in controlling both the size and shape of the nanogold crystals. Reversed-phase high-performance liquid chromatography (HPLC) enabled the isolation and purification of a specific protein with a molecular weight of almost 28 kDa. The isolated protein aided in preparing gold nanoplates with particular shapes, e.g., triangular and hexagonal ones with a high yield. Same authors presented another study in which Ag nanoplates were bio-fabricated via the extract of unicellular green microalgae *Chlorella vulgaris* at room temperature (Xie et al. 2007a). Bioreduction of Ag^+ and shape-controlled synthesis of the prepared AgNPs was attained via the proteins existed within the algal extract. Authors identified the most active functional groups involved in the bioreduction process, and they were the OH groups in tyrosine residues and COOH groups in aspartate and glutamate residues. In 2011, Barwal and coworkers exploited the unicellular microalgae *Chlamydomonas reinhardtii* as a model example for producing spherical- and rectangular-shaped AgNPs at both in vivo and in vitro (Barwal et al. 2011). A comparative study was performed in both cases. In vitro, synthesis was slow and took nearly 13 days, and the formed AgNPs were in a size range of 5 ± 1 to 15 ± 2 nm. Contrary, in vivo synthesis was much faster, taking approximately 10 h, and the produced NPs were in the range of 5 ± 1 to 35 ± 5 nm. TEM micrographs illustrated the existence of the prepared particles in the peripheral cytoplasm and the basal body. Proteins and oxidative reductive machineries were involved in the biosynthesis process of AgNPs. The authors suggested the involvement of certain biomolecules in the biosynthesis of AgNPs including ATP synthase, ferredoxin $NADP^+$ reductase, Ribulose-1,5-bisphosphate (RUBP) carboxylase, sedoheptulose-1,7-bisphosphatase, superoxide dismutase, and oxygen evolving enhancer proteins. Vivek and co-authors presented a study in which AgNPs were biosynthesized using the water extract of red macroalgae *Gelidiella acerosa* (Vivek et al. 2011). UV/Vis spectroscopy, XRD, FTIR, SEM, and TEM were used to further characterize the fabricated particles. The AgNPs were rounded in shape and polydispersed. The biosynthesized AgNPs expressed an antifungal activity against *Mucor indicus* (MTCC 3318), *Humicola insolens* (MTCC 4520), *Fusarium dimerum* (MTCC 6583), and *Trichoderma reesei* (MTCC 3929). Metallic AgNPs as well as ZnO NPs were synthesized via a rapid and microwave-mediated approach as demonstrated by Priyadharshini et al. (2014) via using the extract of macro-algae *Gracilaria edulis* (GE). The formation of GE-derived AgNPs and ZnO NPs was confirmed via different characterization techniques including UV/Vis spectroscopy, EDX, FTIR, XRD, and FESEM. The prepared AgNPs were rounded in shape, while the ZnONPs were rod in shape. The fabricated particles were investigated for their anticancer activity against human PC3 and normal African monkey kidney (VERO) cell lines. AgNPs and ZnONPs exerted an inhibition effect against PC3 cells at concentrations of 39.60 and 28.55 μg/ml, respectively, whereas against PC3 cells, the inhibitory concentrations were 68.49 and 88.05 μg/ml for AgNPs and ZnONPs, respectively, after 48 h of incubation time. These obtained data strongly recommend the use of GE-derived AgNPs and ZnONPs as efficient anticancer agents against PC3 cell lines. AgNPs were successfully tailored using the microalgae *Chlorella* spp. both extracellularly and intracellularly and acted as antibacterial agents as demonstrated by Satapathy et al. (2015) and Annamalai and Nallamuthu (2016) and in wastewater treatment as reported by Karthikeyan et al. (2015) and Aziz et al. (2015). Microalgae *Euglena* spp. (*Euglena gracilis* and *Euglena intermedia*) were employed as potential sources to biosynthesize AgNPs extracellularly and intracellularly for the first time as reported by Li and coworkers (Li et al. 2014). The UV/Vis spectrum displayed a characteristic silver SPR peak at 420 nm. The uptake of silver ions was equal by both strains as revealed by inductively coupled plasma-atomic emission spectrometer. The size of fabricated AgNPs was in the range of 6–24 nm and 15–60 nm in diameter in cases of in vivo and in vitro, respectively, and a spherical-shaped polydispersed of the particles. The prepared particles were polydispersed and exhibited a spherical shape. FTIR illustrated the functional groups related to the bioactive molecules like amines which might have aided in capping and stabilizing the prepared particles. Furthermore, several other green macroalgal species were investigated to biosynthesize AgNPs with different shapes, sizes, and applications like *Nannochloropsis oculata* (Mohseniazar et al. 2011), *Chlorococcum humicola* (Jena et al. 2013), *Scenedesmus* sp. (Jena et al. 2014), *Euglena gracilis* (Li et al. 2015), etc. Additionally, extensive studies have been performed upon macroalgae and their capability to mediate AgNPs production involving *Enteromorpha compressa* (Dhanalakshmi et al. 2012), *Urospora* sp. (Suriya et al. 2012), *Codium capitatum* (Kannan et al. 2013b), *Pithophora oedogonia* (Sinha et al. 2015), and *Spirogyra varians* (Salari et al. 2016). Arsiya et al. (2017) demonstrated the capability of *Chlorella vulgaris* aqueous extract to bioreduce Pd (II) ions into palladium nanoparticles (PdNPs). Biomass of *C. vulgaris* was harvested using a high-speed centrifuge at 4000 rpm for 10 min. Then, the obtained biomass was kept in incubator at 45 °C for 12 hours, till reaching a constant weight. After drying, the sample was grinded in a mortar, and then almost 0.5 g of dry algal material was homogenized in 50 ml double-distilled water and boiled in a sand bath for 15 min. Finally, the crude extract was filtered by vacuum filtration. Afterwards, 10 ml of marine alga crude extract was mixed with 1 mM of palladium chloride ($PdCl_2$). Gradual changes in the color of the reaction mixture from yellow to brown were noticed. UV/Vis spectrophotometer showed the disappearance of Pd (II) ions characteristic peak which occurs at 410 and 420 nm. Additionally, PdNPs distinctive SPR peak was displayed at the range of 370–440 nm. FTIR showed the presence of polyol and amide functional groups within the algal extract. The fabricated PdNPs were monodispersed and spherical shaped with a size in the range of 5–20 nm. XRD confirmed the crystallinity of the biosynthesized particles. A brown macroalgae, namely, *Sargassum siliquosum* was investigated for its potential to biosynthesize AgNPs as reported by Vasquez et al. (2016). The brown seaweed *S. siliquosum* was collected from Bolo Beach, Alaminos, Pangasinan. It was observed that when the reaction mixture was kept at room temperature, poor biosynthesis took place. Whereas, synthesis of AgNPs was confirmed visually from light yellow color to a brown color when the reaction mixture was autoclaved at 121 °C at 15 psi for 15 min. This indicated that the process of AgNPs biosynthesis using polysaccharides from *S. siliquosum* required high temperature as well as high pressure. The prepared AgNPs were analyzed via UV/Vis spectroscopy, FTIR, SEM, EDX, and zeta potential. Toxicity as well as hepatoprotective activity of AgNPs were assessed. It was found that up to 2000 mg/kg BW of AgNPs did not cause any mortality to rats but caused minimum elevation of serum creatinine and blood urea nitrogen. While, low doses of AgNPs were effective in revival of liver enzymes to almost normal levels which indicated the curing effects of *S. siliquosum* derived AgNPs. Ramkumar et al. (2017) reported a simple, one pot and green synthesis of platinum nanoparticles (PtNPs) via the Indian brown seaweed aqueous extract of *Padina gymnospora*. The authors investigated the catalytic and antimicrobial, hemolytic and cytotoxic activities of PtNPs in combination with the polyvinylpyrrolidone (PVP) polymer nanocomposite. The PtNPs derived from *P. gymnospora* were characterized by various characterization techniques like UV/Vis spectrophotometry,

FTIR, XRD, EDX, DLS, zeta potential, FESEM, HRTEM, and SAED. A nanocomposite was prepared by mixing 10 mL of PtNPs (0.01%) and 10 mL of PVP solution (0.1%) to produce polymer/metal nanocomposite (PVP/PtNPs). The prepared PVP/PtNPs was tested for its antibacterial potential against different pathogenic bacterial strains. Maximum inhibition activity was attained against *Escherichia coli* (15.6 mm) followed by *Lactococcus lactis* (14.8 mm) and *Klebsiella pneumoniae* (14.4 mm). Additionally, the effect of PtNPs and PVP/PtNPs nanocomposite upon the integrity of erythrocytes was investigated by in vitro hemolysis. The hemolytic activity was noticed only at higher concentrations (600, 900, and 1200 μg mL^{-1}). Authors also investigated the cytotoxic activity of PVP/PtNPs against *Artemia salina nauplii*. The prepared nanocomposite exerted a cytotoxic effect at 100 ± 4 μg ml^{-1} (LC50). Besides, PVP/PtNPs nanocomposite exhibited an enhanced scavenging activity towards superoxide DPPH, nitric oxide, and hydroxyl radicals. Rao and Pennathur (2017) described a facile synthesis process of CdS NPs using *Chlamydomonas reinhardtii*. Electron microscopy illustrated the morphological structure of the prepared particles as spherical ones with a size of 5 nm. XRD and FTIR confirmed the presence of face centered cubic CdS NPs that were stabilized and capped with *C. reinhardtii* proteins. The photocatalytic degradation potential of the biosynthesized AgNPs against methylene blue dye under UV light was investigated and was found to degrade 90% of the dye within a reaction time of 90 min. Trapping experiments demonstrated that OH groups were the main reactive species behind dye degradation. This study represents an example in using algal cell free extract to fabricate CdS NPs to be used in photocatalytic purposes. Arularasu et al. (2018) described an innovative facile and green synthesis approach for fabricating Fe_3O_4 NPs via the red macroalgae *Kappaphycus alvarezii*. The synthesized Fe_3O_4 NPs were characterized using UV/Vis spectroscopy, XRD, EDX, FTIR, HRTEM, and high-resolution scanning electron microscope (HRSEM) to illustrate the structural, morphological, magnetic, and optical features. The biosynthesized Fe_3O_4 NPs were spherical in shape and highly agglomerated. The photocatalytic degradation capability of the synthesized Fe_3O_4 NPs was evaluated against the cationic textile dye waste (CTDW), e.g., methylene blue. TDW was degraded in presence of Fe_3O_4 NPs due to presence of hydroxyl radicals. Besides, the antibacterial activity of Fe_3O_4 NPs was evaluated against Gram-positive (*Staphylococcus aureus*) and Gram-negative (*Escherichia coli*) bacteria. It was found that the biosynthesized Fe_3O_4 NPs exhibited excellent antibacterial activity. Authors suggested that direct contact of Fe_3O_4 NPs with the tested bacterial cell walls resulted in damaging and loosening of the integrity of the bacterial cell wall along with the liberation of the destructive reactive oxygen species (ROS). Pugazhendhi et al. (2018) investigated the biosynthesis of AgNPs using the marine red macroalgae *Gelidium amansii*. In the present study, *G. amansii* dry powder was mixed with 1, 3, and 5 mM solutions of $AgNO_3$. The initial faint green color of the reaction mixture turned into dark brown. Further confirmation of AgNPs synthesis was affirmed by the excitation of SPR peak at 420 nm. The fabricated algae-mediated AgNPs remained stable for 6 months. FTIR spectrum identified the possible algal bioactive molecules which might have capped and surrounded the synthesized AgNPs. The biosynthesized

AgNPs were spherical in shape with a size ranging from 27 to 54 nm. *G. amansii*-derived AgNPs exhibited antibacterial potential against Gram-negative (*P. aeruginosa*, *V. parahaemolyticus*, *E. coli*, *A. hydrophila*) and Gram-positive (*S. aureus*, *B. pumilus*) biofilm-forming bacteria. Zirconia (ZrO_2) NPs were biosynthesized using brown marine macroalgae *Sargassum wightii* (Kumaresan et al. 2018). XRD pattern confirmed the crystalline nature of the prepared ZrO_2 NPs, and the estimated size was 4.8 nm as calculated by Debye Scherrer equation. TEM micrographs showed the presence of monodispersed and spherical-shaped particles with a size of 5 nm. A strong absorption peak at 277 nm was displayed via the UV/Vis spectrophotometer. Agar well diffusion assay was employed to assess the antibacterial activity of (ZrO_2) NPs against *Bacillus subtilis*, *Escherichia coli*, and *Salmonella typhi*. The fabricated particles exhibited strong antibacterial potential against the tested pathogenic bacteria. Kashyap et al. (2019) reported a facile and novel biosynthesis of Ag/AgCl using the alcoholic microalgal extracts of *Chlorella* sp., *Lyngbya putealis*, *Oocystis* sp., and *Scenedesmus vacuolatus*. Minhas et al. (2018) managed to prepare green synthesized AgNPs via the macroalgal extracts of *Ulva compressa* (L.) Kütz. and (L.) Kütz. Both are filamentous macroalgae which are abundantly distributed in natural water systems. A phytochemical screening analysis was performed to ascertain the active biomolecules which might be involved in biogenesis of AgNPs. It was found that the algal extract was rich with carbohydrates, reducing sugars, flavonoids, terpenoids, tannins, alkaloids, and amino acids. It was observed that upon reaction of the algal extracts with $AgNO_3$, a gradual change in the color of the reaction mixture took place from transparent to brown after 16 h. The fabricated AgNPs were characterized via UV/Vis spectroscopy which revealed the presence of one broad peak at 403 nm. FTIR spectroscopy illustrated the probable biomolecules which might have been involved in the bioreduction process. FTIR bands assigned to hydroxyl and amide groups were detected which might belong to algal polysaccharides. Additionally, bands belonging to amine groups of proteins were observed. XRD demonstrated the distinctive crystalline nature of the prepared AgNPs. The estimated particle sizes for AgNPs were 81.8 and 66.3 nm in case of *C. glomerata* and *U. compressa*, respectively. Morphology and surface roughness were examined via SEM and AFM. A composite membrane was made up of AgNPs with polysulfone (PS). The PS casting solution was synthesized via the dissolving of polymer (15%, w/w) in N-methyl-2 pyrrolidinone and then stirring at room temperature for 24 h to get a homogeneous solution. The synthesized AgNPs/PS nanocomposite membrane was examined for its antimicrobial potential using direct contact test (DCT). It is worth mentioning that the mechanism involved in DCT is mainly dependent upon the close interaction between bacterial culture and tested material for a particular incubation time, and eventually the bacterial number that remained surviving is estimated. Thus, DCT works independently without being influenced by the solubility and diffusibility issues. It was found that AgNPs/PS nanocomposite membrane suppressed the growth of five bacterial pathogens, e.g., *K. pneumoniae*, *P. aeruginosa*, *E. coli*, *E. faecium*, and *S. aureus*. *Ulva lactuca* is commonly referred to as sea lettuce and belongs to green macroalgae. Massironi et al. (2019) exploited the polysaccharides extracted from algal biomass of *U. lactuca* for the first time to biologically synthesize AgNPs in presence of sodium borohydride ($NaBH_4$) as reductant. The fabrication of *U. lactuca*-derived AgNPs was assessed via UV/Vis spectroscopy, XRD, DLS, TEM, and zeta potential. Besides, the antibacterial potential of the synthesized AgNPs was evaluated against Gram-negative bacteria, e.g., *Escherichia coli* ATCC 25922 and *Pseudomonas aeruginosa* ATCC 27853, and against Gram-positive bacteria, e.g., *Staphylococcus aureus* ATCC 33591 and *Staphylococcus epidermidis* ATCC 35984. Princy and Gopinath (2019) presented a study in which marine macroalgae *Padina tetrastromatica* aqueous extract was investigated for its capability to biosynthesize AuNPs. The biosynthesized AuNPs were analyzed via UV/Vis spectroscopy, XRD, TEM, SAED, EDX, and FTIR techniques. A characteristic absorption peak appeared at 532 nm. *P. tetrastromatica*-derived AuNPs exhibited a rounded shape with an average size of 11.43 nm. Nitrophenols (NP) are considered as the most hazardous organic pollutants in wastewater released from industrial and agricultural sectors, involving those generating pesticides, herbicides, insecticides, explosives, and synthetic dyes (Higson 1992). Environmental threats appear because of their inappropriate release into water bodies. The biosynthesized AuNPs acted as a nanocatalyst by reducing anthropogenic organic pollutant, i.e., 4-nitrophenol (4-NP) to 4 aminophenol in presence of $NaBH_4$.

4.4 Mechanism of Algae-Mediated Synthesis of Nanoparticles

Algal extracts are very rich with versatile biomolecules including proteins, carbohydrates, oil, minerals, fats, polyunsaturated fatty acids, as well as a group of biologically active components, for instance, antioxidants (e.g., polyphenols, tocopherols), and pigments such as carotenoids (e.g., carotene, xanthophyll), chlorophylls, and phycobilins (e.g., phycocyanin, phycoerythrin)(Michalak and Chojnacka 2015). According to several published articles, these compounds have been proved to act as efficient reductants and stabilizers Fawcett et al. (2017), Khanna et al. (2019). Algae-derived nanomaterials can be easily fabricated via simple and low-cost procedure involving (i) preparation of algal extract, (ii) reaction with the metal salt precursor, and (iii) incubation of algal extract with metal precursor solution (Sharma et al. 2016). The interaction between the liquid algal extract with the metal salt precursor usually results in change in color of the reaction mixture. The change in coloration is basically the first observable signature for the formation of the desired nanomaterial through reaction nucleation, then followed by growth of NPs in which nucleonic particles union together, hence allowing the preparation of stable NPs with different shapes and sizes (Sharma et al. 2016; Prasad et al. 2016; Fawcett et al. 2017). As in the other biological entities which can be exploited in biogenesis of nanomaterials, parameters affect the synthesis process. Among these factors pH, temperature, shaking rate, metal salt, algal biomass concentration in addition to reaction time (Khanna et al. 2019). Optimizing these factors help to maximize the bio-produced NPs, prevent their agglomeration, and provide more capability to control size and shape of the prepared NPs.

4.5 Applications of Algae-Mediated Nanoparticles

Biologically synthesized NPs via different algal sources have elucidated their capability to have versatile applications (Iravani et al. 2014). They proved to exert antifungal (Azizi et al. 2013), antibacterial (Sharma et al. 2015a) and anticancer activities (Govindaraju et al. 2015), so they compete with other drugs in medical fields. Along with the medicinal applications, algae-mediated NPs possess extensive affinity in electronic, optic, cosmetic, and coating applications (Singh et al. 2014), food packaging, sensing devices, space industries, bioremediation (Iravani et al. 2014), environmental remediation (Husain et al. 2019), mechanics, light emitters, chemical industries, as well as nonlinear optical devices (Khan et al. 2017).

4.5.1 Antimicrobial Potential

AuNPs have been successfully biosynthesized from diverse seaweeds and played a lead role as antibacterial agents against both Gram-positive and Gram-negative pathogenic bacteria (Rajathi et al. 2012; Venkatesan et al. 2014) as well as an antifungal agent against *Humicola insolens* and *Fusarium dimerum* as demonstrated by Varun and Sudha (2014). According to Vijayan et al. (2016), the small size of NPs causes failure to the main functions of cell membrane, i.e., permeability via adherence to cell surface. *Spirulina platensis*-mediated synthesis of AuNPs exhibited high antibacterial activity against Gram-positive bacteria, e.g., *Bacillus subtilis* and *Staphylococcus aureus*. The protein-capped AuNPs succeeded to penetrate through the thick peptidoglycan layer and caused cell damage (Suganya et al. 2015). Rajeshkumar et al. (2013b) elucidated the algal synthesis of AuNPs via the extract of brown algae *Turbinaria conoides*. *Streptococcus* spp. were the most sensitive to the prepared AuNPs, while both *B. subtilis* and pneumonia fever-causing bacteria *Klebsiella pneumoniae* were more resistant. AuNPs synthesized using *Stoechospermum marginatum* were investigated for their antibacterial activity against *Enterobacter faecalis* (Rajathi et al. 2012). Surprisingly, it exerted antibacterial activity higher than that of the positive control "tetracycline," while minimum ZOI was recorded against *K. pneumonia*, and no inhibition was observed against *E. coli*. In a study performed by Abdel-Raouf and co-authors, AuNPs were fabricated from the powder, ethanolic, and ethanolic free extract of *Galaxaura elongata* (Abdel-Raouf et al. 2018). The prepared AuNPs were tested for their antibacterial potential. It was found that maximum ZOI was noticed for AuNPs derived from ethanolic extract against *E. coli*, *K. pneumoniae*, and MRSA, while AuNPs derived from free ethanolic extract of *G. elongata* exhibited high activity only against

MRSA. On the other hand, AuNPs derived from *G. elongata* powder were found to be effective against *K. pneumoniae* and *E. coli.* Ethanolic extract of *Acanthophora spicifera* acted as a reducing agent in fabricating cubic-shaped AgNPs which were efficient in inhibiting a wide range of bacteria including *S. aureus*, *B. subtilis*, *Salmonella* sp., *E. coli*, and *Candida albicans* (Ibm et al. 2016). Sudha et al. (2013) reported the biogenesis of algal-derived AgNPs from *Microcoleus* sp. which exerted a strong antibacterial effect against *E. coli*, *S. aureus*, *Proteus vulgaris*, *Vibrio cholerae*, *Salmonella typhi*, *B. subtilis*, *Streptococcus*, and *Corynebacterium* sp. ZnO NPs were derived from seaweeds isolated from Indian gulf of Mannar as reported by Nagarajan and Kuppusamy (2013). The resulting ZnO NPs were highly stable for a long time approximately 6 months. The prepared particles were efficient antibacterial agents against Gram-positive bacteria, e.g., *Streptococcus mutans* and *Staphylococcus aureus*, and Gram-negative bacteria *Neisseria gonorrhoeae*, *Vibrio cholerae*, and *Klebsiella pneumoniae* in addition to fungal strain *Aspergillus niger* and the yeast strain *Candida albicans*. In another study, red seaweed metabolites of *Gelidiella acerosa* acted as reductants to produce AgNPs which exhibited antifungal activity against different fungal strains involving *Humicola insolens* (MTCC 4520), *Mucor indicus* (MTCC 3318), *Fusarium dimerum* (MTCC 6583), and *Trichoderma reesei* (MTCC 3929) (LewisOscar et al. 2016).

4.5.2 Antifouling and Anti-biofilm Agents

According to several studies, nano-functionalized materials possess the power to inhibit bacterial adhesion and subsequent biofilm formation on surfaces either by coating techniques (Beyth et al. 2008) or impregnation of nanomaterials within biocides (Lellouche et al. 2009). Vijayan et al. (2014) performed a comparative study in which *Turbinaria conoides*-derived AgNPs and AuNPs were investigated for their effectiveness in controlling biofilm formation. AgNPs showed inhibition effectiveness against *E. coli* followed by *Salmonella* sp., *Serratia liquefaciens*, and *Aeromonas hydrophila*, whereas AuNPs did not show any inhibition effect. Furthermore, in the same study, spherical-shaped AgNPs were found to exert a detrimental effect upon brine shrimp *Artemia salina*. The as-prepared AgNPs acted as a strong anti-microfouling agent. Likely, Kumar et al. (2012) synthesized AgNPs from *Sargassum ilicifolium*. The prepared AgNPs displayed a cytotoxic activity towards *Artemia salina* with medial lethal dose (LD_{50}) value of 10 nM/ml. In another report, *Turbinaria ornata* mediated biogenic synthesis of AgNPs (TOAgNPs) (Krishnan et al. 2015).The prepared TOAgNPs was evaluated for its ability to inhibit 15 biofilm-forming bacterial isolates. Although TOAgNPs expressed broad antibacterial spectrum, *Escherichia coli* was the most sensitive a *Micrococcus* sp. was the most resistant. Interestingly, 100% mortality was attained for *Balanu samphitrite* larvae and 56.6% for *Artemia marina* at 250 μg ml^{-1}. Moreover, low toxicity was shown towards non-target species.

4.5.3 Bioremediation

Nanomaterials have been proved to serve as a great platform for treating and remediating toxic waste released via industrial effluents (Khanna et al. 2019). Ramakrishna et al. (2016a) conducted a study in which aqueous extract of brown algae (*Turbinaria conoides* and *Sargassum tenerrimum*) mediated biogenesis of AuNPs. Biosynthesized AuNPs showed effective catalytic activity towards reducing aromatic nitro compounds (4-nitrophenol and p-nitroaniline) and organic dyes (sulforhodamine 101 and rhodamine B). AuNPs derived from *T. conoides* exhibited much higher catalytic potential than that derived from S. *tenerrimum*. *Ulva lactuca* extract was used to biosynthesize AgNPs which in turn degraded methyl orange photocatalytically under visible light illumination (Kumar et al. 2013). AgNPs synthesized from *Microchaete* NCCU-342 exhibited appreciable dye decolorization ability of azo dye methyl red as compared to cyanobacterial extract.

References

Abboud Y, Saffaj T, Chagraoui A et al (2014) Biosynthesis, characterization and antimicrobial activity of copper oxide nanoparticles (CONPs) produced using brown alga extract (*Bifurcaria bifurcata*). Appl Nannosci 4:571–576

Abdel-Raouf N (2017) Green biosynthesis of gold nanoparticles using *Galaxaura elongata* and characterization of their antibacterial activity. Arab J Chem 10:S3029–S3039

Abdel-Raouf N, Al-Enazi NM, Ibraheem IBM et al (2018) Biosynthesis of silver nanoparticles by using of the marine brown alga *Padina pavonia* and their characterization. Saudi J Biol Sci 26:1207–1215

Afzal B, Yasin D, Husain S et al (2019) Screening of cyanobacterial strains for the selenium nanoparticles synthesis and their anti-oxidant activity. Biocatal Agric Biotechnol 21:101307– 101314

Agarwal P, Gupta R, Agarwal N (2019) Advances in synthesis and applications of microalgal nanoparticles for wastewater treatment. J Nanotechnol 2019: 7392713, 9 p. https://doi.org/10.1155/2019/7392713

Alp MT, Ozbay O, Sungur MA (2012) Determination of heavy metal levels in sediment and macroalgae (*Ulva* sp. and *Enteromorpha* sp.) on the Mersin coast. Ekoloji 21:47–55

Annamalai J, Nallamuthu T (2015) Characterization of biosynthesized gold nanoparticles from aqueous extract of *Chlorella vulgaris* and their anti-pathogenic properties. Appl Nanosci 5:603–607

Annamalai J, Nallamuthu T (2016) Green synthesis of silver nanoparticles: characterization and determination of antibacterial potency. Appl Nanosci 6:259–265

Arsiya F, Sayadi MH, Sobhani S (2017) Green synthesis of palladium nanoparticles using *Chlorella vulgaris*. Mater Lett 186:113–115

Arularasu MV, Devakumar J, Rajendran TV (2018) An innovative approach for green synthesis of iron oxide nanoparticles: characterization and its photocatalytic activity. Polyhedron 156:279–290

Arya A, Gupta K, Chundawat TS, Vaya D (2018) Biogenic synthesis of copper and silver nanoparticles using green alga *Botryococcus braunii* and its antimicrobial activity. Bioinorg Chem Appl 2018: 7879403, 9 p. https://doi.org/10.1155/2018/7879403

Ashokkumar T, Vijayaraghavan K (2016) Brown seaweed-mediated biosynthesis of gold nanoparticles. J Environ Biotechnol Res 2:45–50

Aziz N, Faraz M, Pandey R et al (2015) Facile algae-derived route to biogenic silver nanoparticles: synthesis, antibacterial, and photocatalytic properties. Langmuir 31:11605–11612

Azizi S, Namvar F, Mahdavi M et al (2013) Biosynthesis of silver nanoparticles using brown marine macroalga, *Sargassum muticum* aqueous extract. Materials 6:5942–5950

Bakir EM, Younis NS, Mohamed ME, El Semary NA (2018) Cyanobacteria as nanogold factories: chemical and anti-myocardial infarction properties of gold nanoparticles synthesized by *Lyngbya majuscula*. Mar Drugs 16:217–238

Bao Z, Lan CQ (2018) Mechanism of light-dependent biosynthesis of silver nanoparticles mediated by cell extract of *Neochloris oleoabundans*. Colloids Surf B Biointerfaces 170:251–257

Barwal I, Ranjan P, Kateriya S, Yadav SC (2011) Cellular oxido-reductive proteins of *Chlamydomonas reinhardtii* control the biosynthesis of silver nanoparticles. J Nanobiotechnol 9:56–68

Beyth N, Houri-Haddad Y, Baraness-Hadar L et al (2008) Surface antimicrobial activity and biocompatibility of incorporated polyethylenimine nanoparticles. Biomaterials 29:4157–4163

Bhattacharya P, Swarnakar S, Ghosh S et al (2019) Disinfection of drinking water via algae mediated green synthesized copper oxide nanoparticles and its toxicity evaluation. J Environ Chem Eng 7:102867–102877

Bock JH, Norris DO (2016) Additional approaches in forensic plant science. Forensic Plant Sci:129–147

Borowitzka MA (2013) High-value products from microalgae – their development and commercialisation. J Appl Phycol 25:743–756

Brayner R, Barberousse H, Hemadi M et al (2007) Cyanobacteria as bioreactors for the synthesis of Au, Ag, Pd, and Pt nanoparticles via an enzyme-mediated route. J Nanosci Nanotechnol 7:2696–2708

Cannilla C, Bonura G, Frusteri F et al (2014) Development of an ammonia sensor based on silver nanoparticles in a poly-methacrylic acid matrix. J Mater Chem C 2:5778–5786

Castro L, Blázquez ML, Muñoz JA et al (2013) Biological synthesis of metallic nanoparticles using algae. IET Nanobiotechnol 7:109–116

Cepoi L, Rudi L, Chiriac T et al (2014) Biochemical changes in cyanobacteria during the synthesis of silver nanoparticles. Can J Microbiol 61:13–21

Chakraborty N, Banerjee A, Lahiri S et al (2009) Biorecovery of gold using cyanobacteria and an eukaryotic alga with special reference to nanogold formation – a novel phenomenon. J Appl Phycol 21:145–152

Chanda S, Dave R, Kaneria M, Nagani K (2010) Seaweeds: a novel, untapped source of drugs from sea to combat infectious diseases. Curr Res Technol Educ Top Appl Microbiol Microb Biotechnol 1:473–480

Chetia L, Kalita D, Ahmed GA (2017) Synthesis of Ag nanoparticles using diatom cells for ammonia sensing. Sens Bio-Sensing Res 16:55–61

Chlorophyceae PCS, Kannan RRR, Stirk WA, Van Staden J (2013) Synthesis of silver nanoparticles using the seaweed *Codium capitatum*. S Afr J Bot 86:1–4

Chokshi K, Pancha I, Ghosh T et al (2016) Green synthesis, characterization and antioxidant potential of silver nanoparticles biosynthesized from de-oiled biomass of thermotolerant oleaginous microalgae: *Acutodesmus dimorphus*. RSC Adv 6:72269–72274

da Silva Ferreira V, ConzFerreira ME, Lima LMTR et al (2017) Green production of microalgae-based silver chloride nanoparticles with antimicrobial activity against pathogenic bacteria. Enzym Microb Technol 97:114–121

Dahoumane SA, Mechouet M, Alvarez FJ, Agathos SN (2016) Microalgae: an outstanding tool in nanotechnology Microalgas : Una excelente herramienta en nanotecnología. Revista Bionatura 1:196–201

Das T, Pramanik A, Haldar D (2017) On-line ammonia sensor and invisible security ink by fluorescent zwitterionic spirocyclic meisenheimer complex. Sci Rep 7:40465–40477

Davis TA, Volesky B, Mucci A (2003) A review of the biochemistry of heavy metal biosorption by brown algae. Water Res 37:4311–4330

Dhanalakshmi PK, Azeez R, Rekha R et al (2012) Synthesis of silver nanoparticles using green and brown seaweeds. Phykos 42:39–45

Dhas TS, Kumar VG, Karthick V et al (2014) Facile synthesis of silver chloride nanoparticles using marine alga and its antibacterial efficacy. Spectrochim Acta Part A Mol Biomol Spectrosc 120:416–420

Edison TNJI, Atchudan R, Lee YR (2016) Optical sensor for dissolved ammonia through the green synthesis of silver nanoparticles by fruit extract of *Terminalia chebula*. J Clust Sci 27:683–690

El-Gendy NS, Omran BA (2019) Green synthesis of nanoparticles for water treatment. In: Nano and bio-based technologies for wastewater treatment. Wiley, Hoboken, pp 205–263

El-Naggar NEA, Hussein MH, El-Sawah AA (2017) Bio-fabrication of silver nanoparticles by phycocyanin, characterization, in vitro anticancer activity against breast cancer cell line and in vivo cytotxicity. Sci Rep 7:1–20

Fawcett D, Verduin JJ, Shah M, et al (2017) A review of current research into the biogenic synthesis of metal and metal oxide nanoparticles via marine algae and seagrasses. J Nanosci 2017: 8013850, 15 p. https://doi.org/10.1155/2017/8013850

Govindaraju K, Basha SK, Kumar VG, Singaravelu G (2008) Silver, gold and bimetallic nanoparticles production using single-cell protein (*Spirulina platensis*) Geitler. J Mater Sci 43:5115–5122

Govindaraju K, Kiruthiga V, Kumar VG, Singaravelu G (2009) Extracellular synthesis of silver nanoparticles by a marine alga, *Sargassum Wightii* Grevilli and their antibacterial effects. J. Nanosci Nanotechnol 9:5497–5501

Govindaraju K, Krishnamoorthy K, Alsagaby SA et al (2015) Green synthesis of silver nanoparticles for selective toxicity towards cancer cells. IET Nanobiotechnol 9:325–330

Hamouda RA, Hussein MH, Abo-elmagd RA, Bawazir SS (2019) Synthesis and biological characterization of silver nanoparticles derived from the cyanobacterium *Oscillatoria limnetica*. Sci Rep 9:13071–13084

Hauck TS, Jennings TL, Yatsenko T et al (2008) Enhancing the toxicity of cancer chemotherapeutics with gold nanorod hyperthermia. Adv Mater 20:3832–3838

Higson FK (1992) Microbial degradation of nitroaromatic compounds. In: Advances in applied microbiology. Elsevier, pp 1–19

Hoiczyk E, Hansel A (2000) Cyanobacterial cell walls: news from an unusual prokaryotic envelope. J Bacteriol 182:1191–1199

Houck MM, Crispino F, McAdam T (2018) Other types of evidence. Sci Crime Scenes:323–339

Huang X, El-Sayed IH, Qian W, El-Sayed MA (2006) Cancer cell imaging and photothermal therapy in the near-infrared region by using gold nanorods. J Am Chem Soc 128:2115–2120

Hudek L, Rai S, Michalczyk A et al (2012) Physiological metal uptake by Nostoc punctiforme. Biometals 25:893–903

Husain S, Sardar M, Fatma T (2015) Screening of cyanobacterial extracts for synthesis of silver nanoparticles. World J Microbiol Biotechnol 31:1279–1283

Husain S, Afreen S, Yasin D et al (2019) Cyanobacteria as a bioreactor for synthesis of silver nanoparticles-an effect of different reaction conditions on the size of nanoparticles and their dye decolorization ability. J Microbiol Methods 162:77–82

Ibm I, Bee AE, Wf S, Wa F (2016) Green biosynthesis of silver nanoparticles using marine red algae *Acanthophora specifera* and its antimicrobial activity. Nanomed Nanotechnol 7:7–10

Iravani S, Korbekandi H, Mirmohammadi SV, Zolfaghari B (2014) Synthesis of silver nanoparticles: chemical, physical and biological methods. Res Pharm Sci 9:385–406

Jegadeeswaran P, Shivaraj R, Venckatesh R (2012) Green synthesis of silver nanoparticles from extract of *Padina tetrastromatica* leaf. Dig J Nanomater Biostruct 7:991–998

Jena J, Pradhan N, Dash BP et al (2013) Biosynthesis and characterization of silver nanoparticles using microalga *Chlorococcum humicola* and its antibacterial activity. Int J Nanomater Biostruct 3:1–8

Jena J, Pradhan N, Nayak RR et al (2014) Microalga *Scenedesmus* sp.: a potential low-cost green machine for silver nanoparticle synthesis. J Microbiol Biotechnol 24:522–533

Jena J, Pradhan N, Dash BP et al (2015) Pigment mediated biogenic synthesis of silver nanoparticles using diatom *Amphora* sp. and its antimicrobial activity. J Saudi Chem Soc 19:661–666

Kannan RRR, Arumugam R, Ramya D, et al (2013a) Green synthesis of silver nanoparticles using marine macroalga *Chaetomorpha linum*. Appl Nanosci 3:229–233

Kannan RRR, Stirk WA, Van Staden J (2013b) Synthesis of silver nanoparticles using the seaweed *Codium capitatum* PC Silva (Chlorophyceae). South Afr J Bot 86:1–4

Karthikeyan P, Mohan D, Abishek G, Priya R (2015) Synthesis of silver nanoparticles using phytoplankton and its characteristics. Int J Fish Aquat Stud 2:398–401

Kashyap M, Samadhiya K, Ghosh A, Anand V (2019) Screening of microalgae for biosynthesis and optimization of Ag/AgCl nano hybrids having antibacterial effect. RSC Adv 9:25583–25591

Kathiraven T, Sundaramanickam A, Shanmugam N, Balasubramanian T (2015) Green synthesis of silver nanoparticles using marine algae *Caulerpa racemosa* and their antibacterial activity against some human pathogens. Appl Nanosci 5:499–504

Khan MNA, Choi JS, Lee MC et al (2008) Anti-inflammatory activities of methanol extracts from various seaweed species. J Environ Biol 29:465–469

Khan I, Saeed K, Khan I (2017) Nanoparticles: properties, applications and toxicities. Arab J Chem 12:908–931

Khan S, Zada S, Ahmad S et al (2019) Chemosphere concurrent biomineralization of silver ions into Ag0 and Ag_xO by *Leptolyngbya* strain JSC-1 and the establishment of its axenic culture. Chemosphere 215:693–702

Khanehzaei H, Ahmad MB, Kamyar S (2014) Synthesis and characterization of Cu@Cu_2O core shell nanoparticles prepared in seaweed *Kappaphycus alvarezii* media. Int J Electrochem Sci 9:8189–8198

Khanna P, Kaur A, Goyal D (2019) Algae-based metallic nanoparticles: synthesis, characterization and applications. J Microbiol Methods 163:105656–105680

Kim S-K (2011) Handbook of marine macroalgae: biotechnology and applied phycology. Wiley, Hoboken

Kim D, Ganesh R, Shinde S et al (2018) Green synthesis of silver nanoparticles using *Laminaria japonica* extract: characterization and seedling growth assessment. J Clean Prod 172:2910–2918

Krishnan M, Sivanandham V, Hans-Uwe D et al (2015) Antifouling assessments on biogenic nanoparticles: A field study from polluted offshore platform. Mar Pollut Bull 101:816–825

Kröger N, Poulsen N (2008) Diatoms - from cell wall biogenesis to nanotechnology. Annu Rev Genet 42:83–107

Kumar P, Govindaraju M, Senthamilselvi S, Premkumar K (2013) Photocatalytic degradation of methyl orange dye using silver (Ag) nanoparticles synthesized from *Ulva lactuca*. Colloids Surf B Biointerfaces 103:658–661

Kumar B, Smita K, Sánchez E et al (2016) Ecofriendly ultrasound-assisted rapid synthesis of gold nanoparticles using Calothrix algae. Adv Nat Sci Nanosci Nanotechnol 025013–025018

Kumaresan M, Vijai Anand K, Govindaraju K et al (2018) Seaweed *Sargassum wightii* mediated preparation of zirconia (ZrO_2) nanoparticles and their antibacterial activity against gram positive and gram negative bacteria. Microb Pathog 124:311–315

Labib W, Shady AMA, Mohamed LA et al (2015) Ecological studies of macroalgae in Alexandria Mediterranean waters. J Exp Biol 11:169–180

Lellouche J, Kahana E, Elias S et al (2009) Antibiofilm activity of nanosized magnesium fluoride. Biomaterials 30:5969–5978

Lengke MF, Ravel B, Fleet ME et al (2006a) Mechanisms of gold bioaccumulation by filamentous cyanobacteria from gold (III)-chloride complex. Environ Sci Technol 40:6304–6309

Lengke MF, Fleet ME, Southam G (2006b) Synthesis of platinum nanoparticles by reaction of filamentous cyanobacteria with platinum (IV)-chloride complex. Langmuir 22:7318–7323

Lengke MF, Fleet ME, Southam G (2007) Biosynthesis of silver nanoparticles by filamentous cyanobacteria from a silver (I) nitrate complex. Langmuir 23:2694–2699

Levine I (2016) Chapter 1 – algae: a way of life and health. In: Microalgae in health and disease prevention 2018, Elsevier, pp 1–10

LewisOscar F, Vismaya S, Arunkumar M, et al (2016) Algal nanoparticles: synthesis and biotechnological potentials. In: Algae-organisms imminent biotechnology, vol 7, pp 157–182

Li Y, Tang X, Song W et al (2014) Biosynthesis of silver nanoparticles using *Euglena gracilis*, Euglena intermedia and their extract. IET Nanobiotechnol 9:19–26
Li X, Schirmer K, Bernard L et al (2015) Silver nanoparticle toxicity and association with the alga *Euglena gracilis*. Environ Sci Nano 2:594–602
Mahdavi M, Namvar F, Ahmad MB, Mohamad R (2013) Green biosynthesis and characterization of magnetic iron oxide (Fe_3O_4) nanoparticles using seaweed (*Sargassum muticum*) aqueous extract. Molecules 18:5954–5964
Massironi A, Morelli A, Grassi L et al (2019) *Ulvan* as novel reducing and stabilizing agent from renewable algal biomass: application to green synthesis of silver nanoparticles. Carbohydr Polym 203:310–321
Merin DD, Prakash S, Bhimba BV (2010) Antibacterial screening of silver nanoparticles synthesized by marine micro algae. Asian Pac J Trop Med 3:797–799
Metting FB (1996) Biodiversity and application of microalgae. J Ind Microbiol 17:477–489
Michalak I, Chojnacka K (2015) Algae as production systems of bioactive compounds. Eng Life Sci 15:160–176
Minhas FT, Arslan G, Gubbuk IH, Akkoz C (2018) Evaluation of antibacterial properties on polysulfone composite membranes using synthesized biogenic silver nanoparticles with *Ulva compressa* (L.) Kütz. and *Cladophora glomerata* (L.) Kütz. extracts. Int J Biol Macromol 107:157–165
Miyashita K (2009) The carotenoid fucoxanthin from brown seaweed affects obesity. Lipid Technol 21:186–190
Mohamed S, Hashim SN, Rahman HA (2012) Seaweeds: a sustainable functional food for complementary and alternative therapy. Trends Food Sci Technol 23:83–96
Mohseniazar M, Barin M, Zarredar H et al (2011) Potential of microalgae and lactobacilli in biosynthesis of silver nanoparticles. Bioimpacts 1:149–152
Momeni S, Nabipour I (2015) A simple green synthesis of palladium nanoparticles with *Sargassum alga* and their electrocatalytic activities towards hydrogen peroxide. Appl Biochem Biotechnol 176:1937–1949
Mubarak Ali D, Sasikala M, Gunasekaran M, Thajuddin N (2011) Biosynthesis and characterization of silver nanoparticles using marine cyanobacterium, *Oscillatoria willei* ntdm01. Dig J Nanomater Biostruct 6:385–390
MubarakAli D, Arunkumar J, Nag KH et al (2013) Gold nanoparticles from Pro and eukaryotic photosynthetic microorganisms-comparative studies on synthesis and its application on biolabelling. Colloids Surf B Biointerfaces 103:166–173
Murugan K, Samidoss CM, Panneerselvam C et al (2015) Seaweed-synthesized silver nanoparticles: an eco-friendly tool in the fight against *Plasmodium falciparum* and its vector *Anopheles stephensi*? Parasitol Res 114:4087–4097
Nagarajan S, Kuppusamy KA (2013) Extracellular synthesis of zinc oxide nanoparticle using seaweeds of Gulf of Mannar, India. J Nanobiotechnol 11:39–50
Namvar F, Mohamed S, Fard SG et al (2012) Polyphenol-rich seaweed (*Eucheuma cottonii*) extract suppresses breast tumour via hormone modulation and apoptosis induction. Food Chem 130:376–382
Narayanan KB, Sakthivel N (2011) Green synthesis of biogenic metal nanoparticles by terrestrial and aquatic phototrophic and heterotrophic eukaryotes and biocompatible agents. Adv Colloid Interf Sci 169:59–79
Nesakumar T, Immanuel J, Raji E et al (2016) *Caulerpa racemosa*: a marine green alga for eco-friendly synthesis of silver nanoparticles and its catalytic degradation of methylene blue. Bioprocess Biosyst Eng 39:1401–1408
Nishino T, Fukuda A, Nagumo T et al (1999) Inhibition of the generation of thrombin and factor Xa by a fucoidan from the brown seaweed *Ecklonia kurome*. Thromb Res 96:37–49
Pandey S, Goswami GK, Nanda KK (2013) Green synthesis of polysaccharide/gold nanoparticle nanocomposite: an efficient ammonia sensor. Carbohydr Polym 94:229–234
Parial D, Pal R (2011) Green synthesis of gold nanoparticles using cyanobacteria and their characterization. Indian J Appl Res 4:69–72

Parial D, Pal R (2015) Biosynthesis of monodisperse gold nanoparticles by green alga *Rhizoclonium* and associated biochemical changes. J Appl Phycol 27:975–984

Parial D, Patra HK, Dasgupta AKR, Pal R (2012a) Screening of different algae for green synthesis of gold nanoparticles. Eur J Phycol 47:22–29

Parial D, Patra HK, Roychoudhury P et al (2012b) Gold nanorod production by cyanobacteria-a green chemistry approach. J Appl Phycol 24:55–60

Parial D, Gopal PK, Paul S, Pal R (2016) Gold (III) bioreduction by cyanobacteria with special reference to in vitro biosafety assay of gold nanoparticles. J Appl Phycol 28:3395–3406

Pascual García C, Burchardt AD, Carvalho RN et al (2014) Detection of silver nanoparticles inside marine diatom *Thalassiosira pseudonana* by electron microscopy and focused ion beam. PLoS One 9:e96078–e96085

Pasula RR, Lim S (2017) Engineering nanoparticle synthesis using microbial factories. Eng Biol 1:12–17

Patel V, Berthold D, Puranik P, Gantar M (2015) Screening of cyanobacteria and microalgae for their ability to synthesize silver nanoparticles with antibacterial activity. Biotechnol Rep 5:112–119

Pissuwan D, Valenzuela SM, Cortie MB (2008) Prospects for gold nanorod particles in diagnostic and therapeutic applications. Biotechnol Genet Eng Rev 25:93–112

Prasad TNVKV, Kambala VSR, Naidu R (2013) Phyconanotechnology: Synthesis of silver nanoparticles using brown marine algae *Cystophora moniliformis* and their characterisation. J Appl Phycol 25:177–182

Prasad R, Pandey R, Barman I (2016) Engineering tailored nanoparticles with microbes: Quo vadis? Wiley Interdiscip Rev Nanomed Nanobiotechnol 8:316–330

Princy KF, Gopinath A (2019) Spectroscopic investigation on the catalytic efficacy of biofabricated gold nanoparticles using marine macroalgae. Mater Today Proc 9:38–45

Priyadharshini RI, Prasannaraj G, Geetha N, Venkatachalam P (2014) Microwave-mediated extracellular synthesis of metallic silver and zinc oxide nanoparticles using macro-algae (*Gracilaria edulis*) extracts and its anticancer activity against human PC3 cell lines. Appl Biochem Biotechnol 174:2777–2790

Pugazhendhi A, Prabakar D, Jacob JM et al (2018) Synthesis and characterization of silver nanoparticles using *Gelidium amansii* and its antimicrobial property against various pathogenic bacteria. Microb Pathog 114:41–45

Pytlik N, Kaden J, Finger M et al (2017) Biological synthesis of gold nanoparticles by the diatom *Stephanopyxis turris* and in vivo SERS analyses. Algal Res 28:9–15

Rahman A, Kumar S, Bafana A et al (2019) individual and combined effects of extracellular polymeric substances and whole cell components of *Chlamydomonas reinhardtii* on silver nanoparticle synthesis and stability. Molecules 24:956–974

Rajathi FAA, Parthiban C, Kumar VG, Anantharaman P (2012) Biosynthesis of antibacterial gold nanoparticles using brown alga, *Stoechospermum marginatum* (kützing). Spectrochim Acta Part A Mol Biomol Spectrosc 99:166–173

Rajeshkumar S, Malarkodi C, Gnanajobitha G et al (2013a) Seaweed-mediated synthesis of gold nanoparticles using *Turbinaria conoides* and its characterization. J Nanostruct Chem 3:44–51

Rajeshkumar S, Malarkodi C, Vanaja M et al (2013b) Antibacterial activity of algae mediated synthesis of gold nanoparticles from *Turbinaria conoides*. Der Pharma Chem 5:224–229

Ramakrishna M, Babu DR, Gengan RM et al (2016a) Green synthesis of gold nanoparticles using marine algae and evaluation of their catalytic activity. J Nanostruct Chem 6:1–13

Ramakrishna M, Babu DR et al (2016b) Green synthesis of gold nanoparticles using marine algae and evaluation of their catalytic activity. J Nanostruct Chem 6:1–13

Ramkumar VS, Pugazhendhi A, Prakash S et al (2017) Synthesis of platinum nanoparticles using seaweed *Padina gymnospora* and their catalytic activity as PVP/PtNPs nanocomposite towards biological applications. Biomed Pharmacother 92:479–490

Rao MD, Pennathur G (2017) Green synthesis and characterization of cadmium sulphide nanoparticles from *Chlamydomonas reinhardtii* and their application as photocatalysts. Mater Res Bull 85:64–73

Rösken LM, Cappel F, Körsten S et al (2016) Time-dependent growth of crystalline Au0-nanoparticles in cyanobacteria as self-reproducing bioreactors: *Anabaena cylindrica*. Beilstein J Nanotechnol 7:312–327
Salari Z, Danafar F, Dabaghi S, Ataei SA (2016) Sustainable synthesis of silver nanoparticles using macroalgae *Spirogyra varians* and analysis of their antibacterial activity. J Saudi Chem Soc 20:459–464
Salem DMSA, Ismail MM, Aly-Eldeen MA (2019) Biogenic synthesis and antimicrobial potency of iron oxide (Fe_3O_4) nanoparticles using algae harvested from the Mediterranean Sea, Egypt. Egypt J Aquat Res 45:197–204
Satapathy S, Shukla SP, Sandeep KP et al (2015) Evaluation of the performance of an algal bioreactor for silver nanoparticle production. J Appl Phycol 27:285–291
Schröfel A, Kratošová G, Bohunická M et al (2011) Biosynthesis of gold nanoparticles using diatoms-silica-gold and EPS-gold bionanocomposite formation. J Nanopart Res 13:3207–3216
Senapati S, Syed A, Moeez S et al (2012) Intracellular synthesis of gold nanoparticles using alga *Tetraselmis kochinensis*. Mater Lett 79:116–118
Shakibaie M, Forootanfar H, Mollazadeh Moghaddam K et al (2010) Green synthesis of gold nanoparticles by the marine microalga *Tetraselmis suecica*. Biotechnol Appl Biochem 57:71–75
Shankar PD, Shobana S, Karuppusamy I et al (2016) A review on the biosynthesis of metallic nanoparticles (gold and silver) using bio-components of microalgae: Formation mechanism and applications. Enzym Microb Technol 95:28–44
Sharma D, Kanchi S, Bisetty K (2015a) Biogenic synthesis of nanoparticles: a review. Arab J Chem. In Press
Sharma G, Jasuja ND, Kumar M, Ali MI (2015b) Biological synthesis of silver nanoparticles by cell-free extract of spirulina platensis. J Nanotechnol 2015: 132675, 6 p. https://doi.org/10.1155/2015/132675
Sharma A, Sharma S, Sharma K et al (2016) Algae as crucial organisms in advancing nanotechnology: a systematic review. J Appl Phycol 28:1759–1774
Sheehan J, Dunahay T, Benemann J, Roessler P (1998) Look back at the US department of energy's aquatic species program: biodiesel from algae; close-out report. National Renewable Energy Lab, Golden
Singaravelu G, Arockiamary JS, Kumar VG, Govindaraju K (2007) A novel extracellular synthesis of monodisperse gold nanoparticles using marine alga, *Sargassum wightii* Greville. Colloids Surf B Biointerfaces 57:97–101
Singh G, Babele PK, Kumar A et al (2014) Synthesis of ZnO nanoparticles using the cell extract of the cyanobacterium, *Anabaena* strain L31 and its conjugation with UV-B absorbing compound shinorine. J Photochem Photobiol B Biol 138:55–62
Singh AK, Tiwari R, Singh VK et al (2019) Green synthesis of gold nanoparticles from *Dunaliella salina*, its characterization and in vitro anticancer activity on breast cancer cell line. J Drug Deliv Sci Technol 51:164–176
Sinha SN, Paul D, Halder N et al (2015) Green synthesis of silver nanoparticles using fresh water green alga *Pithophora oedogonia* (Mont.) Wittrock and evaluation of their antibacterial activity. Appl Nanosci 5:703–709
Sonker AS, Richa, Pathak J et al (2017) Characterization and in vitro antitumor, antibacterial and antifungal activities of green synthesized silver nanoparticles using cell extract of *Nostoc* sp. strain HKAR-2. Can J Biotechnol 1:26–37
Sudha SS, Rajamanickam K, Rengaramanujam J (2013) Microalgae mediated synthesis of silver nanoparticles and their antibacterial activity against pathogenic bacteria. Indian J Exp Biol 51:393–399
Suganya KSU, Govindaraju K, Kumar VG et al (2015) Blue green alga mediated synthesis of gold nanoparticles and its antibacterial efficacy against Gram positive organisms. Mater Sci Eng C 47:351–356
Suriya J, Raja SB, Sekar V, Rajasekaran R (2012) Biosynthesis of silver nanoparticles and its antibacterial activity using seaweed *Urospora* sp. Afr J Biotechnol 11:12192–12198

Varun S, Sudha S, Kumar PS (2014) Biosynthesis of gold nanoparticles from aqueous extract of *Dictyota bartayresiana* and their anti-fungal activity. Indian J Adv Chem Sci 2:190–193

Vasquez RD, Apostol JG, de Leon JD et al (2016) Polysaccharide-mediated green synthesis of silver nanoparticles from *Sargassum siliquosum* J.G. Agardh: assessment of toxicity and hepatoprotective activity. OpenNano 1:16–24

Venkatesan J, Manivasagan P, Kim S-K et al (2014) Marine algae-mediated synthesis of gold nanoparticles using a novel *Ecklonia cava*. Bioprocess Biosyst Eng 37:1591–1597

Venkatpurwar V, Pokharkar V (2011) Green synthesis of silver nanoparticles using marine polysaccharide: Study of in-vitro antibacterial activity. Mater Lett 65:999–1002

Vijayan SR, Santhiyagu P, Singamuthu M, et al (2014) Synthesis and characterization of silver and gold nanoparticles using aqueous extract of seaweed, *Turbinaria conoides* and their antimicrofouling activity. Sci World J 2014: 938272, 10 p. https://doi.org/10.1155/2014/938272

Vijayan SR, Santhiyagu P, Ramasamy R et al (2016) Seaweeds: a resource for marine bionanotechnology. Enzym Microb Technol 95:45–57

Vivek M, Senthil Kumar P, Steffi S, Sudha S (2011) Biogenic silver nanoparticles by *Gelidiella acerosa* extract and their antifungal effects. Avicenna J Med Biotechnol 3:143–148

Wada K, Nakamura K, Tamai Y et al (2011) Seaweed intake and blood pressure levels in healthy pre-school Japanese children. Nutr J 10:83–90

Xie J, Lee JY, Wang DIC, Ting YP (2007a) Silver nanoplates: from biological to biomimetic synthesis. ACS Nano 1:429–439

Xie J, Lee JY, Wang DIC, Ting YP (2007b) Identification of active biomolecules in the high yield synthesis of single crystalline gold nanoplates in algal solutions. Small 3:672–682

Yee N, Benning LG, Phoenix VR, Ferris FG (2004) Characterization of metal– cyanobacteria sorption reactions: a combined macroscopic and infrared spectroscopic investigation. Environ Sci Technol 38:775–782

Zachleder V (1998) Algae. An introduction to phycology. Photosynthetica vol 4 (ed. van den Hoek C, Mann DG, Jahns HM), pp 506–506

Zinicovscaia I, Rudi L, Valuta A et al (2016) Biochemical changes in nostoc linckia associated with selenium nanoparticles biosynthesis. Ecol Chem Eng S 23:559–569

Chapter 5
Biosynthesized Nanomaterials via Processing of Different Plant Parts (Phytonanotechnology) and Biovalorization of Agro-Industrial Wastes to Nano-Sized Valuable Products

5.1 Introduction

Because of the hazardous toxic effects which might occur during the chemical synthesis of metal nanoparticles, much research has been focused on biogenic synthesis of nanoparticles using plant parts. Extracts of different plant parts such as leaves, roots, stems, barks, seeds, fruit peels, shoots, etc. have been employed to produce nanoparticles (Omran et al. 2017). Such methodology is eco-friendly, rapid, cost-effective, and easily upscalable (El-Gendy and Omran 2019). Scientific literature is filled with great number of investigations regarding the exploitation of different plant parts' extract which have revealed the potential to produce nanoparticles with controlled size and morphology (Küünal et al. 2018). As environmental regulations are striking towards green and clean technologies, phytonanotechnology is arising very rapidly as an alternative for the conventional methodologies which were used in synthesizing nanoparticles. Nonetheless, there are quite a few challenges that need to be overcome before phytosynthesized nanoparticles can completely replace the chemically or physically synthesized ones (Venkat Kumar and Rajeshkumar 2017). Among which the complete determination of the active phytobiomolecules which are involved with reduction and capping processes. This chapter focuses the spot upon using different plant parts as green nanobiofactories, glancing through the biovalorization of agro-industrial and food waste into valuable products such as nanoparticles in addition to illustrating the possible phytoconstituents which might be involved during the synthesis process.

B. A. Omran, *Nanobiotechnology: A Multidisciplinary Field of Science*,
Nanotechnology in the Life Sciences, https://doi.org/10.1007/978-3-030-46071-6_5

5.2 Common Protocols Involved in Plant-Mediated Synthesis of Nanomaterials

Usually certain steps are followed to biologically synthesize nanomaterials using extracts of different plant parts as appended below (Rajeshkumar and Bharath 2017):

- Plant materials are collected from a specific desired site and then washed thoroughly with double distilled water for 2–3 times to remove any debris and unwanted constituents.
- Drying at room temperature or under sunlight.
- Weighing an exact weight of plant materials and then boiled in 100 ml deionized water at temperatures ranging between 60 and 80 °C.
- The obtained solution is filtered with nylon mesh cloth and kept at 4 °C for further processing for synthesis of nanomaterial.
- Then, the collected filtrate was reacted with a specific type of metal salt to produce the desired type of nanomaterial with a specific concentration of the metal salt.
- Optimization of the most influential parameters is carried out like extract, metal salt concentrations, pH, temperature, agitation rate, and reaction duration time.
- The preliminary indication for the biosynthesis of the targeted NPs takes place by the gradual change in color of the reaction mixture.
- Afterwards, the usual spectroscopic and microscopic techniques are employed for further characterization of the prepared nanomaterial.

5.3 Biogenesis of Nanomaterials via Extracts of Plant Parts

A burgeoning branch of nanotechnology is the green synthesis of nanomaterials via the extracts of different plant parts (e.g., leaves, stem, roots, seeds, shoot, aerial parts, flowers, etc.) (Table 5.2). This synthetic approach is referred to as "phytonanotechnology." Plant parts can be exploited in nanoparticle biosynthesis either in their live form or dead/inactive form (Vijayaraghavan and Ashokkumar 2017). Furthermore, plant biomass can either be used in a powder or an extract form (Sorbiun et al. 2018). According to Bali et al. (2006), different plants have the potential to accumulate metals which eventually can be intracellularly to nanoparticles. With the current aspects of nanotechnology, plants are converted superior biosynthesizers compared with other biological entities. This is due to several reasons involving; the availability and the abundance of plant resources in comparison with the other nanoparticle biological synthesizers (Prasad 2014). Plant-mediated synthesis of NPs is much faster (one-pot reaction), safer, simpler, and easier. The reaction takes place at ambient temperature with no need for any harsh conditions (Chung et al. 2016). Furthermore, plant biosynthesis of NPs is more advantageous over the microbial synthesis as it does not require any cell maintenance or culturing (Prasad 2014). Roy and Das (2015) assumed that fabrication of nanoparticles using microbes might not be very appropriate for industrial feasibility due to the requirement for highly aseptic conditions and the required maintenance procedures. Whereas, Kharissova et al. (2013) demonstrated that the microbial mediated synthesis of NPs might be somehow slow as plants carry out the synthesis reaction faster than microbial entities. The reaction rate is relatively high as it may take only several minutes to several hours to be completed in contrast to microbial synthesis methodologies as it takes longer time which may reach two or even several days. Thus, it may be considered as a time-consuming approach (Noruzi 2015). For example, AgNPs and AuNPs were reported to be phytosynthesized using different plant extracts within 2 min (Okafor et al. 2013), 5 min (Saxena et al. 2012; Franco-Romano et al. 2014; Bindhu et al. 2014), 10 min (Kumar et al. 2011; Singh et al. 2016a; Franco-Romano et al. 2014), 20 min (Karuppaiya et al. 2013), 30 min (Krishnaraj et al. 2010; Das et al. 2011; Guo et al. 2015), and 45 min (Singh et al. 2016b). Accordingly, phytonanotechnology possesses outstanding advantages including scalability, biocompatibility, and the capability to be medically applied as the produced NPs are mostly generated using the universal solvent, water (Salouti and Derakhshan 2019). Plant-derived NPs are nontoxic in nature and as a result they can fulfill the high demands for NPs to be applied in biomedical and environmental disciplines (Singh et al. 2016c). For instance, (Mukherjee and Patra 2017) showed the better efficacy of biological synthesis of AgNPs and AuNPs from *Olax scandens* leaf extract in terms of anticancer activity, biocompatibility for drug delivery, and imaging facilitator activity compared with chemically synthesized AgNPs and AuNPs. Furthermore, biologically synthesized NPs showed high anticancer activity in the lung (A549), breast (MCF-7), and colon (COLO 205) cancer cell lines. These results showed significant inhibition of cancer cell proliferation and fluorescence imaging in A549 cancer cells (Mukherjee et al. 2013). Additionally, biological NPs are more biocompatible with the cardio myoblast normal cell line (H9C2), human umbilical vein endothelial cells (HUVEC), and Chinese hamster ovary cells (CHO) than chemically synthesized NPs, which further supports the future applications of biologically fabricated NPs as drug delivery carriers (Makarov et al. 2014).

It is worth noting that, phytosynthesis of NPs is a cost-effective methodology as it is devoid from complex and multistep procedures such as microbial isolation, culturing, and maintenance, and consequently it can be easily scaled up to large scales (Makarov et al. 2014). Interestingly, Shankar et al. (2003) was the first to carry out a study to investigate plant-mediated synthesis of silver nanoparticles (AgNPs) using alfalfa. Different plant parts can be exploited within the biological synthesis of NPs like leaves, roots, stems, bark, shoot, seeds, rhizome, gum, and flowers as their extract was proved to be rich with phytochemicals that could act as both reductants and stabilizers. These bioactive molecules include proteins, amino acids, organic acids, vitamins, and secondary metabolites like flavonoids, terpenoids, alkaloids, polyphenols, carbohydrates, heterocyclic compounds, and polysaccharides (Makarov et al. 2014; Noruzi 2015; Akhtar et al. 2015). The main target of this chapter is to illustrate the role of plants as biological systems for the biosynthesis of metal and metal oxide NPs via plant extracts, demonstrate the parameters that may affect the synthesis process, and glance through the different applications in which they can be employed. El-Gendy and Omran (2019) demonstrated the various classes of plant metabolites which may be involved in nanoparticle synthesis. These bioactive molecules involve alcohol terpenes (linalool), pigments (hydroquinones), phenolic compounds (oleuropein), vitamins (ascorbic acid), polysaccharides (alginic acid), sugar alcohols (mannitol), terpenoids (limonene), furanones (2,4-Dihydroxy-2,5-dimethyl-3(2H)-furan-3-one, sugar derivatives (2, 3-Dihydro-3,5- dihydroxy-6-methyl- 4H-pyran-4-one (DDMP)), dyes (lawsone), saturated fatty acids (tetradecanoic, pentadecanoic, and hexadecanoic acids), monosaccharides (d-allose), flavonoids (kaempferol), phytosterols (sitosterol), glucose derivatives.

Numerous plants have been reported for their capability to accumulate metals and reduce them to their nanoparticle state (Bali et al. 2006). *Hordeum vulgare* is commonly known as barley and is usually available around an oil refinery in Bahregan, Bushehr, Iran. Esmaeili. Saremnia (2016) managed to synthesize NaA zeolite nanoparticles (NaA-ZNPs) derived from *H. vulgare* husks via a hydrothermal methodology. The biosynthesized NaA-ZNPs were characterized via XRD, SEM, BET, and FTIR. Authors investigated its efficacy in removing total petroleum hydrocarbons (TPH) as TPH belongs to the most common groups of persistent organic contaminants. Adsorption of TPH was carried out for both batch and continuous-flow processes. Different parameters were studied in the batch process including time, pH, and adsorbent concentration. The results were analyzed via gas chromatography/mass spectroscopy (GC/MS). One-way analysis of variance (ANOVA) SPSS 21 software was employed to track the removal efficiency of the biosynthesized NaA-ZNPs. In the batch and continuous systems, the highest TPH removal efficiencies under optimal conditions were 92.3% and 87.4%, respectively. Hence, this study introduces a high potential of NaA-ZNPs to be used in treatment or pretreatment of wastewater from oil refineries wastewater containing TPH.

In a study presented by Kuppusamy and colleagues, CuNPs were green synthesized using the aqueous extract of *Commelina nudiflora* (Kuppusamy et al. 2017). Green synthesized CuNPs were rounded in shape with an average size ranging between 45 and 100 nm as measured by FESEM. XRD revealed that the prepared CuNPs were highly pure and crystalline in nature and were found to be highly stable for long time. EDX results showed the presence of intense peaks related to C, O, Si, and Cu in the 0–2.5 keV region of EDX spectra which in turn confirmed the dominance of Cu with a percentage of more than 80% in the tested sample, whereas C (10%) and O (5%) appeared due to presence of bioactive molecules like flavonoids, polyphenols, and reducing sugars within *C. nudiflora* extract. Additionally, authors evaluated the fabricated CuNPs to biologically treat palm oil mill effluent treatment (POME) as POME contains different types of pathogenic microorganisms. The resultant data clearly clarified that POME treated with CuNPs at different concentrations, e.g., 250, 500 and 1000 ppm for 12 h of incubation, decreased bacterial growth of *E. coli*, *Klebsiella* sp., *Pseudomonas* sp., *Aeromonas* sp., and *Bacillus* sp. Since ancient times, *Gardenia resinifera* has been used as a traditional medicinal plant to increase appetite. Malachite green is a dye which has the chemical formula of $C_{23}H_{25}N_2Cl$, and its International Union of Pure and Applied Chemistry (IUPAC) name is 4-[(4- dimethylaminophenyl) phenylmethyl]-*N*, *N*-dimethyl-Amin (Sarmah and Kumar 2011). This dye has several applications in leather, textile, silk, and paper industries due to its high solubility in water (Cheng et al. 2008). But unfortunately, this highly toxic dye cannot be easily removed from water, and it has a carcinogenic effect on humans and represents a major threat to the surrounding water resources. Hence, as a trial, Rup and co-colleagues performed an investigation to overcome such problem via using *Rubus coreanus* derived ZnO NPs (Jahan et al. 2018). ZnO NPs were synthesized from immature fruit extract of *R. coreanus*. The formation of ZnO NPs was indicated by the appearance of a white precipitate powder. For further conformation several spectroscopic and analytical techniques were employed. UV/Vis spectroscopic study revealed the appearance of a maximum absorption peak at 357 nm, thus confirmed the formation of ZnO NPs. Field emission transmission electron microscopy (FETEM) illustrated the size and the shape of the prepared particles as 23.16 nm and leaf-like-shaped particles. FTIR showed the involvement of carbonyl, amine and phenolic groups which might have aided in capping the biofabricated ZnO NPs. ZnO NPs were tested for their catalytic potential. It was observed that malachite green dye was degraded in both dark and light with a degradation percentage of 90% at 4 h of reaction time in both cases. As a result, *R. coreanus* derived ZnO NPs can be used as a safe nanomaterial that can be used in water treatment. Karade et al. (2019) reported the phytosynthesis of superparamagnetic spherical NPs of α-Fe_2O_3 with a size range of 5 nm. Phenolic groups existed within the extract of fresh leaves of *G. Resinifera* were the main functional groups behind the capping of the prepared α-Fe_2O_3 NPs as revealed by FTIR. XRD illustrated the formation of pure form of α-Fe_2O_3 iron oxide phase. VSM revealed the superparamagnetic behavior of the prepared α-Fe_2O_3 NPs with non-saturating MS value of 8.5 emu/g at room temperature (RT). In a study performed by Malini et al. (2019), *Prosopis juliflora* leaf extract was used to biosynthesize AgNPs. The prepared AgNPs was then encapsulated or loaded within chitosan using 1% of chitosan. The prepared nanocapsules were identified using UV/Visible spectroscopy, FTIR, XRD, and SEM. SEM images revealed the presence of spherical-shaped AgNPs in chitosan nanocapsules. Scherer equation showed an estimated size of 30 nm for the biosynthesized AgNPs. Meanwhile, the encapsulated AgNPs showed a significant antibacterial effect against pathogenic *E. coli*. Furthermore, the synthesized nanocapsules proved to exert a biosorption capability and a photocatalytic degradation activity against the toxic rose bengal dye. The biosorption activity was assessed by monitoring the dye concentration at different time intervals.

ZnO NPs were biologically synthesized via the leaf extract of *Artocarpus heterophyllus* as demonstrated by Shahnaz et al. (2019). The addition of zinc nitrate to *A. heterophyllus* leaf extract resulted in formation of a brown color which indicated the formation of ZnO NPs. A maximum absorption peak at 235 nm was observed via UV/Vis spectrophotometer, while FTIR spectroscopy showed that amines and amides were the main lead in stabilizing ZnONPs. EDX confirmed the presence of pure elemental zinc, whereas TEM revealed the formation of rounded particles with a size ranging from 12 to 24 nm. This study showed that these phytosynthesized ZnO NPs exhibited excellent cytotoxic effect against human colon cancer HCT-116 cell line and a low toxicity towards normal Vero cell as was revealed by the morphological changes in tested cells. Aqueous leaf extract of *Passiflora edulis F. Flavicarpa* was used to synthesize AgNPs as presented by Thomas et al. (2019). The study revealed that the prepared AgNPs was a very effective antioxidant agent as well as a catalyst in the photocatalytic degradation of methyl orange and methylene blue dyes. Biogenic AgNPs from three different extracts of *Arisaema flavum* tuber (i.e., ethanolic, ethyl acetate, and water extracts) were synthesized as reported by Ur et al. (2019). UV/Visible spectroscopy revealed the presence of silver SPR between 400 and 430 nm. TEM studies revealed the existence of spherical-shaped particles with diameter ranging between 12 nm and 20 nm. EDX confirmed the presence of strong silver signals. The antibacterial activity of the AgNPs was evaluated against multidrug-resistant bacteria (*E. coli, Bacillus subtilis, BL-21, S. Areous, P. Putida*, and *E. coli* substrain QH4). Results unveiled that the synthesized AgNPs exhibited significant antibacterial activity against all the tested multidrug-resistant bacterial strains and particularly against *E. coli* QH4. *Arisaema flavum* derived AgNPs also displayed photocatalytic activity to reduce methylene blue. Lu et al. (2019) managed to biologically fabricate ZnO NPs using the root extract of *Codonopsis lanceolata* via one-pot coprecipitation reaction. *C. lanceolata* is a perennial herb referred to as "Doedok" in Korea. Roots of *C. lanceolata* were reported to exhibit antioxidant and antimicrobial potential (Lu et al. 2019). UV/Vis spectroscopy displayed an absorption peak at 356 nm which was distinctive to ZnO NPs. XRD, EDX, FETEM, and FTIR were used for further characterization of the prepared ZnO NPs. The *C. lanceolata*-derived ZnO NPs were tested for their catalytic potential by testing the degradation of methylene blue dye (MB) dye under UV light (365 nm). The obtained results showed high degradation of MB dye with a degradation percentage of 90.3% within 40 min by *C. lanceolata*-derived ZnO NPs. A simplified green synthesized AgNPs was biologically fabricated via *Persea americana* seed extract as demonstrated by Girón-vázquez et al. (2019). UV/Vis absorption spectra showed the appearance of AgNPs SPR peak at 430 nm. TEM micrographs showed the formation of spherical-shaped particles. The prepared AgNPs exhibited an antimicrobial potential against *E. coli*. The biogenic AgNPs promoted a significant increase in methylene blue Raman signal, and it has been demonstrated as a suitable substrate for being employed in surface-enhanced Raman scattering (SERS).

Veisi et al. (2019) introduced a study in which PdNPs were biologically synthesized using extract of green tea leaves in the presence of ultrasound waves. Furthermore, Fe_3O_4NPs were synthesized via a coprecipitation method. PdNPs were in situ immobilized onto the magnetic core of the prepared Fe_3O_4NPs. The morphological, structural, and physicochemical features of the prepared nanocatalyst were examined via different analytical techniques including XRD, EDX, TEM, FESEM, ICP, VSM, XPS, and FTIR spectroscopy. It was found that the nanocatalyst Pd/Fe_3O_4NPs was reusable for at least six subsequent reactions. Mohammadalinejhad et al. (2019) carried out a study in which promising, green and one-step approach was performed to fabricate antimicrobial nanohybrids. These nanohybrids were based on organic nanofibers like cellulose (CNF), chitosan (CHNF), and lignocellulose (LCNF) nanofibers which were impregnated with AgNPs. Extract of aerial parts of *Lythrum salicaria* played the dual role of reducing and capping agents. The fabricated AgNPs were spherical in shape with a size ranging between 45 and 60 nm. The average hydrodynamic diameter of AgNPs was found to be 50 nm as demonstrated by DLS. Zeta potential value of AgNPs derived from *L. salicaria* was approximately −20 ± 0.76 mV. It was observed that the amount of attached AgNPs was basically dependent upon the type of the used nanofibers; hence, it was found that LCNF and CHNF supported nanohybrids possessed strong affinity to AgNPs than CNF. XRD confirmed the crystalline nature of the nanofibers after AgNPs immobilization. According to release tests, LCNF displayed more efficacy than the other substrates in controlling the release of AgNPs. Moreover, fabricated nanocomposites manifested antibacterial activity against foodborne pathogenic Gram-negative bacteria *E. coli* and Gram-positive bacteria *S. aureus*. Consequently, authors proposed the potential use of the synthesized nanocomposite to be applied for different purposes for instance, nanofillers in manufacturing of antibacterial food packaging materials and in biological purification of water treatment. Extract of the medicinal plant *Acalypha wilkesiana* was used to fabricate AgNPs via a single-pot green synthesis method (Oluwasogo et al. 2019). Optimization of reaction parameters was carried out and included five main parameters, e.g., pH, reaction time, concentration, volume ratio, and temperature, which might influence the process of AgNPs synthesis. Optimized parameters for synthesizing *Acalypha wilkesiana* silver nanoparticles (AW-AgNPs) were found to be 9 for pH, 90 min for reaction time, 0.001 M for Ag^+ ion concentration, 1:9 for volume ratio (extract: Ag^+ solution), and 90–100 °C for temperature. Phytochemical screening of *A. wilkesiana* extract was performed and was found to contain various bioactive phytomolecules including saponins, phenols, flavonoids, and triterpenes which in turn acted as both capping and stabilizing agents. AW-AgNPs was characterized by UV/Vis spectroscopy, FTIR spectroscopy, SEM, TEM, and EDX. SPR peak of AgNPs was observed at 450 nm which occurred as a result of the collective oscillation of free electrons of AgNPs. SEM and TEM results confirmed the formation of well-dispersed spherical-shaped particles with a size of 10–26 nm. AW-AgNPs exhibited antibacterial potential against both *Escherichia coli* and *Staphylococcus aureus*.

5.4 Biovalorization of Agro-Industrial Wastes into Nanomaterials

One of the major issues that gain the attention of the international communities is food sustainability and security as addressed by the United Nations' Sustainable Development (Poinern and Fawcett 2018). Sustainable food production and excellent monitoring of food resources are the key factors for efficient abundance of foodstuff according to the international policy makers as it is expected for human population to reach 12.3 billion people in 2100 (Gerland et al. 2014). Moreover, according to current and future modeling, the increase in global temperatures, growing demands on energy sources and the deficiency of natural resources as well as the increase in pollution has a great impact upon our environment. Human industrialization and urbanization have an extreme influence in diminishing natural resources.

No doubt, huge amount of food wastes is generated every year. According to Food and Agricultural Organization approximately 1.3 billion tons of all food produced is lost each year (Poinern and Fawcett 2018). As a result, food security involves not only sustainable food production, but also addressing and handling of the high level of generated food waste becomes a must. Conventionally, waste management protocols include treatment, limitation, and prevention strategies to decrease the damaging consequences on environment from disposal methods such as incineration and landfill. The usual waste management strategies that are followed in dealing with food waste involve their use as animal feed, compost manufacturing, incineration, conversion of waste to energy (e.g., anaerobic digestion) and landfill. Unfortunately, some of these strategies have a lethal impact on environment; hence, searching for alternative and environmentally friendly routes have attracted the attention of worldwide researchers. New manufacturing processes and waste valorization procedures are required to make the best use of food waste sources and convert them into more beneficial products. Table 5.1 illustrates five process methodologies that are most important in waste valorization strategies. These products may involve pharmaceuticals, industrial important chemicals, fuel, and biomaterials. Agricultural crop-industrial wastes are also significant candidates to biosynthesize nanomaterials as they are rich with many important phytochemicals that can aid in the bioreduction of metal salt precursors (Shah et al. 2015). Therefore, using agricultural/food wastes are gaining particular momentum in the modern world. Recently, the amounts of agricultural waste are rising quickly worldwide, and their negative impacts on environment have drawn more and more attention. As a result, the need to adopt appropriate approaches to minimize, reuse, recycle and upgrade these wastes is the need of the hour.

Furthermore, food wastes are efficient renewable resources as they are majorly consisted of organic materials, which can be valorized into various valuable products such as natural dyes, bioactive compounds, biofuels, pharmaceuticals, and nanomaterials. Vandermeersch et al. (2014) reported that there is a high motivation in replacing the petrochemical-derived materials with renewable ones, for instance, those derived from food waste as well as the ones produced during food processing. Indeed, the process of waste valorization which refers to the conversion and upgrade of food wastes into more beneficial and higher value products has increasingly attracted much more attention lately (Mirabella et al. 2014). It is worth mentioning that valorizing food waste is not a new strategy, but it has gained a wider appeal because of its ability to produce beneficial, cost-effective, and highly valuable products. Because of the intense global demand upon products like enzymes, biofuels, solvents, pharmaceuticals, and surfactants, this prompted several countries to create new strategies and protocols for developing large-scale facilities to valorize food waste streams into a number of valorized products (Snyder 2015). Though, certain points have to be considered, among which; determination of the most effective type of conversion process, its efficacy (i.e., illustrating degree of waste valorization), its financial viability for commercial operations (Poinern and Fawcett 2018). In order to overcome such massive challenges, a combination effort has to be planned by the collaboration of different scientific disciplines like biochemistry, biotechnology,

Table 5.1 Top strategies employed in waste valorization

Process name	Types	Process description	Examples	Reference
Thermal conversion processes	Hydrothermal carbonization and pyrolysis	Hydrothermal carbonization: low temperature process (180–350 °C) which takes place under autogenous gas pressures	Conversion of urban food wastes in Singapore into high-value hydrochars to be used in removing textile dyes from contaminated water	Parshetti et al. (2014), Heo et al. (2010)
		Pyrolysis: it involves heating biomass at elevated temperatures in absence of oxygen to produce decomposed products and helps to manufacture highly advanced nanometer scale materials, e.g., carbon nanospheres, CNTs, and graphene-like materials	Conversion of sawdust into a bio-oil product	
Solvent extraction processes	–	A process that is employed to separate soluble components from a solid matrix using a suitable solvent. It is usually used to extract a specific component from plant matrix such as antioxidants, organic acids, phenolic compounds, and vitamins from food waste	The use of water and ethanol mixtures for extracting antioxidant compounds from potato waste peels (*Solanum tuberosum*)	Amado et al. (2014)
Chemical and biotechnology processes	Hydrolysis, oxidation reactions, chemoenzymatic, and biotechnological methodologies	These methodologies result in producing sustainable forms of bioenergy like biogas, biodiesel, and bioethanol. It also includes the green synthesis of high-valuable metal and metal oxide NPs	Waste upcycling of sweet orange peels into AgNPs	Omran et al. (2017)

(continued)

Table 5.1 (continued)

Process name	Types	Process description	Examples	Reference
Microwave-assisted processes	Closed and opened processes	It is a process which aids in separating and extracting chemical compounds from food wastes. Microwaves interact with molecules via dipole rotation and ionic conduction; hence, water present within the food waste absorbs microwave energy rapidly till being superheated. The resultant superheated water causes disruption and subsequent release of the inner contents		Ho et al. (2015)
Ultrasound-assisted processes		Ultrasound transducers generate acoustic waves (frequencies greater than 20 kHz) that travel through the solvent (water or organic) causing alternating compression and expansion cycles. The expansion cycle pulls molecules apart to create cavities or bubbles that rapidly grow in the solvent. During the compression cycle there is a suddenly collapse of the bubble releasing large amounts of energy	Extraction of antioxidants from food processing by-products	Pingret (2013), Virot et al. (2010)

environmental science, food production practices, along with governmental legislation and policies. Because all of the mentioned above, waste valorization has come to real by applying different processes that guarantee innovative, environmentally friendly and viable protocols to convert such food waste into high-valuable products like production of NPs (Table 5.2). Figure 5.1 represents a schematic illustration of AgNPs green synthesis using agricultural wastes.

Aqueous peel extract of *Annona squamosa* was found to mediate AgNPs fabrication as revealed by Kumar et al. (2012). The bioreduction of Ag^+ ions were monitored using UV/Vis spectrophotometer, XRD, and TEM. The biosynthesized AgNPs exhibited an irregular spherical shape with a size ranging from 20 to 60 nm. Jaybhaye (2015) managed to exploit the waste vegetable fibers of *Murraya koenigii*, *Moringa oleifera*, and *Luffa acutangula* to biologically synthesize AgNPs. A strong SPR peak of AgNPs was detected via UV/Vis spectroscopy at 422 nm. The average sizes of the prepared AgNPs were 30–40, 30–32, and 30–35 for waste vegetable

Table 5.2 Overview of different plant parts and agro-industrial waste extracts used in the biological synthesis of NPs; size, shape, characterization tools used to determine their characteristics as well as their possible applications

Plant species	Type of NPs produced	Plant part extract	Size (nm)	Shape	Characterization	Application	Reference
Production of metal and metal oxide NPs via plant leaf extracts							
Podophyllum hexandrum	Ag	Leaf	14	Spherical	TEM, XRD, FTIR	Anticancer activity	Jeyaraj et al. (2013)
Magnolia kobus	Cu	Leaf	73–110	Spherical	Inductively coupled plasma spectrometry (ICP), XPS, EDX, HRTEM	Antibacterial activity	Lee et al. (2013)
Lawsonia inermis, *Gardenia jasminoides*	Fe	Leaf	*L. inermis* (21), *G. jasminoides* (32)	Distorted hexagonal	FTIR, TGA, TEM, SEM, AFM, XRD, EDX	Antibacterial activity	Naseem and Farrukh (2015)
Artocarpus altilis	Ag	Leaf	25–43	Spherical	UV/Vis spectroscopy, SEM, TEM, DLS	Antimicrobial and antioxidant activities	Ravichandran et al. (2016)
Lantana camera	Iron oxide	Leaf	10–20	Rod	UV/Vis spectroscopy, XRD, SEM, FTIR, EDX	Antibacterial activity and seed germination effect	Rajiv et al. (2017)
Azadirachta indica	Ag	Leaf	35.4	Spherical	UV/Vis spectroscopy, DLS, zeta potential, FTIR, TEM	Immunomodulatory activity in fish	Rather et al. (2017)
Maclura pomifera (Osage orange)	Ag	Leaf	12	Spherical	UV/Vis spectroscopy, XRD, FTIR, TEM	Antimicrobial activity	Azizian-Shermeh et al. (2017)
Dodonaea viscosa and *Hyptis suaveolens*	Ag	Leaf	40–55	Spherical	UV/Vis spectroscopy, XRD, FTIR, SEM, DLS, zeta potential	Antibiofilm activity	Muthamil et al. (2018)

(continued)

Table 5.2 (continued)

Plant species	Type of NPs produced	Plant part extract	Size (nm)	Shape	Characterization	Application	Reference
Limonia acidissima L.	Ag, ZnO	Leaf	AgNPs (21–42), ZnO NPs (12–53)	Spherical	UV/Vis spectroscopy, FTIR, HRTEM	Antibacterial activity	Patil and Taranath (2018)
Combretum erythrophyllum	Ag	Leaf	13.62	Spherical	UV/Vis spectroscopy, TEM, FTIR, DLS, XRD	Antimicrobial activity	Jemilugba et al. (2019)
Diospyros lotus	Ag	Leaf	10–25	Spherical	UV/Vis spectroscopy, XRD, FTIR, TEM, SEM, atomic absorption spectrometer	Antibacterial and catalytic activities	Hamedi and Shojaosadati (2019)
Agrimoniae herba	Core-shell Cu@Pt	Leaf and shoot	30	Spherical	UV/Vis spectroscopy, FTIR, AFM, SEM, TEM	Antimicrobial activity	Dobrucka and Dlugaszewska (2018)
Passiflora Edulis F. Flavicarpa	Ag	Leaf	3–7	Spherical	UV/Vis spectroscopy, FTIR, XRD, SEM, TEM	Antioxidant and photocatalytic activities	Thomas et al. (2019)
Sida acuta	CuO	Leaf	50	Rod	UV/Vis spectroscopy, FTIR, SEM, TEM, EDX	Antimicrobial and photocatalytic activities	Sathiyavimal et al. (2018)
Alpinia nigra	Au	Leaf	21.52	Spherical	UV/Vis spectroscopy, FTIR, XRD, TEM	Photocatalytic activity	Baruah et al. (2018)
Gloriosa superba	ZrO	Leaf	11.625	Irregular	XRD, SEM, EDX	Sensitized solar cell	Vennila et al. (2018)
Bauhinia tomentosa	ZnO	Leaf	22–94	Spherical, cubic	UV/Vis spectroscopy, TEM, EDX, XRD, FTIR	Antibacterial activity	Sharmila et al. (2018)
Costus pictus D. Don	MgO	Leaf	50	Hexagonal	FTIR, XRD, SEM, TEM	Antimicrobial activity	Suresh et al. (2018)

Atalantia monophylla	ZnO	Leaf	30	Spherical	UV/Vis spectroscopy, fluorescence spectrometer (PL), XRD, TEM, EDX, SEM	Antimicrobial activity	Vijayakumar et al. (2018)
Lantana camara	CuO	Leaf	17	Spheroid	UV/Vis spectroscopy, XRD, FTIR, SEM, EDX	Photocatalytic activity	Arunkumar et al. (2019)
Schinus molle, Equisetum giganteum, Ilex paraguariensis Saint Hilaire	Ag	Leaf	41.0 ± 2.0, 12.0 ± 0.6, 20.0 ± 1.0	Quasi-spherical	UV/Vis spectroscopy, FTIR, TEM	Antimicrobial activity	Barberia-Roque et al. (2019)
Camellia japonica	Au	Leaf	20	Spherical	UV/Vis spectroscopy, FTIR, XRD, SEM	Antimicrobial activity	Sharma et al. (2019b)
Dracocephalum kotschyi	Au	Leaf	5–21	Spherical	UV/Vis spectroscopy, XRD, FTIR, TEM, FESEM	Antibacterial, antioxidant, and catalytic activities	Chahardoli et al. (2019)
Camilla japonica	CuO	Leaf	15–25	Spherical	UV/Vis spectroscopy, FTIR, SEM, TEM, EDX	Antibacterial activity	Rajivgandhi et al. (2019)
(caper)	CuO	Leaf	21–85	Spherical	UV/Vis spectroscopy, TEM, FTIR	Iron sensing	Samari et al. (2019)
Cnici benedicti	Au-CuO, CuO-ZnO	Leaf	Au-CuO (20), CuO-ZnO (30)	Spherical	UV/Vis spectroscopy, FTIR, SEM, TEM, AFM	Antimicrobial, antitumor activities	Dobrucka et al. (2019)
Berberis vulgaris	Ag	Leaf	30–70	Spherical	UV/Vis spectroscopy, XRD, DLS, TEM	Antibacterial activity	Behravan et al. (2019)
Ficus retusa	Ag, Au	Leaf	Ag (15), Au (10–25)	Spherical	UV/Vis spectroscopy, TEM, FTIR, Zetasizer	Detection of heavy metals (Fe^{3+} ions)	Zayed et al. (2019)

(continued)

Table 5.2 (continued)

Plant species	Type of NPs produced	Plant part extract	Size (nm)	Shape	Characterization	Application	Reference
Tinospora cardifolia	Cu	Leaf	178	Spherical	UV/Vis spectroscopy, SEM, TEM, ATR-FTIR, XRD, Raman spectroscopy, AFM, EDX	Antimicrobial potential	Sharma et al. (2019a)
Xanthium strumarium	Pt	Leaf	22	Rectangular	UV/Vis spectroscopy, XRD, FTIR, TEM, cyclic voltammetry analysis, EX	Cytotoxic and antimicrobial activities	Kumar et al. (2019b)
Croton caudatus Geisel	Au	Leaf	20–50	Spherical	UV/Vis spectroscopy, FTIR, XRD, SEM-EDAX, TEM	Free radical scavenging and antifungal activities	Vijaya Kumar et al. (2019)
Simarouba glauca	Au	Leaf	–	Spherical, prism-like	UV/Vis spectroscopy, FTIR, XRD, HRTEM	Antimicrobial activity	Thangamani and Bhuvaneshwari (2019)
Alcea rosea	Au	Leaf	4–95	Triangular, pentagonal, spherical, hexagonal,	UV/Vis spectroscopy, TEM, SEM, EDX, XRD, FTIR, Zetasizer	Antioxidant, catalytic activities	Khoshnamvand et al. (2019)
Costus igneus	ZnO	Leaf	26.55	Hexagonal	UV/Vis spectroscopy, FTIR, XRD, TEM	Antidiabetic, antibacterial and antibiofilm activities	Vinotha et al. (2019)
Mirabilis jalapa	ZnO, Ag	Leaf	ZnO NPs (12.9), AgNPs 32.8	ZnO NPs (needle-like), AgNPs (spherical)	XRD, SEM, EDX, FTIR	Antimicrobial activity	Sumbal et al. (2019)

Ruellia tuberosa	CuO	Leaf	83.23	Rod	UV/Vis spectroscopy, TEM, SAED, DLS, FESEM, EDX	Antimicrobial, photocatalytic activities	Vasantharaj et al. (2019)
Rauvolfia tetraphylla	Ag	Leaf	40	Cubic	UV/Vis spectroscopy, SEM, TEM	Anticancer, antioxidant, and antimitotic activities	Vinay et al. (2019)
Cistus incanus	Au	Leaf	45–85	Popcorn	UV/Vis spectroscopy, TEM, AFM, Multiphoton-excited luminescence (MPL)	–	Klekotko et al. (2019)
Ananas comosus	Pd	Leaf	1.71–16.14	Spherical	UV/Vis spectroscopy, XRD, FTIR, HRTEM, EDX, TGA	Photocatalytic activity	Olajire and Mohammed (2019)
Azadirachta indica	TiO2	Leaf	15–50	Spherical	TEM, SEM, FTIR, XRD	Antibacterial activity	Thakur et al. (2019)
Tecoma castanifolia	ZnO	Leaf	70–75	Spherical	UV/Vis spectroscopy, TEM, EDX, FTIR, XRD	Antioxidant, antibacterial, and anticancer activities	Sharmila et al. (2019)
Parkia speciosa	Ag	Leaf	26–39	Spherical	UV/Vis spectroscopy, XRD, EDX, SEM	Antibacterial, antioxidant, and photocatalytic activities	Ravichandran et al. (2019)
Falcaria vulgaris	Cu	Leaf	20–25	Spherical	UV/Vis spectroscopy, XRD, FTIR, TEM, FESEM	Antioxidant, antifungal, antibacterial, cytotoxic, and cutaneous wound healing activities	Zangeneh et al. (2019)
Murraya koenigii	Ag	Leaf	5–20	Spheroidal	UV/Vis spectroscopy, FTIR, XRD, SEM, TEM	Antimicrobial activity	Qais et al. (2019)

(continued)

Table 5.2 (continued)

Plant species	Type of NPs produced	Plant part extract	Size (nm)	Shape	Characterization	Application	Reference
Production of metal and metal oxide NPs via extracts of other plant parts (gall, flower, rhizome, glove, stem, carb, pod, seeds, aerial parts)							
Pistacia integerrima	Au	Gall extract	20–200	Spherical	UV/Vis spectroscopy, FTIR, SEM	Antifungal activity, antinociceptive, and muscle relaxant properties	Islam et al. (2019)
Chamaemelum nobile	Ag	–	4.2 ± 3.1	Spherical	UV/Vis spectroscopy, TEM, FTIR, XRD, DLS	Antimicrobial activity	Erjaee et al. (2017)
Nyctanthes arbor-tristis	ZnO	Flower	12–32	Spherical	UV/Vis spectroscopy, FTIR, DLS, XRD, TEM	Antifungal activity	Jamdagni et al. (2016)
Callistemon viminalis	Fe_2O_3 NPs	Flower	22–32	Spherical	UV/Vis spectroscopy, FTIR, XRD, HRSEM, HRTEM, EDX	Antileishmanial, anticholinesterase, anticancer, antibacterial, and antioxidant activities	Hassan et al. (2018)
Acorus calamus	Iron oxide, manganese oxide	Rhizome	Iron oxide (20–30)	Iron oxide (spherical), manganese oxide (slimy layer)	UV/Vis spectroscopy, FTIR, XRD, SEM, TEM	Larvicidal, antibacterial, and catalytic activities	Arasu et al. (2019)
Syzygium aromaticum	Cu	Clove (bud)	15	Spherical	UV/Vis spectroscopy, XRD, EDX, FTIR	Antimicrobial activity	Rajesh et al. (2018)

Rhus punjabensis	CeO_2	Stem	23	Spherical	UV/Vis spectroscopy, XRD, FTIR, thermogravimetric analysis (TGA), differential scanning calorimetry (DSC), SEM	Antioxidant, antileishmanial cytotoxic activity	Naz et al. (2019)
Phyllanthus pinnatus	Ag	Stem	>100	Cubic, triangular, spherical	UV/Vis spectroscopy, SEM, XRD, FTIR	Antibacterial activity	Balachandar et al. (2019)
Rhus punjabensis	CeO_2	Stem	20–25	Circular	SEM, XRD, TGA	Antioxidant and Antileishmanial activities	Naz et al. (2019)
Senna alata	Ag	Bark	10–30	Spherical	UV/Vis spectroscopy, FTIR, TEM, EDX	Antimicrobial activity	Ontong et al. (2019)
Arachis hypogaea L. (peanut plant)	Au	Pod	4–6	Spherical	UV/Vis spectroscopy, ICP-AES, HRTEM, EDX, XRD, FTIR, XPS	–	Raju et al. (2012)
Ceratonia siliqua	Iron oxide	Carob pod	7 ± 5	Spherical	TEM, SEM, EDX, XRD, FESEM, Raman spectroscopy	Amoxicillin degradation	Aksu Demirezen et al. (2019)
Centaurea pumilio L.	Ag	Aerial parts	6–8	Spherical	UV/Vis spectroscopy, FTIR, TEM	Effective skin treatment	Mostafa et al. (2019)
Euphorbia maculata	$Ni@Fe_3O_4$, CuO	Aerial parts	$Ni@Fe_3O_4$ (30), CuO (18)	Spherical	UV/Vis spectroscopy, XRD, FTIR, EDX, BET, TGA, FESEM	Photocatalytic activity	Pakzad et al. (2019)
Pongamia pinnata	ZnO	Seed	30.4–40.8	Spherical	UV/Vis spectroscopy, XRD, FTIR, SEM, EDX	Antibacterial, antibiofilm and anticancer activities	Malaikozhundan et al. (2017)

(continued)

Table 5.2 (continued)

Plant species	Type of NPs produced	Plant part extract	Size (nm)	Shape	Characterization	Application	Reference
Elettaria cardamomum	Au	Seed	15.2	Spherical	UV/Vis spectroscopy, XRD, FTIR, TEM	Antioxidant, antibacterial, and anticancer activities	Rajan et al. (2017)
Stevia	Iron oxide	Seed	>25	Spherical	XRD, FESEM, HRTEM, TGA, XPS, VSM, zeta potential	Antioxidant activity	Khatami et al. (2019)
Salvia hispanica L.	Ag	Seed	7	Spherical	UV/Vis spectroscopy, FTIR, FESEM, EDX, TEM	Antibacterial activity	Hernández-Morales et al. (2019)
Cucurbita pepo	TiO_2	Seed	–	–	UV/Vis spectroscopy, XRD, FTIR	–	Abisharani et al. (2019)
Jatropha curcas	Ag	Seed	80–95	Spherical	UV/Vis spectroscopy, zeta potential, SEM, FTIR, XRD	Antibacterial activity	Nayak et al. (2019)
Euclea natalensis	ZrO_2	Root	5.90–8.54	Rounded	XRD, FTIR, TEM	Adsorptive activity	da Silva et al. (2019)
Rhodiola rosea	AgNPs	Root	80–120	Rounded	UV/Vis spectroscopy, SEM, FTIR, DLS, EDX	–	Pawliszak et al. (2019)
Production of metal and metal oxide NPs via agro-industrial waste extracts							
Citrus sinensis	Ag	Peel	35 ± 2 (25 °C), 10 ± 1 (60 °C)	Spherical	UV/Vis spectroscopy, FTIR, EDX, XRD, TEM, FESEM	Antibacterial activity	Kaviya et al. (2011)
Citrus sinensis	Ag	Peel	7.36 ± 8.06	Spherical	UV/Vis spectroscopy, fluorescence emission spectroscopy, XRD, TEM	–	Kahrilas et al. (2013)

Mangifera indica Linn	Ag	Peel	7–27	Quasi-spherical	UV/Vis spectroscopy, FTIR, XRD, TEM	Antibacterial activity	Yang and Li (2013)
Mangifera indica	Au	Peel	6–18	Quasi-spherical	UV/Vis spectroscopy, XRD, TEM, FTIR	Cytotoxic activity	Yang et al. (2014)
Punica granatum	Ag	Peel	5–50	Quasi-spherical	UV/Vis spectroscopy, FTIR, SEM	Antibacterial activity	Shanmugavadivu et al. (2014)
Lansium domesticum	Ag, Au, Au/Ag a lloy	Peel	AuNPs (140 ± 13), Au–AgNPs (128 ± 15)AgNPs (74 ± 6)	AuNPs (triangul AgNPs (spherical), Au–AgNPs (quasi spherical) ar, spherical),	FTIR, DLS, TEM, SAED, zeta potential	Antimicrobial activity	Shankar et al. (2014)
Musa sapientum	Ag	Peel	23.7	Spherical	UV/Vis spectroscopy, XRD, EDX, DLS, TEM, FESEM, SEM	Antibacterial activity	Ibrahim (2015)
Citrus sinensis, Citrus reticulate, Citrus aurantifolia	Ag	Peel	31.0 ± 18.3 (AgNP-Ora), 29.8 ± 18.7 (AgNP-Tan), 18.5 ± 11.6 (AgNP-Lem)	Spherical	UV/Vis spectroscopy, FTIR, HRTEM	Raman probe molecules	de Barros Santos et al. (2015)
Allium cepa L.	Au	Peel	45.42	Spherical, triangular	UV/Vis spectroscopy, FESEM, EDX, XRD, FTIR, DT-TGA	Antibacterial, anticandidal, antioxidant, and proteasome activities	Patra et al. (2016)
Garcinia mangostana	Au	Peel	32.96 ± 5.25	Spherical	UV/Vis spectroscopy, FTIR, XRD, TEM, size particle analyzer	–	Xin Lee et al. (2016)

(continued)

Table 5.2 (continued)

Plant species	Type of NPs produced	Plant part extract	Size (nm)	Shape	Characterization	Application	Reference
Prunus persica L.	AgCl	Peel	15–50	Spherical	UV/Vis spectroscopy, SEM, EDX, FTIR, TGA, XRD	Antibacterial, anticandidal, and antioxidant activities	Patra and Baek (2016)
Citrus sinensis	Ag	Peel	15	Spherical	UV/Vis spectroscopy, XRD, FTIR, TEM, SEM, DLS, zeta potential	–	Omran et al. (2017)
Citrus limon, *Citrus sinensis*, *Citrus limetta*	Ag	Peel	9–64	Spherical, irregular	UV/Vis spectroscopy, TEM, DLS	Antimicrobial, antioxidant and cytotoxic activities	Ahmed et al. (2018)
Mangifera indica	Zero valent iron	Peel	–	Amorphous	UV/Vis spectroscopy, XPS, XRD, FTIR, GC/MS, EDX, SEM	Degradation activity	Desalegn et al. (2019)
Theobroma cacao	Ag	Husk, pulp, seed	6–18	Spherical	UV/Vis spectroscopy, FTIR, XRD, DLS, Zeta potential, TEM, SEM	Antibacterial and photocatalytic activities	Thatikayala et al. (2019)
Garcinia mangostana	Ag, Au, Pt	Rind portion of the fruit	AgNPs (23), AuNPs (20–40), PtNPs (20–25)	Spherical	UV/Vis spectroscopy, FTIR, XRD, HRSEM, HRTEM, zeta potential, EDX	Antibacterial activity	Nishanthi et al. (2019)
Brazilian red propolis	Ag	–	9	Spherical	UV/Vis spectroscopy, DLS, XRD, TEM, FEG-SEM	Antimicrobial activity	Barbosa et al. (2019)
Apple pomace	Ag	Solid residue after apple processing	10–20	Spherical	UV/Vis spectroscopy, TEM, XRD, FTIR	Antibacterial activity	Ren et al. (2019)

Thymus vulgaris L.	ZnO	Waste thyme	10–35	Cubical, rectangle, hexagonal, and rod-radial	UV/Vis spectroscopy, SEM, XRD, FTIR	–	Abolghasemi et al. (2018)
Abelmoschus esculentus	ZnO	Mucilage	29	Microflakes	UV/Vis spectroscopy, FTIR, XRD, XPS, EDX, Zeta potential, HRTEM	Photodegradation of dyes	Prasad et al. (2019)

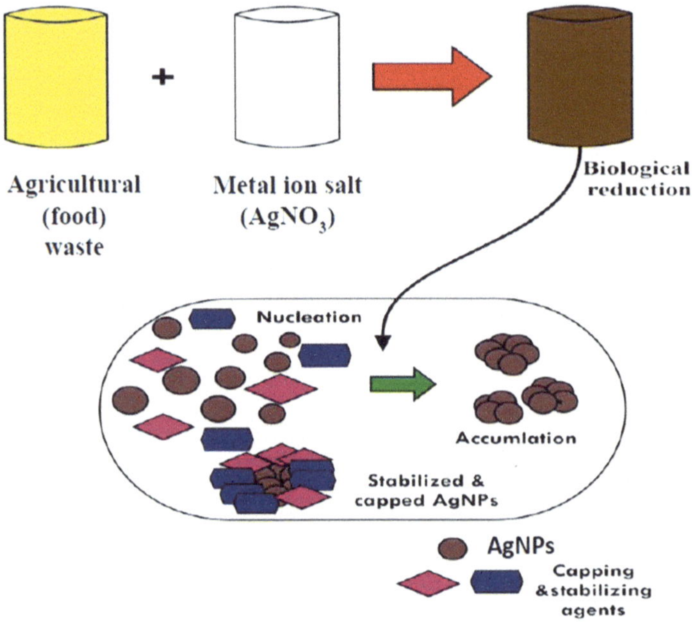

Fig. 5.1 Biological synthesis of NPs (e.g., AgNPs) via agricultural wastes

fibers extract of *M. koenigii*, *M. oleifera*, and *L. acutangula*, respectively. The prepared AgNPs exhibited an antibacterial activity against *Staphylococcus aureus*. A facile synthesis route of highly stable AgNPs using a biopolymer (e.g., xylan) as both reducing and stabilizing agents was documented by Harish et al. (2015). Xylan, a hemicellulose, exists predominantly as a cell wall material in annual plants and agricultural crops (e.g., straw, sorghum, corn stalks, sugar cane, cobs, grasses, cereals, herbs, etc.). It is one of the major constituents of lignocellulosic materials as revealed by Ebringerova and Heinze (2000). Xylan is a heteropolysaccharide compound made up of (1–4) and/or (1–3) xylose residues. Xylan has been employed in different manufacturing disciplines including paper making, film wrapping, textile printing, adhesives, a thickeners as well as additives to plastics. Additionally, it has been used in food industry as an emulsifier and protein foam stabilizer during heating (Ebringerová et al. 1995). In this study, xylan was obtained from waste biomass of wheat bran (WB) via alkaline treatment. The obtained xylan was characterized using Fehling's test, dinitrosalicylic acid assay, FTIR, 1H NMR, and 13C NMR. The synthesized AgNPs were characterized using UV/Vis spectroscopy and TEM. The prepared particles were polydispersed with a size ranging from 20 to 45 nm. The synthesized WB-xylan AgNPs displayed excellent free radical scavenging activity. Furthermore, authors proved that WB xylan AgNPs exhibited a fibrinolytic activity as was evidenced by the zone of clearance in fibrin plate assay. Also, it aided in the dissolution of blood clots. From the obtained results, xylan-derived AgNPs nanocomposite can be beneficial for treating thrombus-related diseases. One of the largest crops produced worldwide is sugarcane. Sugarcane bagasse is generated in high quantities by sugar and alcohol industries. Sugarcane bagasse is a complex substance made up of cellulose, hemicellulose, lignin, and biologically derived silica. It is used in versatile applications, for instance, electricity generation, ethanol production, and an additive of concrete, cement, and mortar mixtures (Faria et al. 2010; Souza et al. 2011; Gusmão et al. 2012). Yet, large quantities of bagasse are not exploited and are being burnt in fields, leading to a massive environmental pollution. As a result, to avoid such negative impact on the environment, it is urgently needed to convert it into value-added products. Recently, silica NPs have received worldwide attention owing to their distinctive biological properties in several biomedical applications, involving their use in separation and adsorption of proteins, detection of nucleic acids, gene therapy, molecular imaging, drug delivery, and scaffolds (An et al. 2010; Liou and Yang 2011). For the aforementioned reasons, Athinarayanan et al. (2015) exploited sugarcane bagasse as a fibrous material for the biogenesis of silica NPs (BSNPs). An autoclave was used for the acid pretreatment of sugarcane in order to eliminate metal ions and promote the hydrolysis of the existed organic constituents. Residues of the acid pretreatment were incinerated at various temperatures to define the role of temperature in forming BSNPs. XRD revealed the crystalline nature of the prepared BSNPs and indicated that the prepared BSNPs exhibited an amorphous nature. Whereas, TEM images showed that BSNPs were irregular in shape with a porous structure. Authors investigated the biocompatibility of BSNPs morphology, viability, mitochondrial, gene expression, and reactive oxygen species (ROS). By employing 3-(4, 5- Dimethylthiazol-2-yl)-2, 5-diphenyltetrazolium bromide assays as well as microscopic techniques, BSNPs were found to exert no effects upon cell viability or morphology. Contrary, BSNPs slightly affected the mitochondrial membrane potential at high concentrations. Hence, it can be concluded that BSNPs were biocompatible which candidate them to be used for biomedical applications.

AgNPs were biologically fabricated for the first time using the seed coat (hull) of *Vigna mungo* as was conducted by Varadavenkatesan et al. (2017). This seed coat is usually discarded in urban life. Utilization of such domestic waste offers a low-cost, safe and natural source for synthesizing AgNPs. UV/Vis spectrophotometry, SEM-EDX, XRD, FTIR, and DLS were employed in characterizing *V. mungo* derived AgNPs. The biogenic AgNPs exhibited anticoagulant as well as antioxidant activities. (Khodadadi et al. 2017) demonstrated the capability of waste peach (*Achillea millefolium* L.) kernel shell extract to mediate AgNPs production. FTIR spectroscopy, UV/Vis spectroscopy, XRD, FESEM, EDX, TEM, and thermogravimetry-differential thermal analysis (TGDTA) were employed to characterize peach kernel shell derived AgNPs. AgNPs/peach kernel shell was found to be a high catalyst in reducing 4- nitrophenol (4-NP), methylene blue (MB), and methyl orange (MO) at room temperature. Furthermore, authors illustrated that AgNPs/peach kernel shell can be recovered and reused for several times without any significant loss in its catalytic activity. Sinsinwar et al. (2018) used the agricultural waste of *Cocos nucifera* (L.) (coconut shell) to produce AgNPs. AgNPs synthesized using coconut shell extract (CSE-AgNPs) was characterized via UV/Vis spectroscopy which revealed the appearance of an absorption peak at 432 nm. TEM displayed the formation of spherical-shaped particles with a size of 14.2–22.96 nm. FTIR illustrated the role of organic molecules in the process of capping and stabilization of the prepared AgNPs by the emergence of bands at 1384, 1609, and 3418 cm^{-1}. CSE-AgNPs exhibited antimicrobial potential against *S. aureus*, *E. coli*, *S. typhimurium*, and *L. monocytogenes* with zone of inhibition (ZOI) of 15, 13, 13, and 10 mm, respectively. Growth curve assay clarified the efficacy of CSE-AgNPs to constrain the growth of selected pathogens when compared with ampicillin and the extract alone. SEM revealed that cell wall degradation might be the possible mechanism behind the antibacterial potential of CSE-AgNPs. Furthermore, authors tested different concentrations of AgNPs (0.078–2.5 mg/ml) against human PBMC cell line and showed no presence of any toxic effects. Deokar and Ingale (2018) investigated for the first time the potential of beetroot waste (peels) aqueous extract to fabricate crystalline AuNPs. Characterization of the prepared nanoplates was monitored via several spectroscopic and microscopic techniques including UV/Vis spectroscopy, FTIR, EDX, XPS, HRTEM, SAED, and XRD. TEM revealed the synthesis of spherical, icosahedral, and triangular shaped particles. Interestingly, the triangular particles possessed a flat surface which was similar to nano-single crystalline plates. Authors assumed that the mechanism of synthesis passed through three main steps. Those steps were described as the following: first the biological synthesis of the nanospheres, second its transformation to icosahedrons, and third, eventually, its fragmentation into triangular nanoplates. The distinctive mentioned mechanism reported by the authors was assessed via HRTEM.

Graphene is a two-dimensional atomic sp^2-hybridized allotrope made up of carbon. It possesses outstanding electrical and thermal conductivity, superior mechanical strength, and chemical features which attracted the attention of worldwide researchers. Because of these excellent optical properties, graphene-based nanomaterials have been extensively investigated for different applications like photoluminescence and fluorescence. To date, graphene oxide (GO) is mostly synthesized by either chemically exfoliation (Yang et al. 2016) and by Hummer's method (Yang et al. 2016). However, both methods release toxic gases. As a result, in a study performed by Tewari et al. (2019), a green, environmentally friendly, low-cost and one-pot hydrothermal route was employed to synthesize potassium-doped graphene oxide (K-doped GO) by using the outer waste layers of *Quercus ilex* fruit (cupule and pericarp). A high percentage of approximately 6.81% of natural doping of potassium was revealed via XPS. The K-doped GO released bright blue photoluminescence (PL) under UV light ($\lambda ex = 365$ nm). Hence, based on these novel results, authors strongly recommended its use as a powerful fluorescence probe in imaging and biosensing with a great promise in biological and analytical fields. Baruah et al. (2019) presented a study in which AgNPs were successfully synthesized using the grinded fruit extract of *Alpinia nigra* after being dried. According to previously published studies, it has been reported that *A. nigra* contains saponins, glycosides, steroids, alkaloids, and rich in polyphenols (Ghosh et al. 2013; Ahmed et al. 2015). Phytochemical screening showed that the total flavonoid and phenol contents of *A. nigra* extract were 718 mg /g extract and 74.9 mg/g extract, respectively. The average particle size of the fabricated AgNPs was 6 nm and was rounded in shape. The biomolecules of *A. nigra* extract exerted a functional dual role in reducing and capping the prepared AgNPs as was evident by FESEM and FTIR. The *A. nigra*-derived AgNPs exhibited promising antimicrobial potential against Gram-negative bacteria, e.g., *Klebsiella pneumoniae*, Gram-positive bacteria, e.g., *Staphylococcus aureus*, and the pathogenic yeast, *Candida albicans*. *K. pneumoniae* was the least resistant to AgNPs among the tested pathogenic strains. Additionally, Baruah and co-authors proved the efficient capability of AgNPs to catalyze the photocatalytic degradation of the anthropogenic dyes (e.g., methyl orange, rhodamine B, and Orange G) in the presence of sunlight.

In a research study introduced by Masoudian et al. (2019), titanium dioxide (TiO_2) nanoparticles (NPs) were loaded onto activated carbon (AC) which was derived from biowaste of watermelon rind (WR) and was referred to as (TiO_2-NPs-ACWR), and was evaluated for its ability to remove phenol red (PhR) and congo red (CR) dyes. TiO_2-NPs-ACWR was characterized via FTIR, FESEM, SEM, EDX, TGA, and BET techniques. Optimization of reaction conditions was carried out. The pH effect was first studied via one-factor-at-a-time technique (OFAT). Whereas, the other reaction parameters such as sonication time, initial dye concentrations, and adsorbent dosage were optimized using central composite design (CCD) combined with desirability function. The optimum conditions were found to be pH 4.3, 0.04 g of TiO_2-NPs-ACWR, 8.22 min of sonication time, and initial concentrations of 45.9 mg L^{-1} and 14.7 mg L^{-1} for Ph R and CR, respectively. A high-removal percentage was attained and reached 100.24% for PhR and 101.45% for CR. Moreover, the adsorption capacities were found to be 55.6 mg g^{-1} for PhR and 17 mg g^{-1} for CR. Dodevska et al. (2019) collected the waste materials of *Rosa damascena* after hydrodistillation from biocertified Mill as a trial to biologically fabricate AgNPs. Four extract types of *R. damascena* waste were prepared, and they were post-distillation water (PW), water extract (WE), and 30% and 70% ethanol extracts (EE). It was shown that the EE was rich in polyphenols and specifically catechin and epicatechin, while the WE was rich in reducing sugars, proteins, and pectic components. The prepared AgNPs from the four extract types displayed irregular shapes with a size of 25.8, 11.5, 11.5, and 8.4 nm for PW, WE, 30% (EE), and 70% (EE), respectively. Authors evaluated the electrocatalytic activity of AgNPs on reducing H_2O_2. The biosynthesized AgNPs were placed onto a spectroscopic graphite electrode (GE), and stabilized using chitosan in order to build modified electrode catalysts. Dodevska and co-authors studied the electrochemical performances of the modified electrodes, i.e., AgNPs/GE and AgNPs/CS/GE using cyclic voltammetry (CV), differential pulse voltammetry (DPV), and chronoamperometry at pH 7.0, to detect their applicability for detection of H_2O_2 and vanillin. It was observed that the synthesized electrodes displayed a sensitive and reproducible response towards quantitative determination of H_2O_2 and to vanillin which offers the excellent potential of *R. damascena* derived AgNPs to be included as an electrochemical sensor in devices to monitor such analytic tests.

5.5 Role of Phytochemicals in Phytosynthesis of Nanoparticles

Detection and determination of the bioactive phytomolecules which might be involved in the bioreduction, capping, and stabilization processes become the major concern which several researchers are focusing upon (Table 5.3). Flavonoids, phenolic acids, polyphenols, terpenoids, proteins, carbohydrates, amino acids, fats, gum, and polysaccharides are among these biomolecules. In 2008, a study intro-

Table 5.3 List of various biomolecules present within plant extracts which act as reducing, capping, and stabilizing agents

Plant species/NPs produced	Biomolecules involved in NP synthesis	Reference
Cola nitida/AgNPs	Crude proteins alkaloids, caffeine (2.8%), theobromine (0.05%), nicotine, and kolatine	Lateef et al. (2016)
Hylocereus undatus/AgNPs	Reducing sugars, proteins	Phongtongpasuk et al. (2016)
Eugenia stipitata/AgNPs	Malic acid, citric acid and carotenoids (lutein, zeaxanthin, zeinoxanthin, α-carotene, β-carotene, β-cryptoxanthin, anhydrolutein, anhydrozeaxanthin)	Kumar et al. (2016)
Musa sapientum/AuNPs	Proteins, sugars (glucose and fructose)	Deokar and Ingale (2016)
Citrus sinensis/AgNPs	Eugenol, with 4-vinyl guaiacol, other phenolic compounds, terpenes, saturated fatty acids, aromatics, and sugar derivatives	Omran et al. (2013, 2017)
Carica papaya/AgNPs	Proteins, vitamins, and natural phenols	Balavijayalakshmi and Ramalakshmi (2017)
Red rise husk/ZnO/rGO nanocomposite	Flavanol, anthocyanin, phytic acid, proanthocyanidin, tocopherol, tocotrienol, ɣ-oryzanol	Alomair and Mohamed (2018)
Juglans regia/iron oxide NPs	Ferulic acid, vanillic acid, myricetin, syringic acid, coumaric acid, juglone	Izadiyan et al. (2018)
Paulownia tomentosa/AgNPs	Flavonoids (anthocyanins, flavones, flavanols, flavanones, and isoflavones)	Ruíz-baltazar et al. (2019)
Thymus vulgaris/ZnO NPs	Phenol, flavonoid, saponins, and thymol (phenolic monoterpenes)	Zare et al. (2019)
Capparis spinosa/CuO NPs	Alkaloids, flavonoids (e.g. kaempferol, quercetin, isorhamnetin, and their *O*-methyl derivative, thomnocitirin, rhamnetin, and rhamnozin), phenols, steroids, glycosides, carbohydrates, saponins, indoles, tannins	Samari et al. (2019)
Culex tritaeniorhynchus/ZnONPs	Proteins, phenolics, amino acids, flavonoids	Velsankar et al. (2019)
Carica papaya/carbon dots	Carbohydrates and fibers	Pooja et al. (2019)

duced by Shukla and colleagues showed that soybeans can act as a phytochemical reservoir for the green synthesis of AuNPs (Shukla et al. 2008). The main phytochemicals were water-soluble proteins (albumins and globulins), carbohydrates (raffinose, sucrose, and stachyose), saponins, isoflavones, and amino acids. Nasrollahzadeh et al. (2015) carried out a high-performance liquid chromatography (HPLC) to the hydroalcoholic extract of *Hippophae rhamnoides* L. The obtained chromatogram showed the qualitative existence of the biomolecules, and they were glycosides A to E, such as isorhamnetin 3-sophoroside-7- rhamnoside, quercetin 3-O-rutinoside, quercetin 3-O-glucoside, isorhamnetin 3-O-rutinoside, and isorhamnetin 3-O-glucoside. These biomolecules were suggested to be involved in the bioreduction of palladium ions to PdNPs. In another study performed by Nasrollahzadeh and Mohammad Sajadi (2015), HPLC analysis of the leaf extract of *Ginkgo biloba* L. displayed the existence of phenolic antioxidants such as quercetin 3-O-b-D- rutinoside, quercetin 3-O-a-L-(b-D-glucopyranosyl)-(1,2)-rhamnopyranoside, kaempferol 3-O-b- D-rutinoside, isorhamnetin 3-O-b-D rutinoside, kaempferol 3-O-a-L-(b-D-glucopyranosyl)-(1,2)- rhamnopyranoside, quercetin 3-O-a- (6-p-coumaroyl glucopyranosyl-b-1,2-rhamnopyranoside), and kaempferol 3-O-a (6-p-coumaroyl glucopyranosyl-b-1,2-rhamnopyranoside). Authors suggested that these phenolic compounds were behind the green synthesis of CuNPs via the leaf extract of *G. biloba* L.

Flavonoids are groups of polyphenolic compounds which involve anthocyanins, chalcones, isoflavonoids, flavones, flavonols, and flavanones (Makarov et al. 2014). Flavonoid main basic structure is aglyconic. Classification of flavonols, flavanones, and their dihydro derivatives is based upon the existence of a condensed benzene ring. Flavonoids are divided into 2-position flavonoids and 3-position isoflavonoids. Jeevanandam et al. (2016) reported that these compounds are able to actively chelate and bioreduce metal ions into NPs. The presence of different functional groups in flavonoids makes them capable of mediating the synthesis of NPs. Makarov et al. (2014) elucidated that flavonoids' tautomeric transformations from the enol-form to the keto-form might release reactive hydrogen atoms which might aid in reducing metal ions to form NPs. In 2010, a study conducted by Ahmed and co-authors postulated that AgNPs was successfully synthesized using the stem and root extracts of *Ocimum basilicum* (Ahmad et al. 2010). Authors suggested that transformation of flavonoids like luteolin and rosmarinic acid from the enol- to the keto-form was responsible for the bioreduction of Ag^{+} to AgNPs. Likewise, Zheng et al. (2013) reported that *Cacumen platycladi*-mediated the synthesis PtNPs by the effect of the existed flavonoid content of the extract. By employing 3.5-dinitrosalicylic acid colorimetric assay, spectrophotometry, phenol-sulfuric acid, and Coomassie brilliant blue assays, it was identified that the metal ion bioreduction took place by ketone conversion to carboxylic acid in flavonoids. Makarov et al. (2014) illustrated the involvement of flavonoids in metal ion chelation like Fe^{2+}, Fe^{3+}, Cu^{2+}, Zn^{2+}, Al^{3+}, Cr^{3+}, Pb^{2+}, and Co^{2+} via carbonyl groups or *p*-electrons. Jeevanandam et al. (2016) assumed that the role of flavonoids is not restricted only to reduction and chelation of metal ions, but they can also be anticipated in nanoparticle formation steps, i.e., nucleation, growth, and stabilization. Beside flavonoids, other phytochemicals are reported to contribute in NPs formation, for instance, phenolic compounds are suggested to be part of the phytochemicals which are employed in reducing metal ions (Ahmad and Sharma 2012; Senthilkumar and Sivakumar 2014).

5.6 Possible Mechanisms Behind Phytosynthesis of Nanoparticles

The exact precise mechanism behind the phytosynthesis of NPs is still somehow mysterious. It depends majorly upon the identification of the exact phytochemical constituents which might act as reducing, capping, and stabilizing agents (Dauthal and Mukhopadhyay 2016; Kumar et al. 2019a). However, several researchers suggested hypothetical mechanisms which might be involved in phytosynthesis of metallic NPs (Rai et al. 2008; Park et al. 2011; Durán and Seabra 2012). For instance, Li and co-authors elucidated that plant extracts contain proteins with functional amino groups, and they were the key behind reducing metal ions (Li et al. 2007). Whereas, Huang et al. (2007) revealed that the presence of functional groups like –C–O–C–, –C– O–, –C=C–, and –C=O– derived from alkaloids, flavones, and anthracenes participated in synthesizing metal NPs. Additionally, Kesharwani et al. (2009) anticipated that quinones and plastohydroquinone molecules existed within the *Datura metel* leaf extract were the main lead behind reduction of metal ions to their NPs state. As a result, it can be noted that phytobiomolecules and heterocyclic compounds were responsible for producing plant extracts derived metal NPs (Jeevanandam et al. 2016). Similar to metal nanoparticles, plant phytochemicals are responsible for biosynthesis, capping, and stabilization of metal oxide NPs. First, metal precursors are reduced by the phytochemicals within the plant extract. Oxygen existed in the atmosphere or released from the degradation of phytochemical bind to the reduced metal ions. The produced metal oxide ions bind to each other via the electrostatic attraction forces and form NPs. The prepared NPs are then stabilized by particular phytochemicals to avoid their aggregation. It is worth noting that phenolic compounds with carboxyl and hydroxyl functional groups suppress the superoxide-driven Fenton reaction, which is an important source for ROS (Iravani 2011). In another study proposed by Makarov et al. (2014), authors assumed that plant extracts not only reduce metal atoms but also encapsulate them with an organic coat in a three phase mechanism to ensure the stability of the fabricated NPs. The three phases include the activation phase in which reduction of metal ions and nucleation of the reduced metal atom take place; then growth phase and stabilization of NPs increases; and eventually, the termination phase which includes tailoring of the final shape of the prepared NPs (Si and Mandal 2007).

Bar et al. (2009) reported the green synthesis of AgNPs using the latex of *Jatropha curcas* and illustrated that cyclic octapeptide (Curcacyclin A and B) and curcain enzyme were the main lead behind the bioreduction as well as the stabilization of AgNPs. Li and co-workers proposed a model which was referred to as "recognition-reduction-limited nucleation and growth" in which AgNPs was green synthesized via *Capsicum annum* extract (Li et al. 2007). Based upon this model, silver ions were captured on the protein surface by means of electrostatic interaction forces. Meanwhile, Ag^+ ions were bioreduced by the proteins existed within the extract of *C. annum*. As a result, silver nuclei were generated and enlarged via further accumulation of elemental silver leading to fabrication of stable spherical shaped AgNPs. In another study introduced by Narayanan and Sakthivel (2011), the authors denied the role of enzymes during the phytosynthesis of AgNPs via the leaf extract of *C. amboinicus*. The reason behind this suggestion was that the extract was heated to a high temperature which reached 90 °C. Meanwhile, authors claimed that other phytochemical constituents such as phenolics, terpenoids, and flavonoids were behind the stabilization and capping of the prepared particles. Likewise, Song et al. (2010) demonstrated that enzymes were not involved in the phytosynthesis of PtNPs from the leaf extract of *Diospyros kaki* as the reaction was performed also at high temperature, e.g., 95 °C. Another proof was the nonexistence of either enzymes or proteins within the FTIR spectrum. Elavazhagan and Arunachalam (2011) carried out an investigation in which both AgNPs and AuNPs were phytosynthesized via the water extract of *Memecylon edule*. Authors suggested that the type of phytochemical constituent responsible for reducing Ag^+ and Au^+ was saponins. In 2011, Ghoreishi et al. (2011) suggested that the hydroxyl groups of flavonoids within the flower extract of *Rosa damascene* were the main keys behind the bioreduction of silver and gold ions. In another investigation, Kasthuri et al. (2009) demonstrated that a flavonoid glycoside referred to apiin existed within the leaf extract of henna (*Lawsonia inermis*) was responsible for synthesizing AuNPs and AgNPs. In a study conducted by Pohlit et al. (2011), authors suggested that phytochemicals like ketones, amides, terpenoids, flavones, aldehydes, and carboxylic acids might be the key constituents behind the biogenesis of AgNPs. Whereas, in a study performed by Doughari (2012), water-soluble phytochemicals such as organic acids, flavones, and quinones were responsible for the immediate bioreduction of silver ions to their nanoparticle state.

Deekonda et al. (2016) suggested a possible mechanism for the biogenic synthesis of AgNPs using carboxymethyl sago pulp (CMSP). The aqueous solution of CMSP reacted with $AgNO_3$. When the CMSP polymer solutions were subjected to electron beam irradiation, free radicals were generated as a result of the excitation and ionization of polymer chains as well as the water molecules. Henceforth, when water is subjected to radiolysis, hydroxyl (OH) radicals, hydrogen atoms, and hydrated electrons are released. The aforementioned hydroxyl radicals produced from water react with CMSP polymer chains and form macroradicals on polymer chains which in turn recombine and lead to the formation of CMSP polymer hydrogel by cross-linking. The produced hydrated electrons along with hydrogen atoms due to the radiolysis aid in the reduction of dissolved Ag^+ to AgNPs. The agglomeration of AgNPs was prevented by the effect of oxygen atoms on the side chain of CMSP which helps AgNPs to be impregnated in the cross-linked CMSP hydrogel. In another study conducted by Suresh and co-authors, hexagonal-shaped magnesium oxide NPs were phytosynthesized using the leaf extract of insulin plant

(*Costus pictus* D. Don) (Suresh et al. 2018). Authors suggested that diosgenin (phytosteroid sapogenin) ($C_{27}H_{42}O_3$) was the main lead behind the formation of MgO NPs. Initially, a metal-diosgenin complex is formed by the combination of diosgenin along with magnesium nitrate via weak hydrogen bonding. Afterwards, the complex solution is retained in a hot air oven for 8 h where it is converted into hydroxide forms. The hydroxide complex undergoes calcination to phytosynthesize the required metal oxide NPs. It is worth mentioning that presence of phenolic groups facilitated the bioreduction process.

Alagesan and Venugopal (2019) proposed a study in which *Parkia speciosa* leaf extract mediated the biological synthesis of AgNPs. Authors suggested that mainly the phenolic and flavonoid compounds acted as effective reducing agents. Both compounds released electrons via the breaking of O–H bond, and the generated electrons were used in reducing Ag^+ to Ag^0. Besides, authors assumed that the existing protein molecule within the extract of *P. speciosa* acted as capping and stabilizing agents. It is worth noting that FTIR introduced an evidence for the involvement of –OH and $–NH_2$ functional groups in synthesizing and capping of *P. speciosa*-derived AgNPs.

5.7 Drawbacks, Suggestions, and Recommendations Concerning Phytonanotechnology

As mentioned earlier, plant-mediated synthesis of metal and metal oxide NPs possess several advantages, yet there are some disadvantages as well. Biological synthesis approach for nanoparticles, including plant-mediated synthesis, is still in its infancy. The major challenges that face researchers are to guarantee long-term stability, prevent agglomeration of the produced NPs, and control crystal growth, size, and morphology (Malik et al. 2014). Henceforth, many plant extracts have been investigated in synthesis of NPs, but the major question among researchers is whether it can be upscaled and whether they can compete with chemical and physical synthesis approaches (Iravani 2011). Therefore, more studies regarding cost analysis, production quantities, biochemical processes, and pathways, which are involved in heavy metal detoxification by plants, are urgently required.

References

Abisharani JM, Devikala S, Kumar R, Dinesh et al (2019) Green synthesis of TiO_2 nanoparticles using *Cucurbita pepo* seeds extract. Mater Today Proc 14:302–307

Abolghasemi R, Haghighi M, Solgi M, Mobinikhaledi A (2018) Rapid synthesis of ZnO nanoparticles by waste thyme (*Thymus vulgaris* L.). Int J Environ Sci Technol 16:6985–6990

Ahmad N, Sharma S (2012) Green synthesis of silver nanoparticles using extracts of *Ananas comosus*. Green Sustain Chem 2:141–147

Ahmad N, Sharma S, Alam MK et al (2010) Rapid synthesis of silver nanoparticles using dried medicinal plant of basil. Colloids Surfaces B Biointerfaces 81:81–86

Ahmed AMA, Sharmen F, Mannan A, Rahman MA (2015) Phytochemical, analgesic, antibacterial, and cytotoxic effects of *Alpinia nigra* (Gaertn.) Burtt leaf extract. J Tradit Complement Med 5:248–252

Ahmed S, Kaur G, Sharma P et al (2018) Fruit waste (peel) as bio-reductant to synthesize silver nanoparticles with antimicrobial, antioxidant and cytotoxic activities. J Appl Biomed 16:221–236

Akhtar M, Swamy MK, Umar A et al (2015) Biosynthesis and characterization of silver nanoparticles from methanol leaf extract of *Cassia didymobotyra* and assessment of their antioxidant and antibacterial activities. J Nanosci Nanotechnol 15:9818–9823

Aksu Demirezen D, Yıldız YŞ, Demirezen Yılmaz D (2019) Amoxicillin degradation using green synthesized iron oxide nanoparticles: kinetics and mechanism analysis. Environ Nanotechnol Monit Manag 11:100219–100228

Alagesan V, Venugopal S (2019) Green synthesis of selenium nanoparticle using leaves extract of *Withania somnifera* and its biological applications and photocatalytic activities. Bionanoscience 9:105–116

Alomair NA, Mohamed HH (2018) Green synthesis of ZnO hollow microspheres and ZnO/rGO nanocomposite using red rice husk extract and their photocatalytic performance. Mater Res Express 5:095012–095022

Amado IR, Franco D, Sánchez M et al (2014) Optimisation of antioxidant extraction from *Solanum tuberosum* potato peel waste by surface response methodology. Food Chem 165:290–299

An D, Guo Y, Zhu Y, Wang Z (2010) A green route to preparation of silica powders with rice husk ash and waste gas. Chem Eng J 162:509–514

Arasu MV, Arokiyaraj S, Viayaraghavan P et al (2019) One step green synthesis of larvicidal, and azo dye degrading antibacterial nanoparticles by response surface methodology. J Photochem Photobiol B Biol 190:154–162

Arunkumar B, Johnson Jeyakumar S, Jothibas M (2019) A sol-gel approach to the synthesis of CuO nanoparticles using *Lantana camara* leaf extract and their photo catalytic activity. Optik (Stuttg) 183:698–705

Athinarayanan J, Periasamy VS, Alhazmi M, Alshatwi AA (2015) Synthesis and biocompatibility assessment of sugarcane bagasse- derived biogenic silica nanoparticles for biomedical applications. J Biomed Mater Res 105B:340–349

Azizian-Shermeh O, Einali A, Ghasemi A (2017) Rapid biologically one-step synthesis of stable bioactive silver nanoparticles using Osage orange (*Maclura pomifera*) leaf extract and their antimicrobial activities. Adv Powder Technol 28:3164–3171

Balachandar R, Gurumoorthy P, Karmegam N et al (2019) Plant-mediated synthesis, characterization and bactericidal potential of emerging silver nanoparticles using stem extract of *Phyllanthus pinnat*us: a recent advance in phytonanotechnology. J Clust Sci 30:1481–1488

Balavijayalakshmi J, Ramalakshmi V (2017) *Carica papaya* peel mediated synthesis of silver nanoparticles and its antibacterial activity against human pathogens. J Appl Res Technol 15:413–422

Bali R, Razak N, Lumb A, Harris AT (2006) The synthesis of metallic nanoparticles inside live plants. In: 2006 International Conference on Nanoscience and Nanotechnology. IEEE

Bar H, Bhui DK, Sahoo GP et al (2009) Green synthesis of silver nanoparticles using latex of *Jatropha curcas*. Colloids Surfaces A Physicochem Eng Asp 339:134–139

Barberia-Roque L, Gámez-Espinosa E, Viera M, Bellotti N (2019) Assessment of three plant extracts to obtain silver nanoparticles as alternative additives to control biodeterioration of coatings. Int Biodeterior Biodegrad 141:52–61

Barbosa VT, Souza JKC, Alvino V et al (2019) Biogenic synthesis of silver nanoparticles using *Brazilian propolis*. Biotechnol Prog 35:e2888–e2897

Baruah D, Goswami M, Yadav RNS et al (2018) Biogenic synthesis of gold nanoparticles and their application in photocatalytic degradation of toxic dyes. J Photochem Photobiol B Biol 186:51–58

Baruah D, Narayan R, Yadav S et al (2019) Biology *Alpinia nigra* fruits mediated synthesis of silver nanoparticles and their antimicrobial and photocatalytic activities. J Photochem Photobiol B Biol 201:111649–111658

Behravan M, Hossein Panahi A, Naghizadeh A et al (2019) Facile green synthesis of silver nanoparticles using *Berberis vulgaris* leaf and root aqueous extract and its antibacterial activity. Int J Biol Macromol 124:148–154

Bindhu MR, Rekha PV, Umamaheswari T, Umadevi M (2014) Antibacterial activities of *Hibiscus cannabinus* stem-assisted silver and gold nanoparticles. Mater Lett 131:194–197

Chahardoli A, Karimi N, Fattahi A, Salimikia I (2019) Biological applications of phytosynthesized gold nanoparticles using leaf extract of *Dracocephalum kotschyi*. J Biomed Mater Res – Part A 107:621–630

Cheng W, Wang S-G, Lu L et al (2008) Removal of malachite green (MG) from aqueous solutions by native and heat-treated anaerobic granular sludge. Biochem Eng J 39:538–546

Chung I-M, Park I, Seung-Hyun K et al (2016) Plant-mediated synthesis of silver nanoparticles: their characteristic properties and therapeutic applications. Nanoscale Res Lett 11:40–54

da Silva AFV, Fagundes AP, Macuvele DLP et al (2019) Green synthesis of zirconia nanoparticles based on *Euclea natalensis* plant extract: optimization of reaction conditions and evaluation of adsorptive properties. Colloids Surfaces A Physicochem Eng Asp 583:123915–123925

Das RK, Gogoi N, Bora U (2011) Green synthesis of gold nanoparticles using *Nyctanthes arbortristis* flower extract. Bioprocess Biosyst Eng 34:615–619

Dauthal P, Mukhopadhyay M (2016) Noble metal nanoparticles: plant-mediated synthesis, mechanistic aspects of synthesis, and applications. Ind Eng Chem Res 55:9557–9577

de Barros Santos E, Madalossi NV, Sigoli FA, Mazali IO (2015) Silver nanoparticles: green synthesis, self-assembled nanostructures and their application as SERS substrates. New J Chem 39:2839–2846

Deekonda K, Muniyandy S, Yan Y, Janarthanan P (2016) Electron beam radiation mediated green synthesis of silver nanoparticles using carboxymethyl sago pulp obtained from sago waste. Polymer 86:147–156

Deokar GK, Ingale AG (2016) Green synthesis of gold nanoparticles (elixir of life) from banana fruit waste extract – an efficient multifunctional agent. RSC Adv 6:74620–74629

Deokar GK, Ingale AG (2018) Unveiling an unexpected potential of beetroot waste in green synthesis of single crystalline gold nanoplates : a mechanistic study. Arab J Chem 11:950–958

Desalegn B, Megharaj M, Chen Z, Naidu R (2019) Heliyon green synthesis of zero valent iron nanoparticle using mango peel extract and surface characterization using XPS and GC-MS. Heliyon 5:e01750–e01759

Dobrucka R, Dlugaszewska J (2018) Antimicrobial activity of the biogenically synthesized core-shell Cu@Pt nanoparticles. Saudi Pharm J 26:643–650

Dobrucka R, Kaczmarek M, Łagiedo M et al (2019) Evaluation of biologically synthesized Au-CuO and CuO-ZnO nanoparticles against glioma cells and microorganisms. Saudi Pharm J 27:373–383

Dodevska T, Vasileva I, Denev P et al (2019) *Rosa damascena* waste mediated synthesis of silver nanoparticles : characteristics and application for an electrochemical sensing of hydrogen peroxide and vanillin. Mater Chem Phys 231:335–343

Doughari JH (2012) Phytochemicals: extraction methods, basic structures and mode of action as potential chemotherapeutic agents. In: Phytochemicals – a global perspective of their role in nutrition and health. InTechOpen, pp 1–35. https://doi.org/10.5772/26052

Durán N, Seabra AB (2012) Metallic oxide nanoparticles: state of the art in biogenic syntheses and their mechanisms. Appl Microbiol Biotechnol 95:275–288

Ebringerova A, Heinze T (2000) Xylan and xylan derivatives–biopolymers with valuable properties, naturally occurring xylans structures, isolation procedures and properties. Macromol Rapid Commun 21:542–556

Ebringerová A, Hromadkova Z, Hřibalová V (1995) Structure and mitogenic activities of corn cob heteroxylans. Int J Biol Macromol 17:327–331

Elavazhagan T, Arunachalam KD (2011) *Memecylon edule* leaf extract mediated green synthesis of silver and gold nanoparticles. Int J Nanomedicine 6:1265–1278

El-Gendy NS, Omran BA (2019) Green synthesis of nanoparticles for water treatment. In: Nano and bio-based technologies for wastewater treatment. Wiley, Hoboken, pp 205–263

Erjaee H, Rajaian H, Nazifi S (2017) Synthesis and characterization of novel silver nanoparticles using *Chamaemelum nobile* extract for antibacterial application. Adv Nat Sci Nanosci Nanotechnol 8:025004–025013

Esmaeili A, Saremnia B (2016) Journal of the Taiwan institute of chemical engineers synthesis and characterization of NaA zeolite nanoparticles from *Hordeum vulgare* L. husk for the separation of total petroleum hydrocarbon by an adsorption process. J Taiwan Inst Chem Eng 61:276–286

Faria KCP, Gurgel RF, Holanda JNF (2010) Characterization of sugarcane bagasse ash for use in ceramic bodies. In: Materials science forum. Trans Tech Publications, Cham, pp 1049–1052

Franco-Romano M, Gil MLA, Palacios-Santander JM et al (2014) Sonosynthesis of gold nanoparticles from a *Geranium* leaf extract. Ultrason Sonochem 21:1570–1577

Gerland P, Raftery AE, Ševčíková H et al (2014) World population stabilization unlikely this century. Science 80:234–237

Ghoreishi SM, Behpour M, Khayatkashani M (2011) Green synthesis of silver and gold nanoparticles using *Rosa damascena* and its primary application in electrochemistry. Phys E Low-Dimens Syst Nanostruct 44:97–104

Ghosh S, Padilla-González GF, Rangan L (2013) Alpinia nigra seeds: a potential source of free radical scavenger and antibacterial agent. Ind Crop Prod 49:348–356

Girón-vázquez NG, Gómez-gutiérrez CM, Soto-robles CA et al (2019) Study of the effect of *Persea americana* seed in the green synthesis of silver nanoparticles and their antimicrobial properties. Results Phys 13:102142–102147

Guo M, Li W, Yang F, Liu H (2015) Controllable biosynthesis of gold nanoparticles from a *Eucommia ulmoides* bark aqueous extract. Spectrochim Acta Part A Mol Biomol Spectrosc 142:73–79

Gusmão KAG, Gurgel LVA, Melo TMS, Gil LF (2012) Application of succinylated sugarcane bagasse as adsorbent to remove methylene blue and gentian violet from aqueous solutions–kinetic and equilibrium studies. Dyes Pigments 92:967–974

Hamedi S, Shojaosadati SA (2019) Rapid and green synthesis of silver nanoparticles using *Diospyros lotus* extract: evaluation of their biological and catalytic activities. Polyhedron 171:172–180

Harish BS, Uppuluri KB, Anbazhagan V (2015) Synthesis of fibrinolytic active silver nanoparticle using wheat bran xylan as a reducing and stabilizing agent. Carbohydr Polym 132:104–110

Hassan D, Khalil AT, Saleem J et al (2018) Biosynthesis of pure hematite phase magnetic iron oxide nanoparticles using floral extracts of *Callistemon viminalis* (bottlebrush): their physical properties and novel biological applications. Artif Cells, Nanomed Biotechnol 46:693–707

Heo HS, Park HJ, Park Y-K et al (2010) Bio-oil production from fast pyrolysis of waste furniture sawdust in a fluidized bed. Bioresour Technol 101:S91–S96

Hernández-Morales L, Espinoza-Gómez H, Flores-López LZ et al (2019) Study of the green synthesis of silver nanoparticles using a natural extract of dark or white *Salvia hispanica* L. seeds and their antibacterial application. Appl Surf Sci 489:952–961

Ho K, Ferruzzi MG, Liceaga AM, San Martín-González MF (2015) Microwave-assisted extraction of lycopene in tomato peels: effect of extraction conditions on all-trans and cis-isomer yields. LWT-Food Sci Technol 62:160–168

Huang J, Li Q, Sun D et al (2007) Biosynthesis of silver and gold nanoparticles by novel sundried *Cinnamomum camphora* leaf. Nanotechnology 18:05104–05115

Ibrahim HMM (2015) Green synthesis and characterization of silver nanoparticles using banana peel extract and their antimicrobial activity against representative microorganisms. J Radiat Res Appl Sci 8:265–275

Iravani S (2011) Green synthesis of metal nanoparticles using plants. Green Chem 13:2638–2650

Islam NU, Jalil K, Shahid M, Muhammad N (2019) *Pistacia integerrima* gall extract mediated green synthesis of gold nanoparticles and their biological activities. Arab J Chem 12:2310–2319

Izadiyan Z, Shameli K, Miyake M et al (2018) Cytotoxicity assay of plant-mediated synthesized iron oxide nanoparticles using Juglans regia green husk extract. Arab J Chem 1151:97–105

Jahan E, Anandapadmanaban G, Mathiyalagan R (2018) Optik synthesis of zinc oxide nanoparticles from immature fruits of *Rubus coreanus* and its catalytic activity for degradation of industrial dye. Opt – Int J Light Electron Opt 172:1179–1186

Jamdagni P, Khatri P, Rana JS (2016) Green synthesis of zinc oxide nanoparticles using flower extract of *Nyctanthes* arbor-tristis and their antifungal activity. J King Saud Univ – Sci 30:168–175

Jaybhaye SV (2015) Antimicrobial activity of silver nanoparticles synthesized from waste vegetable fibers. Mater Today Proc 2:4323–4327

Jeevanandam J, Chan YS, Danquah MK (2016) Biosynthesis of metal and metal oxide nanoparticles. Chem Bio Eng Rev 3:55–67

Jemilugba OT, Sakho EHM, Parani S et al (2019) Green synthesis of silver nanoparticles using *Combretum erythrophyllum* leaves and its antibacterial activities. Colloids Interface Sci Commun 31:100191–100197

Jeyaraj M, Rajesh M, Arun R et al (2013) An investigation on the cytotoxicity and caspase-mediated apoptotic effect of biologically synthesized silver nanoparticles using *Podophyllum hexandrum* on human cervical carcinoma cells. Colloids Surfaces B Biointerfaces 102:708–717

Kahrilas GA, Wally LM, Fredrick SJ et al (2013) Microwave-assisted green synthesis of silver nanoparticles using orange peel extract. ACS Sustain Chem Eng 2:367–376

Karade VC, Parit SB, Dawkar VV et al (2019) Heliyon A green approach for the synthesis of α-Fe2O3 nanoparticles from *Gardenia resinifera* plant and it's in vitro hyperthermia application. Heliyon 5:e02044–e02049

Karuppaiya P, Satheeshkumar E, Chao W-T et al (2013) Anti-metastatic activity of biologically synthesized gold nanoparticles on human fibrosarcoma cell line HT-1080. Colloids Surfaces B Biointerfaces 110:163–170

Kasthuri J, Veerapandian S, Rajendiran N (2009) Biological synthesis of silver and gold nanoparticles using apiin as reducing agent. Colloids Surfaces B Biointerfaces 68:55–60

Kaviya S, Santhanalakshmi J, Viswanathan B et al (2011) Biosynthesis of silver nanoparticles using *Citrus sinensis* peel extract and its antibacterial activity. Spectrochim Acta – Part A Mol Biomol Spectrosc 79:594–598

Kesharwani J, Yoon KY, Hwang J, Rai M (2009) Phytofabrication of silver nanoparticles by leaf extract of *Datura metel*: hypothetical mechanism involved in synthesis. J Bionanoscience 3:39–44

Kharissova OV, Dias HVR, Kharisov BI et al (2013) The greener synthesis of nanoparticles. Trends Biotechnol 31:240–248

Khatami M, Alijani HQ, Fakheri B et al (2019) Super-paramagnetic iron oxide nanoparticles (SPIONs): greener synthesis using *Stevia* plant and evaluation of its antioxidant properties. J Clean Prod 208:1171–1177

Khodadadi B, Bordbar M, Nasrollahzadeh M (2017) *Achillea millefolium* L. extract mediated green synthesis of waste peach kernel shell supported silver nanoparticles: application of the nanoparticles for catalytic reduction of a variety of dyes in water. J Colloid Interface Sci 493:85–93

Khoshnamvand M, Ashtiani S, Huo C et al (2019) Use of *Alcea rosea* leaf extract for biomimetic synthesis of gold nanoparticles with innate free radical scavenging and catalytic activities. J Mol Struct 1179:749–755

Klekotko M, Brach K, Olesiak-Banska J et al (2019) Popcorn-shaped gold nanoparticles: plant extract-mediated synthesis, characterization and multiphoton-excited luminescence properties. Mater Chem Phys 229:56–60

Krishnaraj C, Jagan EG, Rajasekar S et al (2010) Synthesis of silver nanoparticles using *Acalypha indica* leaf extracts and its antibacterial activity against water borne pathogens. Colloids Surfaces B Biointerfaces 76:50–56

Kumar VG, Gokavarapu SD, Rajeswari A et al (2011) Facile green synthesis of gold nanoparticles using leaf extract of antidiabetic potent *Cassia auriculata*. Colloids Surfaces B Biointerfaces 87:159–163

Kumar R, Roopan SM, Prabhakarn A et al (2012) Agricultural waste *Annona squamosa* peel extract: biosynthesis of silver nanoparticles. Spectrochim Acta – Part A Mol Biomol Spectrosc 90:173–176

Kumar B, Smita K, Debut A et al (2016) Extracellular green synthesis of silver nanoparticles using Amazonian fruit Araza (*Eugenia stipitata* McVaugh). Trans Nonferrous Met Soc China 26:2363–2371

Kumar I, Mondal M, Sakthivel N (2019a) Green synthesis of phytogenic nanoparticles. In: Green synthesis, characterization and applications of nanoparticles, micro and nano technologies. Elsevier Inc, Boston, pp 37–73

Kumar PV, Jelastin Kala SM, Prakash KS (2019b) Green synthesis derived Pt-nanoparticles using *Xanthium strumarium* leaf extract and their biological studies. J Environ Chem Eng 7:103146–103154

Kuppusamy P, Ilavenil S, Srigopalram S et al (2017) Treating of palm oil mill effluent using *Commelina nudi* flora mediated copper nanoparticles as a novel bio-control agent. J Clean Prod 141:1023–1029

Küünal S, Rauwel P, Rauwel E (2018) Plant extract mediated synthesis of nanoparticles. In: Emerging applications of nanoparticles and architectural nanostructures, current prospects and future trends, micro and nano technologies. Elsevier Inc, Boston, pp 411–446

Lateef A, Azeez MA, Asafa TB et al (2016) Biogenic synthesis of silver nanoparticles using a pod extract of *Cola nitida* : antibacterial and antioxidant activities and application as a paint additive. Integr Med Res 10:551–562

Lee H, Song JY, Kim BS (2013) Biological synthesis of copper nanoparticles using *Magnolia kobus* leaf extract and their antibacterial activity. J Chem Technol Biotechnol 88:1971–1977

Li S, Shen Y, Xie A et al (2007) Green synthesis of silver nanoparticles using *Capsicum annuum* L. extract. Green Chem 9:852–858

Liou T-H, Yang C-C (2011) Synthesis and surface characteristics of nanosilica produced from alkali-extracted rice husk ash. Mater Sci Eng B 176:521–529

Lu J, Ali H, Hurh J et al (2019) Optik the assessment of photocatalytic activity of zinc oxide nanoparticles from the roots of *Codonopsis lanceolata* synthesized by one-pot green synthesis method. Opt – Int J Light Electron Opt 184:82–89

Makarov VV, Love AJ, Sinitsyna OV et al (2014) "Green" nanotechnologies: synthesis of metal nanoparticles using plants. Acta Nat 6:35–45

Malaikozhundan B, Vaseeharan B, Vijayakumar S et al (2017) Biological therapeutics of *Pongamia pinnata* coated zinc oxide nanoparticles against clinically important pathogenic bacteria, fungi and MCF-7 breast cancer cells. Microb Pathog 104:268–277

Malik P, Shankar R, Malik V, et al (2014) Green chemistry based benign routes for nanoparticle synthesis. J Nanopart, vol. 2014, Article ID 302429, 14 pages. https://doi.org/10.1155/2014/302429

Malini S, Kumar SV, Hariharan R, et al (2019) Materials today : proceedings antibacterial, photocatalytic and biosorption activity of chitosan nanocapsules embedded with *Prosopis juliflora* leaf extract synthesized silver nanoparticles. Mater Today Proc, article in press

Masoudian N, Rajabi M, Ghaedi M (2019) Titanium oxide nanoparticles loaded onto activated carbon prepared from bio-waste watermelon rind for the efficient ultrasonic-assisted adsorption of congo red and phenol red dyes from wastewaters. Polyhedron 173:114105–114114

Mirabella N, Castellani V, Sala S (2014) Current options for the valorization of food manufacturing waste: a review. J Clean Prod 65:28–41

Mohammadalinejhad S, Almasi H, Esmaiili M (2019) Simultaneous green synthesis and in-situ impregnation of silver nanoparticles into organic nano fibers by *Lythrum salicaria* extract: morphological, thermal, antimicrobial and release properties. Mater Sci Eng C 105:110115–110128

Mostafa E, Fayed MAA, Radwan RA, Bakr RO (2019) *Centaurea pumilio* L. extract and nanoparticles: a candidate for healthy skin. Colloids Surfaces B Biointerfaces 182:110350–110358

Muthamil S, Devi VA, Balasubramaniam B et al (2018) Green synthesized silver nanoparticles demonstrating enhanced in vitro and in vivo antibiofilm activity against *Candida* spp. J. Basic Microbiol 58:343–357

Mukherjee S, Vinothkumar B, Prashanthi S et al (2013) Potential therapeutic and diagnostic applications of one-step in situ biosynthesized gold nanoconjugates (2-in-1 system) in cancer treatment. RSC adv 3:2318–2329

Mukherjee S, Patra CR (2017) Biologically synthesized metal nanoparticles: recent advancement and future perspectives in cancer theranostics. Future Sci OA 3:FSO203-FSO207

Narayanan KB, Sakthivel N (2011) Extracellular synthesis of silver nanoparticles using the leaf extract of *Coleus amboinicus* lour. Mater Res Bull 46:1708–1713

Naseem T, Farrukh MA (2015) Antibacterial activity of green synthesis of iron nanoparticles using lawsonia inermis and gardenia jasminoides leaves extract. J Chem, vol. 2015, Article ID 912342, 7 pages, 2015. https://doi.org/10.1155/2015/912342

Nasrollahzadeh M, Mohammad Sajadi S (2015) Green synthesis of copper nanoparticles using *Ginkgo biloba* L. leaf extract and their catalytic activity for the Huisgen [3+2] cycloaddition of azides and alkynes at room temperature. J Colloid Interface Sci 457:141–147

Nasrollahzadeh M, Sajadi SM, Maham M (2015) Green synthesis of palladium nanoparticles using *Hippophae rhamnoides* Linn leaf extract and their catalytic activity for the Suzuki– Miyaura coupling in water. J Mol Catal A Chem 396:297–303

Nayak S, Sajankila SP, Rao CV et al (2019) Biogenic synthesis of silver nanoparticles using *Jatropha curcas* seed cake extract and characterization: evaluation of its antibacterial activity. Energy Sources, Part A Recover Util Environ Eff 1–9

Naz S, Kazmi STB, Zia M (2019) CeO_2 nanoparticles synthesized through green chemistry are biocompatible: in vitro and in vivo assessment. J Biochem Mol Toxicol 33:e22291–e22300

Nishanthi R, Malathi S, John Paul S, Palani P (2019) Green synthesis and characterization of bio-inspired silver, gold and platinum nanoparticles and evaluation of their synergistic antibacterial activity after combining with different classes of antibiotics. Mater Sci Eng C 96:693–707

Noruzi M (2015) Biosynthesis of gold nanoparticles using plant extracts. Bioprocess Biosyst Eng 38:1–14

Okafor F, Janen A, Kukhtareva T et al (2013) Green synthesis of silver nanoparticles, their characterization, application and antibacterial activity. Int J Environ Res Public Health 10:5221–5238

Olajire AA, Mohammed AA (2019) Green synthesis of palladium nanoparticles using *Ananas comosus* leaf extract for solid-phase photocatalytic degradation of low density polyethylene film. J Environ Chem Eng 7:103270–103279

Oluwasogo A, Amoo F, Elizabeth F et al (2019) Silver nanoparticle synthesis by *Acalypha wilkesiana* extract : phytochemical screening, characterization, influence of operational parameters, and preliminary antibacterial testing. Heliyon 5:e02517–e02525

Omran BA, Fatthalah NA, El-Gendy NS et al (2013) Green biocides against sulphate reducing bacteria and macrofouling organisms. J Pure Appl Microbiol 7:2219–2232

Omran BA, Nassar HN, Fatthallah NA et al (2017) Waste upcycling of *Citrus sinensis* peels as a green route for the synthesis of silver nanoparticles. Energy Sources, Part A Recover Util Environ Eff 40:1–10

Ontong JC, Paosen S, Shankar S, Voravuthikunchai SP (2019) Eco-friendly synthesis of silver nanoparticles using *Senna alata* bark extract and its antimicrobial mechanism through enhancement of bacterial membrane degradation. J Microbiol Methods 165:105692–105700

Pakzad K, Alinezhad H, Nasrollahzadeh M (2019) Green synthesis of Ni@Fe_3O_4 and CuO nanoparticles using *Euphorbia maculata* extract as photocatalysts for the degradation of organic pollutants under UV-irradiation. Ceram Int 45:17173–17182

Park Y, Hong YN, Weyers A et al (2011) Polysaccharides and phytochemicals: a natural reservoir for the green synthesis of gold and silver nanoparticles. IET Nanobiotechnol 5:69–78

Parshetti GK, Chowdhury S, Balasubramanian R (2014) Hydrothermal conversion of urban food waste to chars for removal of textile dyes from contaminated waters. Bioresour Technol 161:310–319

Patil BN, Taranath TC (2018) Microbial pathogenesis *Limonia acidissima* L leaf mediated synthesis of silver and zinc oxide nanoparticles and their antibacterial activities. Microb Pathog 115:227–232

Patra JK, Baek K-H (2016) Green synthesis of silver chloride nanoparticles using *Prunus persica* L. outer peel extract and investigation of antibacterial, anticandidal, antioxidant potential. Green Chem Lett Rev 9:132–142

Patra JK, Kwon Y, Baek KH (2016) Green biosynthesis of gold nanoparticles by onion peel extract: synthesis, characterization and biological activities. Adv Powder Technol 27:2204–2213

Pawliszak P, Malina D, Sobczak-kupiec A (2019) *Rhodiola rosea* extract mediated green synthesis of silver nanoparticles supported by nanosilica carrier. Mater Chem Phys 234:390–402

Phongtongpasuk S, Poadang S, Yongvanich N (2016) Environmental-friendly method for synthesis of silver nanoparticles from dragon fruit peel extract and their antibacterial activities. Energy Procedia 89:239–247

Pingret D (2013) Ultrasound-assisted extraction. Royal Society of Chemistry, London, pp 89–112

Pohlit AM, Rezende AR, Baldin ELL et al (2011) Plant extracts, isolated phytochemicals, and plant-derived agents which are lethal to arthropod vectors of human tropical diseases–a review. Planta Med 77:618–630

Poinern GEJ, Fawcett D (2018) Food waste valorization : new manufacturing processes for long-term sustainability. Elsevier, Boston

Pooja D, Singh L, Thakur A, Kumar P (2019) Sensors and actuators B : chemical green synthesis of glowing carbon dots from *Carica papaya* waste pulp and their application as a label-freechemo probe for chromium detection in water. Sensors Actuators B Chem 283:363–372

Prasad R (2014) Synthesis of silver nanoparticles in photosynthetic plants. J Nanoparticles, vol. 2014, Article ID 963961, 8 pages, 2014. https://doi.org/10.1155/2014/963961

Prasad AR, Garvasis J, Oruvil SK, Joseph A (2019) Bio-inspired green synthesis of zinc oxide nanoparticles using *Abelmoschus esculentus* mucilage and selective degradation of cationic dye pollutants. J Phys Chem Solids 127:265–274

Qais FA, Shafiq A, Khan HM, et al (2019) Antibacterial effect of silver nanoparticles synthesized using *Murraya koenigii* (L.) against multidrug-resistant pathogens. Bioinorg Chem Appl, vol. 2019, Article ID 4649506, 11 pages, 2019. https://doi.org/10.1155/2019/4649506

Rai M, Yadav A, Gade A (2008) current trends in phytosynthesis of metal nanoparticles. Crit Rev Biotechnol 28:277–284

Rajan A, Rajan AR, Philip D (2017) *Elettaria cardamomum* seed mediated rapid synthesis of gold nanoparticles and its biological activities. OpenNano 2:1–8

Rajesh KM, Ajitha B, Reddy YAK et al (2018) Assisted green synthesis of copper nanoparticles using *Syzygium aromaticum* bud extract: physical, optical and antimicrobial properties. Optik (Stuttg) 154:593–600

Rajeshkumar S, Bharath LV (2017) Mechanism of plant-mediated synthesis of silver nanoparticles – a review on biomolecules involved, characterisation and antibacterial activity. Chem Biol Interact 273:219–227

Rajiv P, Bavadharani B, Kumar MN, Vanathi P (2017) Synthesis and characterization of biogenic iron oxide nanoparticles using green chemistry approach and evaluating their biological activities. Biocatal Agric Biotechnol 12:45–49

Rajivgandhi G, Maruthupandy M, Muneeswaran T et al (2019) Biologically synthesized copper oxide nanoparticles enhanced intracellular damage in ciprofloxacin resistant ESBL producing bacteria. Microb Pathog 127:267–276

Raju D, Mehta UJ, Ahmad A (2012) Phytosynthesis of intracellular and extracellular gold nanoparticles by living peanut plant (*Arachis hypogaea* L.). Biotechnol Appl Biochem 59:471–478

Rather MA, Bhat IA, Sharma N et al (2017) Synthesis and characterization of *Azadirachta indica* constructed silver nanoparticles and their immunomodulatory activity in fish. Aquac Res 48:3742–3754

Ravichandran V, Vasanthi S, Shalini S et al (2016) Green synthesis of silver nanoparticles using *Atrocarpus altilis* leaf extract and the study of their antimicrobial and antioxidant activity. Mater Lett 180:264–267

Ravichandran V, Vasanthi S, Shalini S et al (2019) Results in physics photocatalytic activity of *Parkia speciosa* leaves extract mediated silver nanoparticles. Results Phys 15:102565–102573

Ren Y, Yang H, Wang T, Wang C (2019) Bio-synthesis of silver nanoparticles with antibacterial activity. Mater Chem Phys 235:121746–121753

Roy S, Das TK (2015) Plant mediated green synthesis of silver nanoparticles-a review. Int J Plant Biol Res 3:1044–1055

Ruíz-baltazar ÁDJ, Maya-cornejo J, Rodríguez-morales AL, Esparza R (2019) Alcoholic extracts from *Paulownia tomentosa* leaves for silver nanoparticles synthesis. Results Phys 12:1670–1679

Salouti M, Derakhshan FK (2019) Phytosynthesis of nanoscale materials, adv phytonanotechnology, from synthesis to application, Elsevier, Boston, p 45–121

Samari F, Balouchi L, Salehipoor H, Yousefinejad S (2019) Controllable phyto-synthesis of cupric oxide nanoparticles by aqueous extract of *Capparis spinosa* (caper) leaves and application in iron sensing. Microchem J 150:104158–104168

Sarmah S, Kumar A (2011) Photocatalytic activity of polyaniline-TiO_2 nanocomposites. Indian J Phys 85:713–726

Sathiyavimal S, Vasantharaj S, Bharathi D et al (2018) Biogenesis of copper oxide nanoparticles (CuONPs) using *Sida acuta* and their incorporation over cotton fabrics to prevent the pathogenicity of Gram negative and Gram positive bacteria. J Photochem Photobiol B Biol 188:126–134

Saxena A, Tripathi RM, Zafar F, Singh P (2012) Green synthesis of silver nanoparticles using aqueous solution of *Ficus benghalensis* leaf extract and characterization of their antibacterial activity. Mater Lett 67:91–94

Senthilkumar SR, Sivakumar T (2014) Green tea (*Camellia sinensis*) mediated synthesis of zinc oxide (ZnO) nanoparticles and studies on their antimicrobial activities. Int J Pharm Pharm Sci 6:461–465

Shah M, Fawcett D, Sharma S et al (2015) Green synthesis of metallic nanoparticles via biological entities. Materials 8:7278–7308

Shahnaz M, Danish M, Hisyamuddin M et al (2019) Anticancer and apoptotic activity of biologically synthesized zinc oxide nanoparticles against human colon cancer HCT-116 cell line-in vitro study. Sustain Chem Pharm 14:100179–100187

Shankar SS, Ahmad A, Pasricha R, Sastry M (2003) Bioreduction of chloroaurate ions by geranium leaves and its endophytic fungus yields gold nanoparticles of different shapes. J Mater Chem 13:1822–1826

Shankar S, Jaiswal L, Aparna RSL, Prasad R (2014) Synthesis, characterization, in vitro biocompatibility, and antimicrobial activity of gold, silver and gold silver alloy nanoparticles prepared from *Lansium domesticum* fruit peel extract. Mater Lett 137:75–78

Shanmugavadivu M, Kuppusamy S, Ranjithkumar R (2014) Synthesis of pomegranate peel extract mediated silver nanoparticles and its antibacterial activity Am J Adv Drug Deliv 2:174–182

Sharma P, Pant S, Dave V et al (2019a) Green synthesis and characterization of copper nanoparticles by *Tinospora cardifolia* to produce nature-friendly copper nano-coated fabric and their antimicrobial evaluation. J Microbiol Methods 160:107–116

Sharma TSK, Selvakumar K, Hwa KY et al (2019b) Biogenic fabrication of gold nanoparticles using *Camellia japonica* L. leaf extract and its biological evaluation. J Mater Res Technol 8:1412–1418

Sharmila G, Muthukumaran C, Sandiya K et al (2018) Biosynthesis, characterization, and antibacterial activity of zinc oxide nanoparticles derived from *Bauhinia tomentosa* leaf extract. J Nanostructure Chem 8:293–299

Sharmila G, Thirumarimurugan M, Muthukumaran C (2019) Green synthesis of ZnO nanoparticles using *Tecoma castanifolia* leaf extract: characterization and evaluation of its antioxidant, bactericidal and anticancer activities. Microchem J 145:578–587

Shukla R, Nune SK, Chanda N et al (2008) Soybeans as a phytochemical reservoir for the production and stabilization of biocompatible gold nanoparticles. Small 4:1425–1436

Si S, Mandal TK (2007) Tryptophan based peptides to synthesize gold and silver nanoparticles: a mechanistic and kinetic study. Chem Eur J 13:3160–3168

Singh P, Kim YJ, Wang C et al (2016a) Biogenic silver and gold nanoparticles synthesized using red ginseng root extract, and their applications. Artif Cells, Nanomedicine, Biotechnol 44:811–816

Singh P, Kim YJ, Yang DC (2016b) A strategic approach for rapid synthesis of gold and silver nanoparticles by *Panax ginseng* leaves. Artif Cells, Nanomedicine, Biotechnol 44:1949–1957

Singh P, Kim YJ, Zhang D, Yang DC (2016c) Biological synthesis of nanoparticles from plants and microorganisms. Trends Biotechnol 34:588–599

Sinsinwar S, Kumar M, Rohit K, Nithyanand P (2018) Microbial pathogenesis use of agricultural waste (coconut shell) for the synthesis of silver nanoparticles and evaluation of their antibacterial activity against selected human pathogens. Microb Pathog 124:30–37

Snyder SW (2015) Commercializing biobased products: opportunities, challenges, benefits, and risks. Royal Society of Chemistry, London

Song JY, Kwon E-Y, Kim BS (2010) Biological synthesis of platinum nanoparticles using *Diopyros kaki* leaf extract. Bioprocess Biosyst Eng 33:159–164

Sorbiun M, Shayegan Mehr E, Ramazani A, Mashhadi Malekzadeh A (2018) Biosynthesis of metallic nanoparticles using plant extracts and evaluation of their antibacterial properties. Nanochem Res 3:1–16

Souza AE, Teixeira SR, Santos GTA et al (2011) Reuse of sugarcane bagasse ash (SCBA) to produce ceramic materials. J Environ Manag 92:2774–2780

Sumbal, Nadeem A, Naz S et al (2019) Synthesis, characterization and biological activities of monometallic and bimetallic nanoparticles using *Mirabilis jalapa* leaf extract. Biotechnol Rep 22:e00338–e00347

Suresh J, Pradheesh G, Alexramani V et al (2018) Green synthesis and characterization of hexagonal shaped MgO nanoparticles using insulin plant (*Costus pictus* D. Don) leave extract and its antimicrobial as well as anticancer activity. Adv Powder Technol 29:1685–1694

Tewari C, Tatrari G, Karakoti M et al (2019) A simple, eco-friendly and green approach to synthesis of blue photoluminescent potassium-doped graphene oxide from agriculture waste for bio-imaging applications. Mater Sci Eng C 104:109970–109977

Thakur BK, Kumar A, Kumar D (2019) Green synthesis of titanium dioxide nanoparticles using *Azadirachta indica* leaf extract and evaluation of their antibacterial activity. S Afr J Bot 124:223–227

Thangamani N, Bhuvaneshwari N (2019) Green synthesis of gold nanoparticles using *Simarouba glauca* leaf extract and their biological activity of micro-organism. Chem Phys Lett 732:136587–136594

Thatikayala D, Jayarambabu N, Banothu V et al (2019) Biogenic synthesis of silver nanoparticles mediated by *Theobroma cacao* extract: enhanced antibacterial and photocatalytic activities. J Mater Sci Mater Electron 30:17303–17313

Thomas B, Vithiya B, Prasad T et al (2019) Antioxidant and photocatalytic activity of aqueous leaf extract mediated green synthesis of silver nanoparticles using *Passiflora edulis f. flavicarpa*. J Nanosci Nanotechnol 19:2640–2648.

Ur A, Ullah A, Yuan Q et al (2019) Tuber extract of *Arisaema flavum* eco-benignly and effectively synthesize silver nanoparticles : Photocatalytic and antibacterial response against multidrug resistant engineered *E. coli* QH4. J Photochem Photobiol B Biol 193:31–38

Vandermeersch T, Alvarenga RAF, Ragaert P, Dewulf J (2014) Environmental sustainability assessment of food waste valorization options. Resour Conserv Recycl 87:57–64

Varadavenkatesan T, Vinayagam R, Selvaraj R (2017) Structural characterization of silver nanoparticles phyto-mediated by a plant waste, seed hull of *Vigna mungo* and their biological applications. J Mol Struct 1147:629–635

Vasantharaj S, Sathiyavimal S, Saravanan M et al (2019) Synthesis of ecofriendly copper oxide nanoparticles for fabrication over textile fabrics: characterization of antibacterial activity and dye degradation potential. J Photochem Photobiol B Biol 191:143–149

Veisi H, Ghorbani M, Hemmati S (2019) Sonochemical in situ immobilization of Pd nanoparticles on green tea extract coated Fe_3O_4 nanoparticles : an efficient and magnetically recyclable nanocatalyst for synthesis of biphenyl compounds under ultrasound irrad. Mater Sci Eng C 98:584–593

Velsankar K, Sudhahar S, Maheshwaran G, Krishna Kumar M (2019) Effect of biosynthesis of ZnO nanoparticles via *Cucurbita* seed extract on *Culex tritaeniorhynchus* mosquito larvae with its biological applications. J Photochem Photobiol B Biol 200:111650–111666

Venkat Kumar S, Rajeshkumar S (2017) Plant-based synthesis of nanoparticles and their impact. In: Nanomaterials in plants, algae, and microorganisms, concepts and controversies: volume 1. Elsevier Inc, New York, pp 33–57

Vennila R, Kamaraj P, Arthanareeswari M et al (2018) Biosynthesis of ZrO nanoparticles and its natural dye sensitized solar cell studies. Mater Today Proc 5:8691–8698

Vijaya Kumar P, Mary Jelastin Kala S, Prakash KS (2019) Green synthesis of gold nanoparticles using *Croton Caudatus* Geisel leaf extract and their biological studies. Mater Lett 236:19–22

Vijayakumar S, Mahadevan S, Arulmozhi P et al (2018) Green synthesis of zinc oxide nanoparticles using *Atalantia monophylla* leaf extracts: characterization and antimicrobial analysis. Mater Sci Semicond Process 82:39–45

Vijayaraghavan K, Ashokkumar T (2017) Plant-mediated biosynthesis of metallic nanoparticles: a review of literature, factors affecting synthesis, characterization techniques and applications. J Environ Chem Eng 5:4866–4883

Vinay SP, Udayabhanu NG et al (2019) *Rauvolfia tetraphylla* (devil pepper)-mediated green synthesis of Ag nanoparticles: applications to anticancer, antioxidant and antimitotic. J Clust Sci 30:1545–1564

Vinotha V, Iswarya A, Thaya R et al (2019) Synthesis of ZnO nanoparticles using insulin-rich leaf extract: anti-diabetic, antibiofilm and anti-oxidant properties. J Photochem Photobiol B Biol 197:111541–111553

Virot M, Tomao V, Le Bourvellec C et al (2010) Towards the industrial production of antioxidants from food processing by-products with ultrasound-assisted extraction. Ultrason Sonochem 17:1066–1074

Xin Lee K, Shameli K, Miyake M, et al (2016) Green synthesis of gold nanoparticles using aqueous extract of *Garcinia mangostana* fruit peels. J Nanomaterials, vol. 2016, Article ID 8489094, 7 pages, 2016. https://doi.org/10.1155/2016/8489094

Yang N, Li W-H (2013) Mango peel extract mediated novel route for synthesis of silver nanoparticles and antibacterial application of silver nanoparticles loaded onto non-woven fabrics. Ind Crop Prod 48:81–88

Yang N, Li W, Hao L (2014) Biosynthesis of Au nanoparticles using agricultural waste mango peel extract and its in vitro cytotoxic effect on two normal cells. Mater Lett 134:67–70

Yang S, Lohe MR, Müllen K, Feng X (2016) New generation graphene from electrochemical approaches: production and applications. Adv Mater 28:6213–6221

Zangeneh MM, Ghaneialvar H, Akbaribazm M et al (2019) Novel synthesis of *Falcaria vulgaris* leaf extract conjugated copper nanoparticles with potent cytotoxicity, antioxidant, antifungal, antibacterial, and cutaneous wound healing activities under in vitro and in vivo condition. J Photochem Photobiol B Biol 197:111556–111569

Zare M, Namratha K, Thakur MS, Byrappa K (2019) Biocompatibility assessment and photocatalytic activity of bio-hydrothermal synthesis of ZnO nanoparticles by *Thymus vulgaris* leaf extract. Mater Res Bull 109:49–59

Zayed MF, Eisa WH, El-kousy SM et al (2019) *Ficus retusa*-stabilized gold and silver nanoparticles: controlled synthesis, spectroscopic characterization, and sensing properties. Spectrochim Acta – Part A Mol Biomol Spectrosc 214:496–512

Zheng B, Kong T, Jing X et al (2013) Plant-mediated synthesis of platinum nanoparticles and its bioreductive mechanism. J Colloid Interface Sci 396:138–145

Chapter 6
Versatile Applications of Biosynthesized Nanoparticles, Global Safety Issues, Grand Challenges, and Future Perspectives Regarding Nanobiotechnology

6.1 Introduction

Nowadays, the whole world is seeking for synthesizing novel materials or systems with exceptional characteristics and performance. Research in modern biotechnology is mainly focused upon exploiting the distinctive features and functions of living organisms and microorganisms. The main vision of this technology is very broad including gene technology (e.g., "cloning"); bioinformatics; optical, electrical, or mechanical information; nanobiotechnology; etc. Nanobiotechnology is the science which is mainly concerned with the study of fine structures originated from biological macromolecules, and their behavior in the nanoscale range to be applied in different disciplines. The magnificent integration between nanotechnology and biotechnology formulated the modern science of nanobiotechnology which will soon become an important scientific discipline in human's life. It has very promising potential applications in different sectors involving biomedical, electronics, materials science, environmental science, pharmaceutical, solar energy, bioremediation, etc. Research in nanobiotechnology is majorly spotted upon two main aspects: the first is to exploit the already existed biological molecules, for instance, DNA, proteins, enzymes, and ion channels, and the second aspect involves the use of processed nanomaterials such as metal, metal oxide nanoparticles, quantum dot particles (QDs), and magnetic nanoparticles. In this context, this chapter glances through the versatile applications of biosynthesized nanomaterials, safety, and toxicity of the prepared nanomaterials to human health and to the surrounding environment in addition to the future perspectives which need to be taken into account.

B. A. Omran, *Nanobiotechnology: A Multidisciplinary Field of Science*,
Nanotechnology in the Life Sciences, https://doi.org/10.1007/978-3-030-46071-6_6

6.2 Versatile Applications of Biologically Fabricated Nanoparticles

6.2.1 Applications in Agriculture (Agro-nanotechnology)

Nanobiotechnology has tremendous applications in the agricultural sector including the development of novel nanofertilizers, nanoherbicides, nanosensors, and nanopesticides (de Oliveira et al. 2014; Liu and Lal 2015; Grillo et al. 2016). Figure 6.1 represents the possible applications of biologically fabricated NPs in agriculture. Singh et al. (2019b) reported the employment of nanotechnology in the development of healthy seeds, thereby improving plant germination, plant breeding, growth, yield, and quality. Moreover, it has the potential to increase the storage period for both fruits and vegetables. AgNPs have shown great capability to get rid of a myriad of several phytopathogens (fungal and bacterial pathogens) which cause various plant diseases. Mishra et al. (2014) proposed an efficient application of biologically synthesized NPs derived from *Serratia* sp. BHU-S4. It was found that they were effective against *Bipolaris sorokiniana* which causes spot blotch disease in

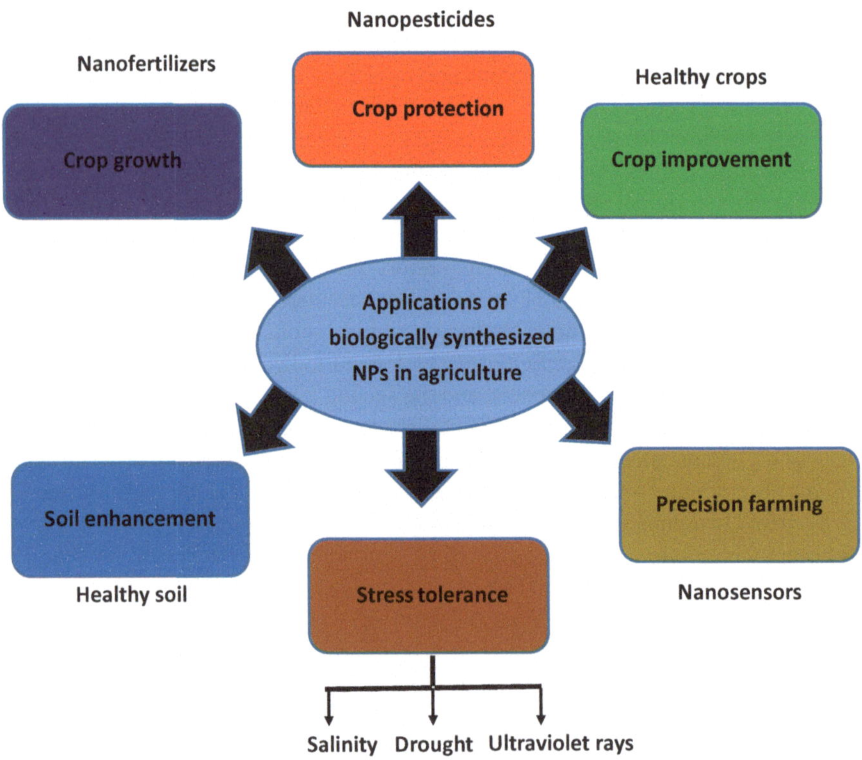

Fig. 6.1 Different applications of biologically fabricated NPs in agriculture

wheat. Vanti et al. (2019) demonstrated the successful biological synthesis of AgNPs using the stem extract of *Gossypium hirsutum* (cotton plant). Authors investigated the antibacterial efficacy of the phytosynthesized AgNPs against the plant pathogenic bacteria *Xanthomonas axonopodis* pv. *malvacearum* and *Xanthomonas campestris* pv. *campestris*. The resultant data showed that AgNPs exhibited antibacterial potential against both pathogens. Two concentrations of AgNPs (50 and 100 μg/ml) exhibited zone of inhibition (ZOI) of 11.0 ± 1.0 and 12.3 ± 0.5 mm in case of *X. axonopodis pv. malvacearum* and 9.7 ± 0.6 and 15.33 ± 1.0 mm for *X. campestris* pv. *campestris*, respectively. Ndeh and co-workers managed to biosynthesize AuNPs via a green approach using the aqueous leaf extract of *Tiliacora triandra* (Ndeh et al. 2017). AuNPs were further evaluated for their effect on germination of rice roots and shoot roots. Different concentrations of AuNPs were employed (0, 10, 100, 500, 1000, 2000 mg l^{-1}) for 1 week. AuNPs exhibited superior effect upon germination rate with a percentage ranging between 95 and 98.38%. However, a slight decrease in the length of both root and shoot was observed compared with the control. Authors also carried out phytotoxicity studies, and the obtained results revealed that the plant-derived AuNPs were of minimal toxicity to rice seedlings. Besides, increases in cell death, H_2O_2 formation, and lipid peroxidation in both roots and shoots were noticed. Yet, such increases were not statistically significant.

While nanotechnology possesses several positive aspects particularly in agriculture and food industries, few published data explains the consequences regarding the risk to humans. For instance, Tsuji et al. (2005) illustrated that when rats were exposed to titanium dioxide NPs via inhalation, this led to microvascular dysfunction. Moreover, inhalation of titanium dioxide engineered nanomaterials (ENM) affected the systematic circulation in rats as reported by Nurkiewicz et al. (2008). On the contrary, Radomski et al. (2005) demonstrated that exposure to NPs enhanced platelet aggregation as well as vascular thrombosis in the tested rats. As a result, it is extremely essential to obtain more reliable data regarding the effect of nanomaterials on the ecosystems of plants and animals before being applied in the agricultural sector.

6.2.2 Applications in Environmental Pollution Management

The rapid global industrialization and population increased the percentage of exposure to several toxic materials like polyaromatic hydrocarbons, heavy metals, etc. Consequently, an urgent need for clean technology has become a worldwide demand. Textile industry is one of the major industries that consume large volumes of water annually. Dyestuff has been recognized as the first contaminant of water pollution. The discharge of dyes in water is one of the greatest concerns that cause severe environmental threats and worries upon human health as well as the aquatic life ecosystem. As a result, the removal of dyes, pollutant chemicals, and pathogenic microorganisms and treatment of wastewater became very urgent

(Omran et al. 2018, 2019a). Researchers around the world made severe efforts regarding the use of biologically synthesized NPs to degrade environmental pollutants including dyes.

Fulekar et al. (2018) managed to isolate microorganisms from the root zone of the rhizosphere of sorghum plant. The isolated microorganisms were identified using 16 srDNA as *Micrococcus lylae* (MF1), *Micrococcus aloeverae* (MF2), and *Cellulosimicrobium* sp. (MF3). They had the potential to mediate the synthesis of TiO_2 NPs. Authors illustrated that TiO_2 NPs derived from MF1, MF2 and MF3 had the potential to degrade methyl orange (MO) within 195, 180, and 105 min with ~99% degradation, respectively. Extract of immature fruits of *Rubus coreanus* acted as a reducing agent for the biogenic synthesis of ZnO NPs as demonstrated by Jahan et al. (2018). Authors noticed the development of a white precipitate. Using the analytical spectroscopic and microscopic techniques, it was revealed that the white precipitate was ZnO NPs.

UV/Vis spectroscopy revealed a maximum absorption spectrum at 375 nm. The catalytic activity of ZnO NPs was determined against malachite green (MG). MG is also referred to as basic green or Victoria green. Malachite green has the chemical formula of $C_{23}H_{25}N_2Cl$ and its IUPAC name is 4-[(4-dimethulamonophenyl)-phenyl methyl]-N, N-dimethyl-Amin (Sarmah and Kumar 2011). This dye has a high affinity regarding solubility in water. According to Srivastava et al. (2004) and Cheng et al. (2008), MG crystal powder is widely used in industries such as leather, textile, silk, and paper. But, unfortunately, this dye is highly toxic and extremely water soluble hence cannot be easily removed from water. This dye has carcinogenic effects on human and also causes threats upon the aquatic life ecosystem. As a result, *Rubus coreanus* zinc oxide nanoparticles (Rc-ZnO NPs) were tested for their degradation capability against the organic pollutant MG in both the dark and light (indoor light) conditions. In the presence of light source, authors assumed that the light source possess a band gap higher than that of the biologically fabricated ZnO NPs. Then, the light is absorbed via the electrons in the valence band of the NPs and moves to the conduction band, resulting in the formation of an electron hole. The hydroxide free radicals are then formed by the reaction with the dissolved oxygen, leading to formation of superoxide which in turn generates hydroxide free radicals. These [–OH] radicals have strong potential towards MG degradation. UV/Vis spectroscopy revealed the decrease in the absorption peak of MG at 618 nm after being treated with Rc-ZnO NPs. Moreover, Rc-ZnO NPs exerted a high catalytic activity (approximately 90%) in both the light and dark conditions. Hence, there is no further need to any light source either solar or UV light to activate the catalyst. Accordingly, Rc-ZnO NPs are strongly recommended as environmentally friendly agents for treating water contaminated with dye effluents particularly from MG.

Morinda tinctoria leaf extract was reported to mediate the biogenic synthesis of AgNPs (Vanaja et al. 2014). The photocatalytic activity of the green synthesized AgNPs was determined against the MB dye under sunlight irradiation. The green synthesized AgNPs were found to effectively degrade the dye after 72 h of exposure time with a degradation efficiency of approximately 95%. The main absorption peak of MB at 660 nm gradually reduced which confirmed the photocatalytic degradation of MB dye. Hollow microspheres of ZnO and ZnO hollow microspheres/reduced graphene oxide (ZnO/rGO) nanocomposites were successfully synthesized for the first time via a totally green method using the extract of red rice husk (Alomair and Mohamed 2018). Red rice husk served as an efficient oxidizing and a chelating agent. The green synthesized ZnO and ZnO/rGO nanocomposite were characterized. Raman and XRD analyses revealed that the fabricated ZnO nanostructures exhibited wurtzite hexagonal phase with high crystallinity. The photocatalytic activity of the green synthesized ZnO microspheres was evaluated for the degradation of malachite green dye. Authors observed that the photocatalytic activity increased in the presence of ZnO/rGO nanocomposite and attributed this result to the role played by graphene oxide. Graphene oxide participated in capturing the photo-generated electrons and in reducing electron-hole pair recombination. Palladium NPs were biologically fabricated via the extract of *Saccharomyces cerevisiae* (Sriramulu and Sumathi 2018). The photocatalytic potential of the yeast-derived PdNPs was evaluated for the reduction of direct blue 71 dye under the UV light. The decrease in the intensity of the characteristic absorption peak of direct blue 71 (5091 nm) was observed. It was noticed that 98% of the textile dye was degraded within 60 min by using the yeast-derived PdNPs.

An innovative, one-pot, and green synthesis of magnetic iron oxide NPs (MIONPs) via the waste fruit extract of *Cynometra ramiflora* was proposed by Bishnoi et al. (2018). Bishnoi and co-authors evaluated the photocatalytic activity of *Cynometra ramiflora*-derived MIONPs against methylene blue (MB) as a model pollutant under sunlight irradiation. It was observed that MIONPs accelerated the degradation of MB as a function of time. The absorption peak characteristic to MB at 663 nm gradually reduced as the irradiation time was increased. It eventually disappeared after 110 min. It is worth mentioning that the reusability of the photocatalyst plays a major role in the photodegradation process. In this study, the recovered MIONPs were reused for five consecutive times. AgNPs were mycosynthesized via the mycelial cell free filtrate of *Aspergillus brasiliensis* ATCC 16404 (Omran et al. 2018). The mycosynthesized AgNPs was investigated for its capability to inhibit the growth of some pathogenic microorganisms including *Bacillus subtilis* ATCC 6633 and *Staphylococcus aureus* ATCC 35556, Gram-negative *Pseudomonas aeruginosa* ATCC 10145 and *Escherichia coli* ATCC 23282, and a yeast strain *Candida albicans* IMRU3669. The obtained results proved the good biocidal activity of *A. brasiliensis*-derived AgNPs against the tested pathogenic microorganisms which can cause some water-related diseases and health problems to local residents. Palladium nanoparticles (PdNPs) were biologically fabricated via the gum ghatti (*Anogeissus latifolia*) as investigated by Kora and Rastogi (2018). UV/Vis spectrophotometry, DLS, TEM, and XRD were employed to characterize the generated PdNPs. The appearance of the intense brown color gave a preliminary indication for PdNPs formation. The prepared NPs exhibited a spherical shape with an average particle size of 4.8 ± 1.6 nm. The catalytic activity of PdNPs was evaluated against Coomassie Brilliant Blue (CBB) G-250, MB, MO, and nitrophenolic compound (e.g., 4-nitrophenol with sodium borohydride). The catalytic degradation potential against CBB by PdNPs was assessed by monitoring the characteristic absorption peak of CBB at 588 nm. Within 2 min of reaction, the color of CBB was changed from blue to colorless. Additionally, the disappearance of MB characteristic absorption peak at 664 nm was noted within 2 min of reaction, thus indicating the complete reduction of the MB dye to leuco methylene blue. Besides, the characteristic absorption peak of MO was observed at 462 nm and was found to disappear within 2 min of reaction as well. Moreover, a characteristic peak at 294 nm appeared in the presence of Pd nanocatalyst which indicated the reduction of 4-nitrophenol (4-NP) to 4-aminophenol (4-AP). Change in color was also observed from yellow to colorless. Copper oxide NPs (CuO NPs) were successfully phytosynthesized using *Solanum nigrum* as was demonstrated by Lakshmi et al. (2019). The degradation efficiency of the phytosynthesized CuO NPs was evaluated against a wide range of industrial dyes such as methyl orange (MO), methylene blue (MB), safranin O, Congo red (CR) and Coomassie Brilliant Blue R-250 (CBB). Among the tested dyes, the biologically fabricated CuO NPs exhibited a better degradation efficacy against CBB, CR, and methylene blue. The degradation efficiency was found to reach 75.32 ± 2.07%, 89.33 ± 0.739%, and 64.448 ± 1.1411% for CBB, methylene blue, and CR, respectively. Furthermore, authors investigated the effect of different parameters that might influence the degradation efficiency which involved irradiation time, catalyst load, dye concentration, and pH. Besides, authors proposed a mechanism to elucidate the role played by the phytosynthesized CuO NPs for dye degradation. First, electrons consume the energy in the valence band and get excited to the conduction band when the sunlight falls on the surface of the nanocatalyst CuO NPs, leading to generation of electron-hole pair. Afterwards, the holes generated in the valence band react with the hydroxyl groups on the surface, leading to their reduction to more reactive hydroxyl radicals [$OH^{\bullet}$]. Electrons in the conduction band react with oxygen molecules adsorbed on the catalyst surface and oxidize them into superoxide anion radicals [$O^{\bullet}_2$]. As a consequence, the generated free radicals become highly reactive and react with the dye molecules, leading to the formation of Co_2, H_2O, and other by-products.

In a study proposed by Xu et al. (2019), *Shewanella oneidensis* MR-1 mediated the biological synthesis of novel, Pd-Pt alloy NPs. Data showed the successful synthesis of Pd-Pt alloy NPs both intracellularly and extracellularly. The size of the biosynthesized Pd-Pt alloy NPs was in the range from 3 to 40 nm. The biologically synthesized Pd-Pt alloy NPs were evaluated for their catalytic activity towards methyl red (MR), MO, and acid red 14 (AR 14) and 4-NP reduction reaction. 4-NP solution displayed a peak centered near 318 nm, and after the addition of $NaBH_4$ solution, it was shifted to 400 nm. After the treatment with the nanocatalyst Pd-Pt alloy NPs, the absorption peak at 400 nm started to decrease, and an absorption peak at 300 nm started to appear, indicating the transformation of 4-NP to 4-AP. The reaction was terminated within 4 min. In cases of MO, MR, and AR 14, characteristic absorption peaks appeared at 465, 431, and 506 nm, respectively, due to the presence of azo groups. It was observed that after the addition of Pd-Pt alloy nanocatalyst, the absorbance of azo groups was reduced with reaction time, which took place within 2, 6, and 13 min for MO, MR, and AR 14, respectively. In a study performed by Altikatoglu Yapaoz and Attar (2019), TiO_2 NPs were biologically fabricated using the extract of *Salvia officinalis*. The photocatalytic decomposition performance of TiO_2 NPs was evaluated against three organic dyes including Reactive Blue 19, Brilliant Blue R, and Reactive Black 5. The obtained results indicated that the biologically fabricated TiO_2 NPs exhibited a notably high photocatalytic efficacy towards the degradation of the three tested azo dyes. Decolorization percentage increased by increasing the reaction time. Authors elucidated a possible mechanism which might be responsible for dye degradation in the presence of *S. officinalis*-derived TiO_2 NPs. This mechanism revealed that when the dye is in the ground state, it becomes excited under the effect of the visible light which provided TiO_2 NPs with electrons which in turn caused semi-oxidation to the target dyes.

6.2.3 Applications in Medical Sector

Biologically fabricated NPs have several medical applications as revealed in Fig. 6.2.

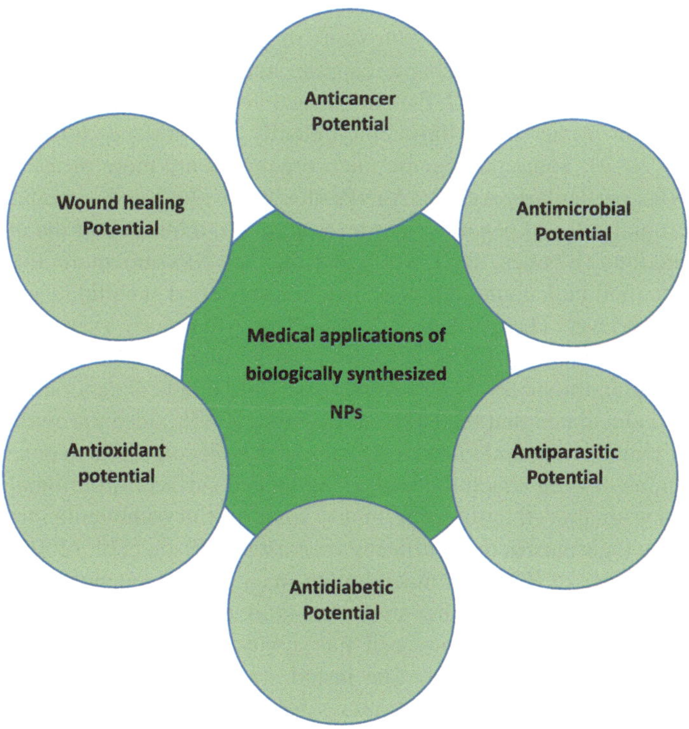

Fig. 6.2 Versatile medical applications of NPs in the medical field

6.2.3.1 Anticancer Activity

One of the deadliest diseases that occurred in the modern world is cancer. Cancer resulted in the death of millions of people around the world. However, it can be prevented and can be curable if detected at an early stage and the proper treatment is given at the right time. It has been reported that cancer has more than 100 types according to the National Cancer Institute (NCI), United States (Barabadi et al. 2017). Additionally, different estimations anticipate that incidence of cancer worldwide will reach 21.7 million by 2030. Besides, the death rate will reach approximately 13 million death cases due to the increase in population growth as well as aging. Radiotherapy, chemotherapy, and surgery are among the usually employed cancer treatments. However, the side effects which usually accompany these treatments are massive (Lokina et al. 2014; Rao et al. 2016; Anwar et al. 2017). Despite the advancements in cancer control, incidence and death rates are highly increased. Lately, nanobiotechnology showed novel strategies for diagnosis and treatment of cancer. Table 6.1 shows a list of some of the green synthesized nanomaterials which proved to be highly significant in cancer treatment.

Cytotoxicity studies are useful initial steps in determining the potential toxicity of a particular substance. Minimal to no toxicity is essential for the successful development of a pharmaceutical or cosmetic preparation, and in this regard, cellular toxicity studies play a crucial role. Studying the toxicity issues related to NPs is a very important aspect. Cytotoxicity of NPs is majorly influenced by different factors including size, shape, capping agent, and concentration of NPs (Roy et al. 2019). However, NPs synthesized from non-biological routes are basically more toxic than those biologically synthesized (Gahlawat and Choudhury 2019). Some pathogenic microorganisms are more prone to certain types of NPs, particularly AgNPs. AgNPs slowly envelope and surround the targeted microbial cell and penetrate inside, resulting in inhibition of the main metabolic functions. Besides, it is worth noting that NPs are more toxic when compared to their bulk counterparts as they become lethal at cellular, subcellular, and molecular levels (Jayasree et al. 2006). The cytotoxicity exerted by NPs is assumed to be generated via reactive oxygen species (ROS) which lead to a reduction in glutathione levels and an increase in free radical generation. Omran et al. (2018) elucidated that the large surface area of NPs aids in providing a better contact with the targeted microorganisms as they become able to penetrate the cell membrane or even attach themselves to the cell surface. Additionally, as the size becomes smaller, they become highly toxic to the pathogenic strains, and consequently their antibacterial efficacy increases. Carlson et al. (2008) demonstrated that release of ROS was much higher in case of 15 nm sized AgNPs than to 55 nm sized AgNPs. In another study, Liu et al. (2010) investigated the effect of AgNPs size upon four human cell lines, which were referred to as A549, HePG2, MCF-7, and SGC-7901. The tested sizes of AgNPs were 5, 20, and 50 nm. Authors performed several assays which involved observation of cell

Table 6.1 List of biologically fabricated nanomaterials with anticancer potential

Biological entity	Biologically synthesized NPs	Size (nm)	Shape	Type of cancer cell line	Dose	Exposure time	IC50	References
Albizia adianthifolia (leaf extract)	AgNPs	4–35	Spherical, near spherical	Lung cancer cells (A549)	10 μg/ml	6 h	–	Gengan et al. (2013)
Sesbania grandiflora (leaf extract)	AgNPs	22	Spherical	Breast cancer cell line (MCF-7)	0, 10, 20, 30, 40, 50 μg/ml	24 h	20 μg/ml	Jeyaraj et al. (2013)
Vitex negundo L. (leaf extract)	AgNPs	5–47	Spherical	Human colon cancer cell line Hct15	10, 20, 50, 70, 100 μg/ml	48 h	20 μg/ml	Prabhu et al. (2013)
Acalypha indica (leaf extract)	CuO NPs	26 ± 4	Spherical	Breast cancer cell line (MCF-7)	100 μg/ml	48 h	56.16 μg/ml	Sivaraj et al. (2014)
Syzygium cumini (fruit extract)	AgNPs	10–15	Spherical	Dalton lymphoma (DL) cell lines	50, 100, 250, 500 μg/ml	4–6 h	66.38 μg/ml	Mittal et al. (2014)
Sargassum swartzii	AuNPs	35	Spherical	Human cervical carcinoma (HeLa) cells	15.63, 31.25, 62.5, 125, 250, 500 μg/ml	24 h	41.10 μg/ml	Dhas et al. (2014)
Brown seaweed (*Sargassum muticum*)	Fe_3O_4 NPs	–	–	Human cell lines for leukemia (Jurkat cells), breast cancer (MCF-7 cells), cervical cancer (HeLa cells), and liver cancer (HepG2 cells)	–	72 h	HepG2 (23.83 ± 1.1 μg/ml), MCF- 7 (18.75 ± 2.1 μg/ml), HeLa (2.5 ± 1.7 μg/ml), Jurkat (6.4 ± 2.3 μg/ml)	Namvar et al. (2014)

(continued)

Table 6.1 (continued)

Biological entity	Biologically synthesized NPs	Size (nm)	Shape	Type of cancer cell line	Dose	Exposure time	IC50	References
Areca catechu nut	AuNPs	13.7	Spherical	HeLa cervical cancer cell line (NCCS Pune)	100 μl	24 h	25.17 μl	Rajan et al. (2015)
Clerodendrum serratum (leaf extract)	AgNPs	5–30	Spherical	Ehrlich ascites carcinoma cells	5, 10, 20, 30, 40, 50, and 60 μg/ml	24 h	–	Priyadharshini Raman et al. (2015)
Antigonon leptopus	AuNPs	13–28	Spherical, triangular	Human adenocarcinoma breast cancer (MCF-7) cells	31.25, 62.5, 125, 250, 500, 1000 μg/ml	48 h	257.8 μg/ml	Balasubramani et al. (2015)
Cucurbita maxima (petal extract), *Moringa oleifera* (leaf extract), and *Acorus calamus* (rhizome extract)	AgNPs	30–70	Roughly spherical, cuboidal	Epidermoid A431 carcinoma	10, 20, 40, 60, 80, 100, 150 μg/ml	24 h	82.39 ± 3.1, 83.57 ± 3.9 and 78.58 ± 2.7 μg/ml	Nayak et al. (2015)
Solanum trilobatum (fruit extract)	AgNPs	12.50–41.90	Spherical	Human breast cancer cell line MCF-7	5, 10, 20, 30, 40, and 50 μg/ml	24 h	–	Ramar et al. (2015)

Eclipta alba (leaf extract)	AgNPs	310–400	Spherical, cuboid	RAW 264.7 (mouse macrophage cells), MCF-7 (human breast cancer cells), Caco-2 (human adenocarcinoma cells)	0–1000 ng/ml	48 h	–	Premasudha et al. (2015)
Rosa indica (petal extract)	AgNPs	23.52–60.83	Spherical	HCT15 cell line	5, 10, 20, 30, 40, 50, and 60 μg/ml	24 h	–	Manikandan et al. (2015)
Butea monosperma (leaf extract)	AgNPs, AuNPs	AgNPs (20–80), AuNPs (10–30, 50–75, 30–100, 15–35)	AgNPs (spherical, triangular), AuNPs (spherical, rod, triangular, hexagonal)	B16F10, MCF-7, A549, and HNGC-2 cells	–	48 h	–	Patra et al. (2015)
Erythrina indica Lam (root extract)	AgNPs	20–118	Spherical	MCF-7 and HEPG2 cell lines	25–0.625 μg/ml	24 h	–	Rathi Sre et al. (2015)
Annona squamosa L. (peel extract)	SnO_2 NPs	25 ± 5	Spherical	Hepatocellular carcinoma cell line (HEPG2)	25–500 μg/ml	24 h	148 μg/ml	Roopan et al. (2014)
Cassytha filiformis	AgNPs	16–66	Spherical	Human, colon, colorectal carcinoma cell line (HCT116)	10 μg/ml	24 h	–	Jena et al. (2016)

(continued)

Table 6.1 (continued)

Biological entity	Biologically synthesized NPs	Size (nm)	Shape	Type of cancer cell line	Dose	Exposure time	IC50	References
Hibiscus sabdariffa	AuNPs	10–60	Spherical	U87 glioblastoma cell line	2.0 ng/ml	48 h	–	Mishra et al. (2016)
Mimosa pudica (leaf extract)	AuNPs	12.5	Spherical	Breast cancer cell line (MDA-MB-231 and MCF-7)	MDA-MB-231 (4 μg/ml), MCF-7 (6 μg/ml)	48 h	–	Suganya et al. (2016)
Ficus benghalensis, *A. indica* (bark extract)	AgNPs	*F. benghalensis* (40), *A. indica* (50)	Spherical	MG-63 osteosarcoma cell line	10, 20, 40, 60, 80, 100, 150, 200 μg/ml	48 h	*F. benghalensis* (81.8 ± 2.6 μg/ml), *A. indica* (75.5 ± 2.4 μg/ml)	Nayak et al. (2016)
Rubus glaucus Benth (leaf extract)	AgNPs	12–50	Quasi-spherical	Hepatic cancer (HepG2) cell line	0.01, 0.1, 0.2, 0.5, and 1.0 μM	2 h	–	Kumar et al. (2016a)
Artemisia marschalliana Sprengel (aerial part extract)	AgNPs	5–50	Spherical	Human gastric carcinoma (AGS) and human embryonic kidney (HEK293) cell lines	(3.125, 6.25, 12.5, 25, 50, 100 μg/ml)	4 h	21.05 μg/ml	Salehi et al. (2016)

Sargassum muticum (seaweed water extract)	Hyaluronan/zinc oxide (nanocomposite)	10.2 ± 1.5	Hexagonal	Pancreatic adenocarcinoma (PANC-1), ovarian adenocarcinoma (CaOV-3), colonic adenocarcinoma (COLO205), and acute promyelocytic leukemia (HL-60) cells	–	4 h	PANC-1 (10.8 ± 0.3 μg/ml), CaOV-3 (15.4 ± 1.2 μg/ml), COLO205 (12.1 ± 0.9 μg/ml), HL-60 cells (6.25 ± 0.5 μg/ml)	Namvar et al. (2016)
Rosa canina (fruit extract)	ZnO NPs	Less than 50	30–80	Spherical	0.05, 0.1, 0.25, 0.5 mg/ml	24 h	–	Jafarirad et al. (2016)
Taxus yunnanensis (callus extract)	AgNPs	6.4–27.2	Spherical	Human colon adenocarcinoma cells (LS174T), human lung adenocarcinoma cells (A549), human breast cancer cells (MCF-7), human hepatoma cells (SMMC-7721), and human liver cells (HL-7702)	10, 20, 30, 40, and 50 μg/ml	24 h	SMMC-7721 cells (27.75 μg/ml), HL- 7702 (81.39 μg/ml)	Xia et al. (2016)
Marine sponge *(Haliclona exigua)*	AgNPs	100–120	Flower like	Human oral cancer cell line	–	48 h	1.23 μg/ml	Inbakandan et al. (2016)

(continued)

Table 6.1 (continued)

Biological entity	Biologically synthesized NPs	Size (nm)	Shape	Type of cancer cell line	Dose	Exposure time	IC50	References
Dimocarpus longan Lour. (peel extract)	AgNPs	9–32	Spherical	Human prostate cancer pc-3 cells	2–30 μg/ml	72 h	10 μg/ml	He et al. (2016)
Helicteres isora (stem bark extract)	AgNPs	16–95	Spherical	Oral carcinoma (KB) line	0, 20, 40, 60, 80, 100, 120, 140, 160, 180, and 200 μg/ml	24 h	70 μg/ml	Bhakya et al. (2016)
Borago officinalis (leaf extract)	AgNPs	30–80	Spherical	Lung cancer cell lines (A549) and cervical cancer cell line (HeLa)	A549 (5 μg/ml), Hela (2 μg/ml)	24 h	–	Singh et al. (2017)
Lavandula vera (leaf extract)	Zn NPs	30–80	Spherical	A549 (lung cancer), MCF-7 (breast cancer), HT-29 (colorectal cancer), and Caco-2 (colorectal cancer)	10–160 μg/ml	48 h	A549 (22.3 ± 1.1 μg/ml), MCF-7 (86 ± 3.7 μg/ml), HT- 29 (10.9 ± 0.5 μg/ml), Caco- 2 (56.2 ± 2.8 μg/ml)	Salari et al. (2017)
Costus pictus D. Don	ZnO NPs	–	Hexagonal, elongated, rod shaped	Dalton's ascites cells (DLA)	10, 20, 50 μg/ml	3 h	–	Suresh et al. (2018)

Streptomyces naganishii (MA7)	AgNPs	5–50	Spherical	HeLa cervical cancer cell line	0.1, 1, 10, 100, 300 µg/ml	48 h	24.39 µg	Shanmugasundaram et al. (2013)
Aspergillus flavus	AgNPs	33.5	–	Human promyeloid leukemia cells (HL-60)	5–10 µg/ml	24 h	–	Sulaiman et al. (2015)
Fusarium oxysporum	Tb_2O_3 NPs	10	Spherical	Human osteosarcoma cell lines, Saos- 2 and MG-63	0.023–0.373 µg/ml	24 h	0.102 µg/ml	Iram et al. (2016)
Beauveria bassiana	AgNPs	20.44 ± 34.16	Spherical	Human cervical cancer (HeLa) cell line	50 µg/ml	7 days	28.8 ± 2.5 µg/ml	Prabakaran et al. (2016)

morphology, cell viability, integrity of cell membrane, cell cycle progression, and oxidative stress. It was found that 5 and 20 nm sized AgNPs exhibited lethal effects by elevating the levels of the released ROS. ICP-MS was employed to assess the influence of AgNPs on cells, and the resultant data revealed that smaller NPs can penetrate cells more easily than larger ones. Thus, a higher toxic effect is exhibited. Coating NPs with natural biocompatible agents not only increases the stabilization of NPs and prevents their agglomeration but also helps to be suitable for biomedical applications (Roy et al. 2019).

AgNPs were green synthesized from the aqueous leaf extract of pomegranate (*Punica granatum*) as demonstrated by Sarkar and Kotteeswaran (2018). The biogenic synthesis of AgNPs was depicted by the change in color from golden yellowish to dark brownish. UV/Vis spectroscopy, zeta potential, FTIR, SEM, EDX, and XRD were employed to further confirm the green synthesis of AgNPs. In this study, the green synthesized AgNPs were evaluated on viability of human cervical cancer cells using 3-(4,5-Dimethylthiazol-2-Yl)-2,5-Diphenyltetrazolium (MTT) assay. The resultant data depicted that the viability of human cervical cancer cells was significantly reduced by AgNPs via a dose-dependent manner. Inhibition of the growth potential of HeLa cells took place at different concentrations of AgNPs (i.e., 50, 100, 150, 200, and 250 $\mu g\ ml^{-1}$). Additionally, lactate dehydrogenase (LDH) cell cytotoxicity assay was employed to determine the cytotoxic effect of the biologically derived AgNPs from *P. granatum* leaf extract. Generally, LDH is present inside all normal mammalian cells, but it is released when cells are damaged. Thus, the estimation of the amount of released LDH can facilitate the tracking of the toxic effect of the tested NPs. It was found that the quantity of the released LDH was elevated upon increasing the concentration of AgNPs. Hence, this result confirmed that AgNPs exhibited a cytotoxic effect on HeLa cell line as was confirmed by the release of lactate dehydrogenase in the media.

Among the severe cancer types is the liver cancer, which is considered as the sixth most common cancer type and the second deadliest cancer type in males and the sixth in females. Unfortunately, it is responsible for approximately 745,517 death cases (Iqbal et al. 2019). Consumption of heavy alcohols is the main reason behind the increasing risk of its occurrence in addition to infections by hepatitis C virus (HCV) and hepatitis B virus (HBV) as well as aflatoxins. Iqbal et al. (2019) tested the capability of CoO NPs derived from *Geranium wallichianum* leaf extract to combat HepG2 (hepatocellular carcinoma) cell line. Cancer cells were exposed to different concentrations of CoO NPs at different concentration ranging from 3.9 to 500 μg/ml for 24 h. Results showed that cancer cell inhibition was dose dependent. The IC50 value was found to reach 31.4 μg/ml. However, authors observed that the anticancer potential exerted by the biogenic CoO NPs decreased by lowering the concentration of the used CoO NPs.

Nickel oxide nanoparticles (NiO NPs) were reported to be biologically fabricated using *Geranium wallichianum* extract as demonstrated by Abbasi et al. (2019a). *G. wallichianum* served as a reducing and capping agent. Formation, morphological structure, crystalline structure, elemental composition, surface charge, and stability were investigated via different characterization techniques involving UV/Vis spectroscopy, Raman spectroscopy, XRD, SEM, TEM, FTIR, EDX, and DLS. The cytotoxic potential of the prepared NiO NPs was investigated against HepG2 (hepatocellular carcinoma) cells (RCB1648) cancer cell line. It was found that treatment of cancer cells treated with different concentrations of NiO NPs (500 to 3.9 μg/ml) for 24 h resulted in inhibition of cancer cells (dose dependent). The resultant data revealed the strong anticancer potential for the bioinspired NiO NPs. The calculated IC50 value was 37.84 μg/ml.

6.2.3.2 Antimicrobial Potential

Bacterial and fungal infections represent severe public health problem owing to the levels of morbidity and mortality. Effective antibiotics are the main used drugs to combat such microbial infections. However, certain problems have arisen regarding their use; the most common is the acquired resistance of microbes against such antibiotics. Hence, researchers around the world are exerting much effort to find alternatives to antibiotics as well as to minimize the risk of spreading such infectious diseases (Iqbal et al. 2017). Thus, the great advancement in nanobiotechnology provides new tools to formulate innovative biologically originated NPs with antimicrobial potential (Rizzello et al. 2013). Biologically fabricated CuNPs were green synthesized from the leaf extract of *Magnolia kobus* as elucidated by Lee et al. (2013). The synthesized CuNPs coated latex foams in order to investigate their antibacterial power. Latex foams were thoroughly washed, sterilized, and dried before use. The copper colloid solution coated the latex foam and was placed in a shaking incubator for 1 h at 37 °C, and then they were allowed to dry in an oven for 24 h at 50 °C. It is worth noting that the incorporation of CuNPs to the latex foams resulted in changing the color of the foam from white to brown. The antibacterial potential of the copper-coated foams was evaluated against *Escherichia coli* strain (ATCC 25922). It was observed that the colony count of the untreated latex foam was 22,400 CFU ml^{-1} and it decreased to 190 CFU ml^{-1} upon treatment with CuNPs. It is worth mentioning that the antibacterial activity was indirectly proportional to the size of the prepared CuNPs. As the size got smaller, a higher antimicrobial activity was observed. Extracellular cell free filtrate of *Penicillium oxalicum* was exploited in the biogenic synthesis of AgNPs (Bhattacharjee et al. 2017). The biosynthesized AgNPs was tested against *B. subtilis* (MTCC-619), *E. coli* (MTCC-40), *P. aeruginosa* (MTCC-424), and *S. aureus* (MTCC-96). Both AgNPs and AgNPs combined with streptomycin displayed significant inhibition zone against all the tested four bacterial strains. Bhattacharjee and colleagues noticed that Gram-negative bacteria were more susceptible to AgNPs rather than the Gram-positive ones.

Mycosynthesis of AuNPs was performed using the aqueous mycelial extract of *Trichoderma hamatum* (Abdel-Kareem and Zohri 2018). The bioinspired AuNPs were identified via UV/Vis spectrometry, TEM, and FTIR. A characteristic SPR peak appeared at 530 nm. The prepared AuNPs exhibited spherical, pentagonal, and hexagonal shapes with a size ranging from 5 to 30 nm. The biosynthesized AuNPs displayed an antimicrobial potential against four pathogenic bacterial strains including *Bacillus subtilis* ACCB 133, *Staphylococcus aureus* ACCB 136, *Pseudomonas aeruginosa* ACCB 156, and *Serratia* sp. ACCB 178. Results showed that AuNPs displayed an antibacterial activity against all the tested pathogenic bacteria, but the clear zone was smaller than that of streptomycin. Green and low-cost methodology for the biosynthesis of cobalt oxide NPs via the leaf extract of *Geranium wallichianum* was demonstrated by Iqbal et al. (2019). CoO NPs were synthesized by the effect of the natural phytochemical constituents which included stigmasterol, β-sitosterol, β-sitosterol, herniarin, galactoside, and ursolic acid. The prepared CoO NPs were tested in diverse applications.

The antimicrobial activity of *G. wallichianum*-derived CoO NPs was assessed against five different bacterial strains involving *Pseudomonas aeruginosa*, *Staphylococcus* aureus, *Bacillus subtilis*, *Klebsiella pneumonia*, and *Escherichia coli*. It was noticed that by increasing the concentration of CoO NPs, the antibacterial activity increased as well. Both *P. aeruginosa* and *K. pneumonia* were found to be more resistant with MIC value of 175 μg/ml. Whereas, *B. subtilis* was found to be the least resistant to the biogenic CoO NPs with MIC value of 21.875 μg/ml. Moreover, it was tested against *Candida albicans*, *Mucor racemosus*, *Aspergillus niger*, *Aspergillus flavus*, and *Fusarium solani*. Authors also observed a direct relationship between inhibiting fungus growth and concentration of the employed CoO NPs. The highest resistant strain was *A. flavus* with MIC value of 175 μg/ml and the least resistant strain was *M. racemosus* with MIC of 10.937 μg/ml.

Plant-mediated biogenic synthesis of iron oxide nanoparticles (IONPs) via the leaf extract of *Rhamnus virgata* (Roxb.) was proposed by Abbasi et al. (2019b). The physicochemical features of the synthesized IONPs were characterized via UV/Vis spectroscopy, Raman spectroscopy, DLS, XRD, FTIR, TEM, and SEM. The antimicrobial activity of biosynthesized IONPs was assessed against Gram-positive bacterial strains, e.g., *Staphylococcus aureus* (ATCC 25923) and *Bacillus subtilis* (ATCC 6633), and against Gram-negative bacterial strains including *Pseudomonas aeruginosa* (ATCC 9721), *Escherichia coli* (ATCC 15224), and *Klebsiella pneumoniae* (*ATCC* 4617). Interestingly, most of the tested bacteria were susceptible to the biologically fabricated IONPs. Additionally, in a previous study performed by Behera et al. (2012), the chemically synthesized IONPs did not exhibit any antimicrobial potential towards *P. aeruginosa* even at high concentrations reaching 50 mg/ml. Authors mentioned in details that both *B. subtilis* and *E. coli* were the most susceptible to IONPs with MIC values of 31.25 μg/ml for both microorganisms. On the other hand, *P. aeruginosa* and *Klebsiella pneumonia* were more resistant, and the MIC values recorded 125 μg/ml. Authors assumed that generation of reactive oxygen species (ROS) is the main leading cause behind the cellular damage caused by the employed NPs. Besides, authors evaluated for the first time the antifungal properties of the *R. virgate*-inspired IONPs against five fungal strains including *Aspergillus flavus* (FCBP 0064), *Aspergillus niger* (FCBP 0918), *Fusarium solani* (FCBP 0291), *Mucor racemosus* (FCBP 0300), and *Candida albicans* (FCBP 478). The obtained results illustrated that *A. flavus* was the most resistant strain with MIC value of 125 μg/ml, while *M. racemosus*, *A. niger*, and *F. solani* were the least resistant strains with MIC value of 31.25 μg/ml. Biologically fabricated NiO NPs from the extract of the medicinal plant *Geranium wallichianum* was evaluated for its antibacterial and antifungal potential (Abbasi et al. 2019a). Antibacterial assay was performed against *Bacillus subtilis* ATCC 6633 and *Staphylococcus aureus* ATCC 25923, *Escherichia coli* ATCC15224, *Pseudomonas aeruginosa* ATCC9721, and *Klebsiella pneumonia* ATCC4617. Moreover, the antifungal activity of NiO NPs was assessed against different fungal strains *Candida albicans* FCBP 478, *Mucor racemosus* FCBP 0300, *Aspergillus niger* FCBP 0918, *Fusarium solani* FCBP 0291 and *Aspergillus flavus* FCBP 0064. Authors noticed that most of the studied bacteria were susceptible to NiO NPs. Yet, the most susceptible bacterial strain to biogenic NiO NPs was *B. subtilis* (MIC value 21.875 μg/ml). Both *P. aeruginosa* and *K. pneumonia* were found to be more resistant with MIC values of 175 μg/ml. In the case of the antifungal potential of NiO NPs, *A. flavus* was the most resistant strain with MIC value of 175 μg/ml, while *F. solani* and *M. racemosus* were the least resistant strains with MIC value of 21.875 μg/ml. Authors elucidated that the interference of NiO NPs with the fungal hyphae and spores as well as the generation of ROS were the main reasons behind the antifungal potential of NiO NPs. Co_3O_4-NPs were successfully mycosynthesized via the MCFF of *Aspergillus brasiliensis* ATCC 16404 (Omran et al. 2019a). Authors proved that up till now, the exact real mechanisms for investigating the biocidal activity of metal NPs are still not clear, yet three hypothetical mechanisms are usually introduced (Singh et al. 2019a). These mechanisms are mentioned below:

1. Lok et al. (2006) reported that when metal ions penetrate inside the cells, degradation of intracellular adenosine triphosphate (ATP) and DNA duplication becomes prohibited.
2. Kim et al. (2007) demonstrated that metal NPs are able to generate ROS which in turn have a noxious impact on cellular structures. ROS could be either radical or non-radical. Radical ROS include superoxide, singlet oxygen, hydroxyl, hydroperoxyl, carbon dioxide radical, carbonate, peroxyl, and alkoxyl radicals. Wheresas, the non-radical ROS involve ozone, hydrogen peroxide, nitric oxide, hypochlorous acid, hypochlorite, hypobromous acid, organic peroxides, peroxynitrite, peroxynitrate, peroxynitrous acid and peroxomonocarbonate (Wu et al. 2014).
3. McQuillan (2010) depicted that NPs usually accumulate and remain dissolved in the bacterial membrane; hence this results in alterations of membrane permeability and driving away the proton motive forces (PMF).

It is worth mentioning that oxidation reactions exert diverse hazardous effect concerning regulation of cell survival, death, differentiation, signaling, and generation of ROS under stress conditions (Mueller et al. 2005). Reactions of ROS are usually mediated and catalyzed by enzymes such as superoxide dismutases (SOD) (Johnson and Giulivi 2005). Superoxide is produced via nicotinamide adenine dinucleotide phosphate (NADPH) oxidase (NOX). It inactivates several enzymes, and

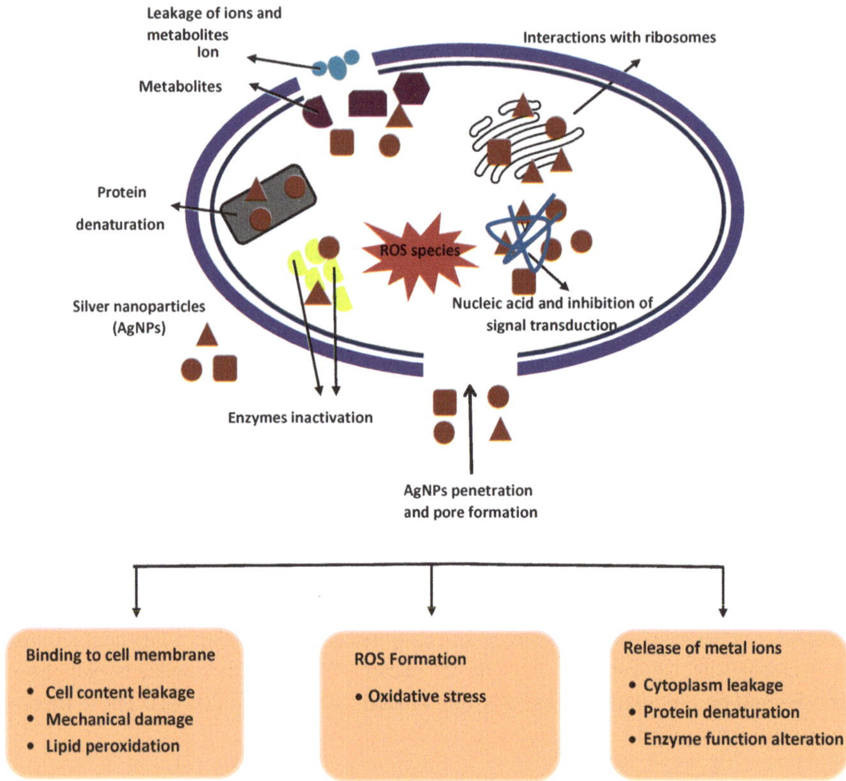

Fig. 6.3 A suggested mode of action of AgNPs towards Gram-positive bacteria

lipid peroxidation of cellular membranes takes place (Brand 2010). Singh et al. (2019a) pointed out that ROS production induced by NPs plays a major role in genotoxicity either directly or indirectly. Figure 6.3 illustrates the mode of action of AgNPs against Gram-positive bacteria.

6.2.3.3 Antiparasitic (Antileishmanial) Potential

Leishmaniasis is a tropical illness whose main causative agent is the genus *Leishmania*. The bite of the female phlebotomine sand fly is responsible for disease transmittance (Ahmad et al. 2015). According to reports of the World Health Organization (WHO), leishmaniasis is listed as a category 1 disease (most emerging and uncontrollable disease). It affects nearly 88 countries with annual incidence of cutaneous leishmaniasis in approximately 1.5 million people worldwide. Leishmaniasis rate has been reported to increase around the world because of the possible increase in disease vectors due to global warming. Hence, there is an urgent need for more efficient therapeutic agents with broad-spectrum antileishmanial potential. Thivaharan et al. (2018) demonstrated that metal NPs have the ability to generate ROS via a process called "respiratory burst mechanism," leading to the inhibition of pathogenic microbes. *Leishmania* spp. are highly susceptible to ROS. As a result, NPs can be powerful candidates as antileishmanial drugs. *Leishmania tropica* is the main causative agent of leishmaniasis as demonstrated by Kaye and Scott (2011). Unfortunately, the usual drugs employed in the treatment of leishmanial parasites are not very efficient, toxic, and somehow expensive. Ahmad et al. (2015) conducted a study in which the aqueous leaf extract of *Euphorbia prostrata* was used for the biogenic synthesis of AgNPs and TiO_2 NPs. The potential of the biologically synthesized NPs to act as antileishmanial agents was investigated. The synthesized AgNPs were found to be more lethal against *Leishmania* parasites after 24 h exposure, with IC50 value of 14.94 μg/ml and 3.89 μg/ml in promastigotes and intracellular amastigotes, respectively. *Sargentodoxa cuneate* mediated the biological synthesis of AgNPs as well as AuNPs as depicted by Ahmad et al. (2015). The antileishmanial activity of AgNPs and AuNPs was performed against *Leishmania tropica* promastigotes. Hemocytometer was employed to count the parasites in both the control and the treated samples at different incubation times, i.e., 24, 48, 72, and 96 h, and the activity was expressed as percent of inhibition. It has been noticed that the number of cells significantly decreased within the first 24 h when compared with the control group after the exposure to AgNPs and AuNPs. AgNPs and AuNPs showed 90 and 62% inhibition, respectively. After 48 h of exposure, the number of cell count was further decreased in the treated samples. However, AgNPs expressed an excellent antileishmanial activity with maximum inhibition of 95.45% after 48 h of incubation. Authors noticed that after 48 h, AgNPs showed a negligible decline in cell number, while an increase in leishmanicidal activity was noticed for AuNPs (77.5%).

A novel green approach for synthesis of AgNPs was carried out using the extract of *Isatis tinctoria* as illustrated by Ahmad et al. (2016). Under visible light irradiation, biogenic AgNPs displayed significant activity towards *Leishmania tropica* with an IC50 value of 4.2 μg/ml. Furthermore, authors depicted that the leishmanicidal potential of *I. tinctoria*-derived AgNPs was majorly enhanced by conjugation with amphotericin B with IC50 value of 2.43 μg/ml. The leaf extract of *Anethum graveolens* mediated the bioreduction of AgNPs as reported by Kalangi et al. (2016). The biosynthesized AgNPs alone with a concentration of 50 μM exerted antileishmanial effect against the promastigote stages of *Leishmania* parasite. However, the combination of AgNPs along with miltefosine resulted in an increase in the leishmanicidal effect of miltefosine by approximately twofolds. Scanning electron microscope (SEM) observations illustrated the morphological aberration and fragmentation of genomic DNA in promastigotes which further assured the enhanced effect of miltefosine in association with AgNPs. In a study performed by Iqbal et al. (2019), biogenic CoO NPs exerted an outstanding antileishmanial activity against amastigotes in comparison with promastigote. In another study, Abbasi et al. (2019a) investigated the cytotoxic potential of the biologically prepared NiO NPs via the extract of *Geranium wallichianum* against the promastigote and amastigote cultures of *Leishmania tropica* 'KWH23 strain collected from Khyber Medical Centre,

Peshawar Pakistan. Abbasi and co-authors observed that the antileishmanial potential increased by increasing the concentration of NiO NPs. The IC50 value recorded 22.12 μg/ml.

6.2.3.4 Antidiabetic Potential

The medicinal plant *Solanum nigrum* is traditionally used in treating diabetes mellitus. In a study conducted by Sengottaiyan et al. (2016), AgNPs were green synthesized using the leaf extract of *S. nigrum*. The phytosynthesized AgNPs were tested for their antidiabetic potential in alloxan- induced diabetic rats. It was found that AgNPs significantly enhanced the dyslipidemic condition and reduced the blood glucose level over the period of treatment. Additionally, the improvement in body weight was also a clear evidence for *S. nigrum* extract-mediated AgNPs to possess a potential antidiabetic agent against alloxan-induced diabetic rats. Balan et al. (2016) demonstrated the phytosynthesis of AgNPs via the aqueous leaf extract of *Lonicera japonica*. The antidiabetic capability of *L. japonica*-mediated AgNPs was evaluated by the effective inhibition towards carbohydrate digestive enzymes like α-amylase and α-glucosidase with IC50 values of 54.56 and 37.86 μg/ml, respectively. A facile one-pot phytosynthesis of AgNPs was carried out using the aqueous leaf extract of *Cinnamomum tsoi*. A significant inhibition was observed of α-amylase and α-glucosidase enzymes upon treatment with the biologically fabricated AgNPs. Authors suggested that the enhanced in vitro antidiabetic activity of the AgNPs might be due to the oxidized polyphenol constituents that capped the surface of AgNPs. Prabhu et al. (2018) presented a study in which the green synthesized AgNPs using aqueous leaf extract of *Pouteria sapota* was investigated to evaluate the in vitro and in vivo antidiabetic properties. The in vitro antidiabetic activity of the biosynthesized AgNPs was assured by different assays including non-enzymatic glycosylation of hemoglobin, glucose uptake by yeast cells followed by exposure of cells to 5 or 10 mmol/L glucose solution, and inhibition of α-amylase. Further, in vivo antidiabetic activity was assessed in streptozotocin-induced rats. Rats were treated with AgNPs (10 mg/kg) for 28 days. Following treatment, rats were examined for any biochemical and histopathological alterations in kidney and liver samples. Additionally, decrease in blood sugar levels was observed in rats treated with AgNPs. Hence, Prabhu and co- authors postulated that the bioinspired AgNPs possessed an antidiabetic activity.

6.2.3.5 Antioxidant Potential

An antioxidant is referred to as the molecule which hinders the oxidation of other molecules (Choudhary and Madhuri 2018). On the other hand, several pathological diseases are caused by the influence of free radicals varying from cancer to aging, diabetes, atherosclerosis, Alzheimer's, cardiovascular diseases, etc. Levels of free radicals (e.g., $O^{\bullet}_2$, $O^{\bullet}$, $HO^{\bullet}$, $NO^{\bullet}$) and antioxidants are usually balanced in normal metabolism. Nonetheless, free radical over generation leads to oxidative damage and consequently to different chronic diseases. Yet, the use of synthetic antioxidants is limited because of their adverse toxicity. As a result, research is now pushed towards naturally occurring antioxidants (Kumar et al. 2016b). Searching for green synthesized NPs for treating such free radical-related medical disorders is one of the important applications of the biologically fabricated NPs (Table 6.2). Dauthal and Mukhopadhyay (2016) assumed that the intake of antioxidant guarantees protection towards the damage caused by free radicals.

The aqueous extract of the rind of water melon *Citrullus lanatus* was used in the biogenic synthesis of AuNPs as reported by Patra and Baek (2015). AuNPs exhibited strong antioxidant potential via exerting 2,2- diphenyl-1-picrylhydrazyl (DPPH) radical scavenging (24.69%), nitric oxide scavenging (25.62%), and ABTS scavenging (29.42%) activities. The presence of several constituents like citrulline, lycopene, vitamins, and a variety of phenolic compounds governed the antioxidant potential of biosynthesized AuNPs. In 2016, Markus and colleagues reported the novel biological synthesis of AuNPs using *Lactobacillus kimchicus* (DCY51T) isolated from Korean kimchi. The biosynthesized AuNPs exerted outstanding and superior antioxidant properties against DPPH. Authors manifested that the scavenging activity is dose dependent, and it increased by increasing the concentration of the AuNPs. Kumar et al. (2016b) reported the use of lavender (*Lavandula angustifolia*) as a reductant and a stabilizing agent for biosynthesis of AuNPs. Authors compared the inhibition percentage of DPHH of both AuNPs and lavender leaf extract (LLE). It was found that better DPPH quenching activity took place at low concentration of AuNPs in comparison with LLE. Furthermore, authors depicted that DPPH scavenging activity of AuNPs decreased by increasing its concentration, whereas a contrasting trend took place by LLE, owing to the less solubility of AuNPs. AgNPs were biologically synthesized using the aqueous extract of fresh onion (*Allium cepa*) extract as revealed by Gomaa (2017). The fabricated AgNPs exhibited a powerful radical scavenging potential against the lethal disruption caused by the free radicals. It is worth noting that Gomaa and co-authors used three

Table 6.2 List of biologically synthesized NPs possessing antioxidant activity

Biological entity	Type of NPs	References
Streptomyces naganishii (MA7)	AgNPs	Shanmugasundaram et al. (2013)
Antigonon leptopus (leaf extract)	AuNPs	Balasubramani et al. (2015)
Areca catechu nut	AuNPs	Rajan et al. (2015)
Aspergillus flavus	AgNPs	Sulaiman et al. (2015)
Rubus glaucus Benth. (leaf extract)	AgNPs	Kumar et al. (2016a)
Artemisia marschalliana (aerial part extract)	AgNPs	Salehi et al. (2016)
Lavandula vera (leaf extract)	Zn NPs	Salari et al. (2017)
Gossypium hirsutum (stem extract)	AgNPs	Vanti et al. (2019)
Citrus maxima (peel extract)	AgNPs	Huo et al. (2019)
Ceratonia siliqua (leaf extract)	CeO2 NPs	Javadi et al. (2019)

methods to determine the antioxidant properties of the biogenic AgNPs. These methods included (i) potassium ferricyanide reduction method (acts as indicator of potential antioxidant activity); (ii) phosphor-molybdenum method (determines the total antioxidant capacity depending upon the reduction of Mo (VI) to Mo (V) by the tested antioxidant compound); and (iii) DPPH scavenging capacity test (used to evaluate the antioxidant activity, characterized by its high speed, ease, and stable radical formation strategy).

6.2.3.6 Wound Healing Applications

Wound healing refers to the natural response of the body towards tissue injury. It takes place via a complex cascade of cellular and biochemical processes which includes restoration, resurfacing, and reformation of the injured skin (Choudhary and Madhuri 2018). Generally, when the skin is injured, four phases usually occur including inflammation, proliferation, maturation, and hemostasis, which eventually end up with a scar. For wound healing, fibrin production is extremely essential, and it serves as a primary component for the wound matrix at which both of cells and plasma proteins migrate. During the inflammatory phase which takes place in the first 2–4 days of healing, macrophages and neutrophils (viz., inflammatory cells) protect the injured tissue and the skin from further infection and aid in releasing mitogenic and chemotactic factors (Midwood et al. 2004). During the proliferative phase, collagen is produced from fibroblasts from the surrounding tissue and begins to proliferate onto the fibrin matrix. Afterwards, newly formed collagen molecules cross-link with the present collagen and protein molecules, and this step represents the maturation phase which helps in elevating the tensile strength of the scar. At the end of the second week, the healing process starts to proceed and continues for a period of time (Ziv-Polat et al. 2010). Table 6.3 represents a list of biologically synthesized NPs with wound healing capabilities.

6.2.4 Applications of Biologically Synthesized Nanoparticles as Antibiofoulers

Biofouling refers to the undesirable accumulation or colonization of plants, animals, and microorganisms either on artificial or natural surfaces when immersed in aquatic environments (Choudhary and Madhuri 2018). Biofouling involves microbiofouling which occurs due to the attachment of bacteria and other microorganisms on surfaces, while macrobiofouling occurs due to the attachment of larvae of higher organisms, like mussels, barnacles, etc., leading to biofilm formation (Omran et al. 2013). Among the most common treatment approaches is the use of chemical antifouling agents to combat biofouling caused by micro- and macro-foulers. Shanmugasundaram et al. (2013) investigated the capability of AgNPs to inhibit the

Table 6.3 List of biologically synthesized NPs with wound healing capabilities

Biological entity	Type of NPs	References
Coleus forskohlii (root extract)	AgNPs/AuNPs	Naraginti et al. (2016)
Delonix elata (leaf extract)	AgNPs	Wang et al. (2018)
Prosopis farcta (seed extract)	Ag, ZnO, and Ag/ZnO NPs	Khatami et al. (2018)
Barleria gibsoni (leaf extract)	ZnO NPs	Shao et al. (2018)
Euphorbia milii (leaf extract)	AgNPs	Gong et al. (2018)
Ocimum sanctum	AgNPs	Sood and Chopra (2018)
Ziziphora clinopodioides lam (leaf extract)	MnNPs	Mahdavi et al. (2019)
Fenugreek (leaf extract)	AgNPs	Ying et al. (2019)
Raphanus sativus (white radish) (root extract)	ZnO NPs	Kiran Kumar et al. (2019)
Lippia citriodora	AgO NPs	Li et al. (2019)
Prosopis juliflora (leaf extract)	AgNPs	Arya et al. (2019)
Chamaecostus cuspidatus	AuNPs	Ponnanikajamideen et al. (2019)
Falcaria vulgaris L. (leaf extract)	CuNPs	Zangeneh et al. (2019)
Allium eriophyllum Boiss	TiNPs	Seydi et al. (2019)
Brassica oleracea (capitate extract)	AgNPs	Ahsan and Farooq (2019)
Tridax procumbens (leaf extract)	AgNPs	Ravindran et al. (2019)

biofouling activity of ten bacterial strains, e.g., P*seudomonas* sp. P1, *Aeromonas* sp. P26, *Bacillus* sp. P31, *Bacillus* sp. P46, *Alcaligenes* sp. P47, *Micrococcus* sp. P56, *Staphylococcus* sp. PP3, *Micrococcus* sp. PP5, *Aeromonas* sp. PP6, and *Alcaligenes* sp. PP8. Cultures were grown overnight and were diluted to 1:100 fresh LB medium containing 50% sea water. Different concentrations of AgNPs were tested (1, 5, and 10 mg/ml) and were added to the bacterial cells and incubated at 28 °C for 48 h. Afterwards, the medium was removed, walls of the test tubes were exhaustively washed with 1X phosphate-buffered saline (PBS), and 100 L of 0.2% (w/v) crystal violet was added and incubated for 20 min. The crystal violet was then removed and washed thoroughly with 1X PBS. Measurements carried out by the crystal violet were in absolute ethanol, and later the absorbance was measured at 570 nm. Authors observed that all the ten bacterial strains had the potential to form a biofilm. Results obtained after 48 h showed a decrease in biofilm formation of more than 10–40% at 1 mg/ml, 40–70% at 5 mg/ml, and 50–90% at 10 mg/ml when treated with AgNPs. This would candidate the use of biologically derived AgNPs as an alternative to antifouling compounds. Krupa and Raghavan demonstrated the capability of the fruit extract of *Aegle marmelos* to mediate the biogenic synthesis of AgNPs (Krupa and Raghavan 2014). Authors investigated the potential of the AgNPs to prevent/control the biofilm-forming bacterial communities via conducting antibiofouling studies. Biofilm samples were collected from ship hulls anchored at Royapuram Harbor. Twelve different bacterial strains were isolated and screened for their ability to form biofilms. Among the 12 isolates, five isolates exhibited maximum fouling activity, and they were identified as *Pseudomonas otitidis* strain NV1, *Pseudomonas aeruginosa* strain NV2, *Enterobacter cloacae* strain NV3, *Microbacterium* sp. NV4, and *Staphylococcus hominis* strain NV5. Authors employed the microtiter plate method to carry out their antibiofilm studies as it facilitates the direct quantification of the attached bacteria. Biofilms were stained with crystal violet and the optical density was measured at 570 nm. Obtained results showed that the optical density value for the AgNPs loaded wells were lower than the control wells, i.e., in absence of NPs. Additionally, it was noticed that AgNPs affect the production of extracellular polymeric substances (EPS), thereby aiding in preventing biofilm formation. Aqueous extract of the seaweed *Turbinaria conoides* mediated the biogenic synthesis of AgNPs and AuNPs (Vijayan et al. 2014). The antibiofilm activity of both AgNPs and AuNPs was tracked against four different marine biofilm-forming bacterial strains, namely, *Salmonella* sp. (JN596113), *E. coli* (JN585664), *S. liquefaciens* (JN596115), and *A. hydrophila* (JN561697). After 24 h of incubation, CLSM images illustrated the weak adherence and disintegrated biofilm architecture of the four tested bacterial strains. Omran and co-authors presented a study in which AgNPs were mycosynthesized via the mycelial cell free filtrate of *Trichoderma longibrachiatum* DSMZ 16517 (Omran et al. 2019b). The mycosynthesized AgNPs were found to possess a high biocidal effect against a halotolerant mixed culture of sulfate-reducing bacteria (SRB) which are known to be responsible for the anaerobic microbial corrosion in oil industries. HRTEM analysis displayed a clear evidence of the alterations in cell morphology, cell membrane disruption, cell wall lysis, and cytoplasmic extraction after treatment with AgNPs.

6.2.5 Applications of Biologically Synthesized Nanoparticles in Food Industry

One of the important steps which are essential in food industry is the process of food packaging. Food packaging has to guarantee the best quality and stability of processed food till reaching the consumer (Tripathi et al. 2018). Since there is a global public awareness about diseases transmitted by microorganisms in food, there has been a tremendous demand for manufacturing antibacterial packaging materials to help in inhibiting food contamination by bacteria (Park and Zhao 2004). Yam (2010) reported that several conventional food packaging materials are made up of polymeric films like polypropylene, polystyrene, and polyethylene terephthalate. These materials provide gas barrier and mechanical properties but unfortunately they lack biodegradability. Employing biologically driven polymers such as polysaccharide (starch) polymers which have gained much popularity because of their biodegradability, but problems regarding their humidity have limited their commercial use (Lagaron et al. 2008). Henceforth, tremendous research focused on employing nanomaterials as ideal food packaging materials by dispersing the nanoscaled fillers within the polymer matrices. SiO_2 nanoparticles (Vladimirov et al. 2006; Jia et al. 2007), graphene (Ramanathan et al. 2008), starch nanocrystals (Chen et al. 2008), and cellulose-based nanofibers (Podsiadlo et al. 2005) have been employed as nano-fillers to enhance the properties of the used food packaging polymers. These nano-fillers were found to upgrade different food packaging properties like flame resistance, thermal features, wettability, and hydrophobicity; unfortunately they did not prevent contamination of food products by microorganisms. Another methodology was employed by incorporating nanomaterials such as silver, titanium dioxide, or copper NPs into the polymer films to prepare anti-food packaging materials (Kim et al. 2003; Tankhiwale and Bajpai 2009). In a study presented by Tripathi et al. (2018), authors developed a biodegradable a nanocomposite film by fabricating PVA along with biogenic AgNPs to be used as an antibacterial packaging material. The AgNPs were biosynthesized via the leaf extract of *Ficus benghalensis*. Tripathi and co-authors investigated the antibacterial activity of the nanocomposite film against *Salmonella typhimurium*. It was found that the ZOI increased by increasing the concentration of biogenic AgNPs in the film. Figure 6.4 shows the possible applications of biologically fabricated NPs in food industry. They can serve as anti-caking agents (i.e., agents which enhance the consistency of the desired material and prevent lump formation), gelating agents (i.e., agents that improve food textile), nanoadditives (i.e., agents that improve the nutritional value of the food), and nano-carriers (i.e., agents that protect the aroma and flavor of food ingredients). Additionally, they can have a major role in food packaging via improving food physical performance (improved packaging), acting as antimicrobial agents (active packaging), and serving as nano-bio-sensors to detect pathogens in food (smart packaging).

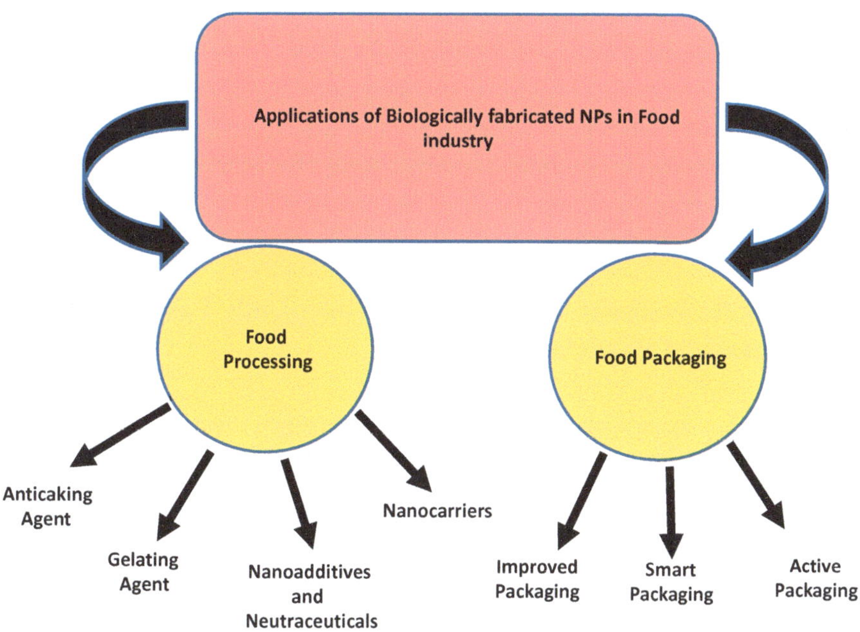

Fig. 6.4 Different applications of biologically synthesized NPs in food industry

6.3 Routes for Maximizing the Productivity of the Biosynthesized Nanomaterials

In order to maximize the efficiency of biosynthesized nanomaterials, a number of parameters which control the nucleation and results in formation of stabilized NPs have to be optimized. These factors involve pH, shaking speed, dark/illumination, reactant concentrations, reaction time, and temperature as reported by Omran et al. (2017, 2018, 2019b). Apart from optimizing these factors, the use of biofilms is another important approach for efficient biosynthesis of nanoparticles (Tanzil et al. 2016). According to Ikuma et al. (2015), biofilms have been postulated as the most dynamic growth mode of bacteria. Biofilms possess more advantages than their planktonic counterparts as they exhibit superior properties like catalytic activity, high reducing matrix, and capability to control electrochemical reactions (Kalathil et al. 2011). Besides, the entire process is usually free of contamination because of the high protective nature of biofilms as it limits the diffusion of outside materials, thus making it a promising perspective for biosynthesizing NPs in aqueous systems (Moon et al. 2016). Highly efficient and large-scale production is guaranteed via biofilms because of the high biomass concentration as well as the large surface areas. Teitzel and Parsek (2003) elucidated that biofilms can resist and tolerate heavy metals up to 600 times higher than their planktonic counterparts. Biofilms catalyze electrochemical redox reactions in an appropriate environment by the effect of natural reducing agents like proteins, peptides, and heterocyclic compounds for metal reduction to nanoparticles (Khan et al. 2013, 2014). However, till now few studies investigated nanoparticle synthesis in biofilms (Khan et al. 2014; Gahlawat and Choudhury 2019). Additionally, little information is available regarding the mechanisms of synthesis, capping, and stabilization. Hence, it is suggested to perform thorough studies related to the molecular mechanism of NPs synthesis in biofilms and in their planktonic counterparts. This will facilitate future researchers to develop microbial systems that can robust rapid biosynthesis of NPs with the desired size and shape.

6.4 Risk Assessment Regarding the Biologically Fabricated Nanosized Particles

The release of nanoparticulates in their original or modified structures usually takes place via their production, use, and disposal (Jeevanandam et al. 2018). The human skin represents the first defense against any foreign substances. On the contrary, organs are susceptible to foreign substances such as the lungs and gastrointestinal tract. NPs that are inhaled can reach the bloodstream and other organs within the human body like the liver, heart, or blood cells. It is significant to mention that the toxicity of NPs is related to their structure, size, and shape. The smaller the size of NPs, the easier their translocation through organism barriers such as skin, lung, body tissues, and organs. As a result, organelle damage, asthma, irreversible oxidative stress, and cancer may take place by NPs if they were not handled properly. Additionally, generation of ROS and protein denaturation usually take place.

Chanda and Baravalia (2011) demonstrated that brine shrimp lethality assay (BLSA) is one of the most employed assays to confine biological potency of any compound. Iqbal et al. (2019) investigated the potential of *Geranium wallichianum*-derived CoO NPs against brine shrimps at different concentrations ranging from 1 to 200 μg/ml. The median lethality dose (IC50) was found to be 18.12 μg/ml. Abbasi and co-authors tracked the cytotoxic effect of bioinspired IONPs derived from the leaf extract of *Rhamnus virgate* (Abbasi et al. 2019b). Authors employed BLSA for evaluating the cytotoxic potential of IONPs. The IC50 value was found to be 32.41 μg/ml. In a study presented by Bhakya et al. (2016), AgNPs were biologically prepared by the stem bark extract of *Helicteres isora*. A significant mortality rate was observed against *Artemia* with an IC50 concentration of 70 μg/ml after 108 h of exposure time, so they were cytotoxic at high concentrations and after prolonged exposure time.

6.5 Challenges, Conclusions, and Future Perspectives

Nanobiotechnology is one of the interesting scientific disciplines and its main building blocks are the nanoparticles. Green synthesis is one of the best eco-friendly approaches for the synthesis of metal and metal oxide nanoparticles. The reduced form of metals is characterized by its high reactivity and large surface area which allow their applications in several fields ranging from biomedical applications including antimicrobial (antibacterial and antifungal), anticancer, antioxidants, anti-coagulants, wound healing, antileishmanial, and anti-inflammatory activities, drug delivery to several environmental applications concerned with their catalytic activities towards toxic dye effluents, phenolic compounds, heavy metals and organic pollutant removal. The activities of NPs are chiefly dependent upon the surface area as well as particles' size. According to the reported literature, the smaller the size of the prepared NPs, the higher they get dispersed and their capability to be applied in biomedical applications. However, a great concern regarding the possible adverse effects of biologically synthesized nanoparticles on humans' health as well as on the environment is currently spotted upon. Yet, the potential toxicological effects of nanomaterials have received little attention so far. Assessment of the long-term toxic effects on the surrounding environment, aquatic organisms, and humans will help to figure out the complex interactions and mechanisms that may take place.

References

Abbasi BA, Iqbal J, Mahmood T et al (2019a) Plant-mediated synthesis of nickel oxide nanoparticles (NiO) via *Geranium wallichianum*: characterization and different biological applications. Mater Res Express 6:0850a7–0850a24

Abbasi BA, Iqbal J, Mahmood T et al (2019b) Biofabrication of iron oxide nanoparticles by leaf extract of *Rhamnus virgata*: characterization and evaluation of cytotoxic, antimicrobial and antioxidant potentials. Appl Organomet Chem 33:1–15

Abdel-Kareem MM, Zohri AA (2018) Extracellular mycosynthesis of gold nanoparticles using *Trichoderma hamatum*: optimization, characterization and antimicrobial activity. Lett Appl Microbiol 67:465–475

Ahmad A, Syed F, Shah A et al (2015) Silver and gold nanoparticles from *Sargentodoxa cuneata*: synthesis, characterization and antileishmanial activity. RSC Adv 5:73793–73806

Ahmad A, Wei Y, Syed F et al (2016) *Isatis tinctoria* mediated synthesis of amphotericin B-bound silver nanoparticles with enhanced photoinduced antileishmanial activity: a novel green approach. J Photochem Photobiol B Biol 161:17–24

Ahsan A, Farooq MA (2019) Therapeutic potential of green synthesized silver nanoparticles loaded PVA hydrogel patches for wound healing. J Drug Deliv Sci Technol 54:101308–101318

Alomair NA, Mohamed HH (2018) Green synthesis of ZnO hollow microspheres and ZnO/rGO nanocomposite using red rice husk extract and their photocatalytic performance. Mater Res Express 5:095012–095021

Altikatoglu Yapaoz M, Attar A (2019) *Salvia officinalis*-derived rutile TiO_2 NPs: production, characterization, antibacterial evaluation and its effect on decolorization. Mater Res Express 6

Anwar A, Ovais M, Khan A, Raza A (2017) Docetaxel-loaded solid lipid nanoparticles: a novel drug delivery system. IET Nanobiotechnol 11:621–629

Arya G, Kumari RM, Sharma N et al (2019) Catalytic, antibacterial and antibiofilm efficacy of biosynthesised silver nanoparticles using *Prosopis juliflora* leaf extract along with their wound healing potential. J Photochem Photobiol B Biol 190:50–58

Balan K, Qing W, Wang Y et al (2016) Antidiabetic activity of silver nanoparticles from green synthesis using *Lonicera japonica* leaf extract. RSC Adv 6:40162–40168

Balasubramani G, Ramkumar R, Krishnaveni N et al (2015) Structural characterization, antioxidant and anticancer properties of gold nanoparticles synthesized from leaf extract (decoction) of *Antigonon leptopus* Hook. & Arn. J Trace Elem Med Biol 30:83–89

Barabadi H, Ovais M, Shinwari ZK, Saravanan M (2017) Anti-cancer green bionanomaterials: present status and future prospects. Green Chem Lett Rev 10:285–314

Behera SS, Patra JK, Pramanik K et al (2012) Characterization and evaluation of antibacterial activities of chemically synthesized iron oxide nanoparticles. World J Nano Sci Eng 2:196–200

Bhakya S, Muthukrishnan S, Sukumaran M et al (2016) Antimicrobial, antioxidant and anticancer activity of biogenic silver nanoparticles-an experimental report. RSC Adv 6:81436–81446

Bhattacharjee S, Debnath G, Das AR et al (2017) Characterization of silver nanoparticles synthesized using an endophytic fungus, *Penicillium oxalicum* having potential antimicrobial activity. Adv Nat Sci Nanosci Nanotechnol 8:045008–045014

Bishnoi S, Kumar A, Selvaraj R (2018) Facile synthesis of magnetic iron oxide nanoparticles using inedible *Cynometra ramiflora* fruit extract waste and their photocatalytic degradation of methylene blue dye. Mater Res Bull 97:121–127

Brand MD (2010) The sites and topology of mitochondrial superoxide production. Exp Gerontol 45:466–472

Carlson C, Hussain SM, Schrand AM et al (2008) Unique cellular interaction of silver nanoparticles: size-dependent generation of reactive oxygen species. J Phys Chem B 112:13608–13619

Chanda S, Baravalia Y (2011) Brine shrimp cytotoxicity of *Caesalpinia pulcherrima* aerial parts, antimicrobial activity and characterisation of isolated active fractions. Nat Prod Res 25:1955–1964

Chen Y, Cao X, Chang PR, Huneault MA (2008) Comparative study on the films of poly (vinyl alcohol)/pea starch nanocrystals and poly (vinyl alcohol)/native pea starch. Carbohydr Polym 73:8–17

Cheng W, Wang S-G, Lu L et al (2008) Removal of malachite green (MG) from aqueous solutions by native and heat-treated anaerobic granular sludge. Biochem Eng J 39:538–546

Choudhary R, Madhuri R (2018) Eco-friendly noble metal nanoparticles for therapeutic applications: present and future scenario. In: Green Metal nanoparticles: synthesis, characterization and their applications, pp 629–665

Dauthal P, Mukhopadhyay M (2016) Noble metal nanoparticles: plant-mediated synthesis, mechanistic aspects of synthesis, and applications. Ind Eng Chem Res 55:9557–9577

de Oliveira JL, Campos EVR, Bakshi M et al (2014) Application of nanotechnology for the encapsulation of botanical insecticides for sustainable agriculture: prospects and promises. Biotechnol Adv 32:1550–1561

Dhas TS, Kumar VG, Karthick V et al (2014) Biosynthesis of gold nanoparticles using *Sargassum swartzii* and its cytotoxicity effect on HeLa cells. Spectrochim Acta A 133:102–106

Fulekar J, Dutta DP, Pathak B, Fulekar MH (2018) Novel microbial and root mediated green synthesis of TiO_2 nanoparticles and its application in wastewater remediation. J Chem Technol Biotechnol 93:736–743

Gahlawat G, Choudhury AR (2019) A review on the biosynthesis of metal and metal salt nanoparticles by microbes. RSC Adv 9:12944–12967

Gengan RM, Anand K, Phulukdaree A, Chuturgoon A (2013) A549 lung cell line activity of biosynthesized silver nanoparticles using *Albizia adianthifolia* leaf. Colloids Surf B Biointerfaces 105:87–91

Gomaa EZ (2017) Antimicrobial, antioxidant and antitumor activities of silver nanoparticles synthesized by *Allium cepa* extract : a green approach. J Genet Eng Biotechnol 15:49–57

Gong CP, Li SC, Wang RY (2018) Development of biosynthesized silver nanoparticles based formulation for treating wounds during nursing care in hospitals. J Photochem Photobiol B Biol 183:137–141

Grillo R, Abhilash PC, Fraceto LF (2016) Nanotechnology applied to bio-encapsulation of pesticides. J Nanosci Nanotechnol 16:1231–1234

He Y et al (2016) Biosynthesis, Antibacterial activity and anticancer effects against prostate cancer (PC-3) cells of silver nanoparticles using *Dimocarpus Longan* Lour. peel extract. Nanoscale Res Lett 11:300–310

Huo C, Khoshnamvand M, Liu P et al (2019) Eco-friendly approach for biosynthesis of silver nanoparticles using *Citrus maxima* peel extract and their characterization, catalytic, antioxidant and antimicrobial characteristics. Mater Res Express 6:015010–015023

Ikuma K, Decho AW, Lau BLT (2015) When nanoparticles meet biofilms—interactions guiding the environmental fate and accumulation of nanoparticles. Front Microbiol 6:591–597

Inbakandan D, Kumar C, Bavanilatha M et al (2016) Ultrasonic-assisted green synthesis of flower like silver nanocolloids using marine sponge extract and its effect on oral biofilm bacteria and oral cancer cell lines. Microb Pathog 99:135–141

Iqbal J, Abbasi BA, Mahmood T et al (2017) Plant-derived anticancer agents: a green anticancer approach. Asian Pac J Trop Biomed 7:1129–1150

Iqbal J, Abbasi BA, Batool R et al (2019) Biogenic synthesis of green and cost effective cobalt oxide nanoparticles using *Geranium wallichianum* leaves extract and evaluation of *in vitro* antioxidant, antimicrobial, cytotoxic and enzyme inhibition properties. Mater Res Express 6:115407–115423

Iram S, Khan S, Ansary AA et al (2016) Biogenic terbium oxide nanoparticles as the vanguard against osteosarcoma. Spectrochim Acta A 168:123–131

Jafarirad S, Mehrabi M, Divband B, Kosari-Nasab M (2016) Biofabrication of zinc oxide nanoparticles using fruit extract of *Rosa canina* and their toxic potential against bacteria: a mechanistic approach. Mater Sci Eng C 59:296–302

Jahan E, Anandapadmanaban G, Mathiyalagan R (2018) Synthesis of zinc oxide nanoparticles from immature fruits of *Rubus coreanus* and its catalytic activity for degradation of industrial dye. Optik 172:1179–1186
Javadi F, Taghavizadeh Yazdi ME, Baghani M, Es-Haghi A (2019) Biosynthesis, characterization of cerium oxide nanoparticles using *Ceratonia siliqua* and evaluation of antioxidant and cytotoxicity activities. Mater Res Express 6:065408–065414
Jayasree L, Janakiram P, Madhavi R (2006) Characterization of *Vibrio* spp. associated with diseased shrimp from culture ponds of *Andhra Pradesh* (India). J World Aquac Soc 37:523–532
Jeevanandam J, Barhoum A, Chan YS et al (2018) Review on nanoparticles and nanostructured materials: history, sources, toxicity and regulations. Beilstein J Nanotechnol 9:1050–1074
Jena S, Singh RK, Panigrahi B et al (2016) Photo-bioreduction of Ag^+ ions towards the generation of multifunctional silver nanoparticles: mechanistic perspective and therapeutic potential. J Photochem Photobiol B Biol 164:306–313
Jeyaraj M, Sathishkumar G, Sivanandhan G et al (2013) Biogenic silver nanoparticles for cancer treatment: an experimental report. Colloids Surf B Biointerfaces 106:86–92
Jia X, Li Y, Cheng Q et al (2007) Preparation and properties of poly (vinyl alcohol)/silica nanocomposites derived from copolymerization of vinyl silica nanoparticles and vinyl acetate. Eur Polym J 43:1123–1131
Johnson F, Giulivi C (2005) Superoxide dismutases and their impact upon human health. Mol Aspects Med 26:340–352
Kalangi SK, Dayakar A, Gangappa D et al (2016) Biocompatible silver nanoparticles reduced from *Anethum graveolens* leaf extract augments the antileishmanial efficacy of miltefosine. Exp Parasitol 170:184–192
Kalathil S, Lee J, Cho MH (2011) Electrochemically active biofilm-mediated synthesis of silver nanoparticles in water. Green Chem 13:1482–1485
Kaye P, Scott P (2011) Leishmaniasis: complexity at the host–pathogen interface. Nat Rev Microbiol 9:604–615
Khan MM, Kalathil S, Han TH et al (2013) Positively charged gold nanoparticles synthesized by electrochemically active biofilm-a biogenic approach. J Nanosci Nanotechnol 13:6079–6085
Khan MM, Ansari SA, Lee JH et al (2014) Electrochemically active biofilm assisted synthesis of Ag@CeO_2 nanocomposites for antimicrobial activity, photocatalysis and photoelectrodes. J Colloid Interface Sci 431:255–263
Khatami M, Varma RS, Zafarnia N et al (2018) Applications of green synthesized Ag, ZnO and Ag/ZnO nanoparticles for making clinical antimicrobial wound-healing bandages. Sustain Chem Pharm 10:9–15
Kim B, Kim D, Cho D, Cho S (2003) Bactericidal effect of TiO_2 photocatalyst on selected food-borne pathogenic bacteria. Chemosphere 52:277–281
Kim JS, Kuk E, Yu KN et al (2007) Antimicrobial effects of silver nanoparticles. Nanomedicine Nanotechnol Biol Med 3:95–101
Kiran Kumar ABV, Saila ES, Narang P et al (2019) Biofunctionalization and biological synthesis of the ZnO nanoparticles: the effect of *Raphanus sativus* (white radish) root extract on antimicrobial activity against MDR strain for wound healing applications. Inorg Chem Commun 100:101–106
Kora AJ, Rastogi L (2018) Green synthesis of palladium nanoparticles using gum ghatti (*Anogeissus latifolia*) and its application as an antioxidant and catalyst. Arab J Chem 11:1097–1106
Krupa NDA, Raghavan V (2014) Biosynthesis of silver nanoparticles using aegle marmelos (Bael) fruit extract and its application to prevent adhesion of bacteria: a strategy to control microfouling. Bioinorg Chem Appl 2014, Article ID 949538, 8 pages, 2014. https://doi.org/10.1155/2014/949538
Kumar B, Smita K, Seqqat R et al (2016a) In vitro evaluation of silver nanoparticles cytotoxicity on Hepatic cancer (Hep-G2) cell line and their antioxidant activity: green approach for fabrication and application. J Photochem Photobiol B Biol 159:8–13

Kumar B, Smita K, Vizuete KS, Cumbal L (2016b) Aqueous phase Lavender leaf mediated green synthesis of gold nanoparticles and evaluation of its antioxidant activity. Biol Med 8:3–7

Lagaron J, Sanchez-Garcia M, Gimenez E (2008) Thermoplastic nanobiocomposites for rigid and flexible food packaging applications. In: Environmentally compatible food packaging. Elsevier, pp 63–89

Lakshmi A, Venkata K, Anthony SP (2019) Synthesis of *Solanum nigrum* mediated copper oxide nanoparticles and their photocatalytic dye degradation studies. J Chem Technol Biotechnol 2013 88:1971–1977

Lee H, Song JY, Kim BS (2013) Biological synthesis of copper nanoparticles using *Magnolia kobus* leaf extract and their antibacterial activity. J Chem Technol Biotechnol 88:1971

Li R, Chen Z, Ren N et al (2019) Biosynthesis of silver oxide nanoparticles and their photocatalytic and antimicrobial activity evaluation for wound healing applications in nursing care. J Photochem Photobiol B Biol 199:111593

Liu R, Lal R (2015) Potentials of engineered nanoparticles as fertilizers for increasing agronomic productions. Sci Total Environ 514:131–139

Liu W, Wu Y, Wang C et al (2010) Impact of silver nanoparticles on human cells: effect of particle size. Nanotoxicology 4:319–330

Lok C-N, Ho C-M, Chen R et al (2006) Proteomic analysis of the mode of antibacterial action of silver nanoparticles. J Proteome Res 5:916–924

Lokina S, Stephen A, Kaviyarasan V et al (2014) Cytotoxicity and antimicrobial activities of green synthesized silver nanoparticles. Eur J Med Chem 76:256–263

Mahdavi B, Paydarfard S, Zangeneh MM, et al (2019) Assessment of antioxidant, cytotoxicity, antibacterial, antifungal, and cutaneous wound healing activities of green synthesized manganese nanoparticles using *Ziziphora clinopodioides* Lam leaves under in vitro and in vivo condition. Appl Organomet Chem e5248–e5253

Manikandan R, Manikandan B, Raman T et al (2015) Biosynthesis of silver nanoparticles using ethanolic petals extract of *Rosa indica* and characterization of its antibacterial, anticancer and anti-inflammatory activities. Spectrochim Acta A Mol Biomol Spectrosc 138:120–129

McQuillan J (2010) Bacterial-nanoparticle interactions, PhD thesis

Midwood KS, Williams LV, Schwarzbauer JE (2004) Tissue repair and the dynamics of the extracellular matrix. Int J Biochem Cell Biol 36:1031–1037

Mishra S, Singh A, Keswani C, Singh HB (2014) Nanotechnology: exploring potential application in agriculture and its opportunities and constraints. Biotech Today 4:9–14

Mishra P, Ray S, Sinha S et al (2016) Facile bio-synthesis of gold nanoparticles by using extract of *Hibiscus sabdariffa* and evaluation of its cytotoxicity against U87 glioblastoma cells under hyperglycemic condition. Biochem Eng J 105:264–272

Mittal AK, Bhaumik J, Kumar S, Banerjee UC (2014) Biosynthesis of silver nanoparticles: elucidation of prospective mechanism and therapeutic potential. J Colloid Interface Sci 415:39–47

Moon JW, Phelps TJ, Fitzgerald CL et al (2016) Manufacturing demonstration of microbially mediated zinc sulfide nanoparticles in pilot-plant scale reactors. Appl Microbiol Biotechnol 100:7921–7931

Mueller CFH, Laude K, McNally JS, Harrison DG (2005) Redox mechanisms in blood vessels. Arterioscler Thromb Vasc Biol 25:274–278

Namvar F, Rahman HS, Mohamad R et al (2014) Cytotoxic effect of magnetic iron oxide nanoparticles synthesized via seaweed aqueous extract. Int J Nanomedicine 9:2479–2488

Namvar F, Azizi S, Rahman HS et al (2016) Green synthesis, characterization, and anticancer activity of hyaluronan/zinc oxide nanocomposite. Onco Targets Ther 9:4549–4559

Naraginti S, Kumari PL, Das RK et al (2016) Amelioration of excision wounds by topical application of green synthesized, formulated silver and gold nanoparticles in albino Wistar rats. Mater Sci Eng C 62:293–300

Nayak D, Pradhan S, Ashe S et al (2015) Biologically synthesised silver nanoparticles from three diverse family of plant extracts and their anticancer activity against epidermoid A431 carcinoma. J Colloid Interface Sci 457:329–338

Nayak D, Ashe S, Rauta PR et al (2016) Bark extract mediated green synthesis of silver nanoparticles: evaluation of antimicrobial activity and antiproliferative response against osteosarcoma. Mater Sci Eng C 58:44–52

Ndeh NT, Maensiri S, Maensiri D (2017) The effect of green synthesized gold nanoparticles on rice germination and roots. Adv Nat Sci Nanosci Nanotechnol 8:0352008–0035018

Nurkiewicz TR, Porter DW, Hubbs AF et al (2008) Nanoparticle inhalation augments particle-dependent systemic microvascular dysfunction. Part Fibre Toxicol 5:1–13

Omran BA, Fatthalah NA, El-Gendy NS et al (2013) Green biocides against sulphate reducing bacteria and macrofouling organisms. J Pure Appl Microbiol 7:2219–2232

Omran BA, Nassar HN, Fatthallah NA et al (2017) Waste upcycling of *Citrus sinensis* peels as a green route for the synthesis of silver nanoparticles. Energy Sources A Recover Util Environ Eff 40:1–10

Omran BA, Nassar HN, Fatthallah NA et al (2018) Characterization and antimicrobial activity of silver nanoparticles mycosynthesized by *Aspergillus brasiliensis*. J Appl Microbiol 125: 370–382

Omran BA, Nassar HN, Younis SA et al (2019a) Novel mycosynthesis of cobalt oxide nanoparticles using *Aspergillus brasiliensis* ATCC 16404: Optimization, characterization and antimicrobial activity. J Appl Microbiol 128:438–457

Omran BA, Nassar HN, Younis SA et al (2019b) Physiochemical properties of T*richoderma longibrachiatum* DSMZ 16517 synthesized silver nanoparticles for the mitigation of halotolerant sulphate reducing bacteria. J Appl Microbiol 126:138–154

Park S, Zhao Y (2004) Incorporation of a high concentration of mineral or vitamin into chitosan-based films. J Agric Food Chem 52:1933–1939

Patra JK, Baek K-H (2015) Novel green synthesis of gold nanoparticles using *Citrullus lanatus* rind and investigation of proteasome inhibitory activity, antibacterial, and antioxidant potential. Int J Nanomedicine 10:7253–7264

Patra S, Mukherjee S, Barui AK et al (2015) Green synthesis, characterization of gold and silver nanoparticles and their potential application for cancer therapeutics. Mater Sci Eng C 53:298–309

Podsiadlo P, Choi S-Y, Shim B et al (2005) Molecularly engineered nanocomposites: layer-by-layer assembly of cellulose nanocrystals. Biomacromolecules 6:2914–2918

Ponnanikajamideen MI, Rajeshkumar S, Vanaja M, Annadurai G (2019) In vivo type 2 diabetes and wound-healing effects of antioxidant gold nanoparticles synthesized using the insulin plant *Chamaecostus cuspidatus* in Albino Rats. Can J Diabetes 43:82–89.e6

Prabakaran K, Ragavendran C, Natarajan D (2016) Mycosynthesis of silver nanoparticles from: *Beauveria bassiana* and its larvicidal, antibacterial, and cytotoxic effect on human cervical cancer (HeLa) cells. RSC Adv 6:44972–44986

Prabhu D, Arulvasu C, Babu G et al (2013) Biologically synthesized green silver nanoparticles from leaf extract of *Vitex negundo* L. induce growth-inhibitory effect on human colon cancer cell line HCT15. Process Biochem 48:317–324

Prabhu S, Vinodhini S, Elanchezhiyan C, Rajeswari D (2018) Evaluation of antidiabetic activity of biologically synthesized silver nanoparticles using *Pouteria sapota* in streptozotocin- induced diabetic rats. J Diabetes 10:28–42

Premasudha P, Venkataramana M, Abirami M et al (2015) Biological synthesis and characterization of silver nanoparticles using *Eclipta alba* leaf extract and evaluation of its cytotoxic and antimicrobial potential. Bull Mater Sci 38:965–973

Priyadharshini Raman R, Parthiban S, Srinithya B et al (2015) Biogenic silver nanoparticles synthesis using the extract of the medicinal plant *Clerodendron serratum* and its in-vitro antiproliferative activity. Mater Lett 160:400–403

Radomski A, Jurasz P, Alonso-Escolano D et al (2005) Nanoparticle induced platelet aggregation and vascular thrombosis. Br J Pharmacol 146:882–893

Rajan A, Vilas V, Philip D (2015) Studies on catalytic, antioxidant, antibacterial and anticancer activities of biogenic gold nanoparticles. J Mol Liq 212:331–339

Ramanathan T, Abdala AA, Stankovich S et al (2008) Functionalized graphene sheets for polymer nanocomposites. Nat Nanotechnol 3:327–331

Ramar M, Manikandan B, Marimuthu PN et al (2015) Synthesis of silver nanoparticles using *Solanum trilobatum* fruits extract and its antibacterial, cytotoxic activity against human breast cancer cell line MCF 7. Spectrochim Acta A Mol Biomol Spectrosc 140:223–228

Rao PV, Nallappan D, Madhavi K, et al (2016) Phytochemicals and biogenic metallic nanoparticles as anticancer agents. Oxid Med Cell Longev 2016, Article ID 3685671, 15 pages, 2016. https://doi.org/10.1155/2016/3685671

Rathi Sre PR, Reka M, Poovazhagi R et al (2015) Antibacterial and cytotoxic effect of biologically synthesized silver nanoparticles using aqueous root extract of *Erythrina indica* lam. Spectrochim Acta A Mol Biomol Spectrosc 135:1137–1144

Ravindran J, Arumugasamy V, Baskaran A (2019) Wound healing effect of silver nanoparticles from Tridax procumbens leaf extracts on *Pangasius hypophthalmus*. Wound Med 27:100170–100175

Rizzello L, Cingolani R, Pompa PP (2013) Nanotechnology tools for antibacterial materials. Nanomedicine 8:807–821

Roopan SM, Kumar SHS, Madhumitha G, Suthindhiran K (2014) Biogenic-production of SnO_2 nanoparticles and its cytotoxic effect against hepatocellular carcinoma cell line (HepG2). Appl Biochem Biotechnol 175:1567–1575

Roy A, Bulut O, Some S et al (2019) Green synthesis of silver nanoparticles: biomolecule-nanoparticle organizations targeting antimicrobial activity. RSC Adv 9:2673–2702

Salari Z, Ameri A, Forootanfar H et al (2017) Microwave-assisted biosynthesis of zinc nanoparticles and their cytotoxic and antioxidant activity. J Trace Elem Med Biol 39:116–123

Salehi S, Sadat Shandiz SA, Ghanbar F et al (2016) Phytosynthesis of silver nanoparticles using *Artemisia marschalliana* sprengel aerial part extract and assessment of their antioxidant, anticancer, and antibacterial properties. Int J Nanomedicine 11:1835–1846

Sarkar S, Kotteeswaran V (2018) Green synthesis of silver nanoparticles from aqueous leaf extract of Pomegranate (*Punica granatum*) and their anticancer activity on human cervical cancer cells. Adv Nat Sci Nanosci Nanotechnol 9:025014–025024

Sarmah S, Kumar A (2011) Photocatalytic activity of polyaniline-TiO_2 nanocomposites. Indian J Phys 85:713–727

Sengottaiyan A, Aravinthan A, Sudhakar C et al (2016) Synthesis and characterization of Solanum nigrum-mediated silver nanoparticles and its protective effect on alloxan-induced diabetic rats. J Nanostruct Chem 6:41–48

Seydi N, Saneei S, Jalalvand AR et al (2019) Synthesis of titanium nanoparticles using *Allium eriophyllum* Boiss aqueous extract by green synthesis method and evaluation of their remedial properties. Appl Organomet Chem 33:1–18

Shanmugasundaram T, Radhakrishnan M, Gopikrishnan V et al (2013) A study of the bactericidal, anti-biofouling, cytotoxic and antioxidant properties of actinobacterially synthesised silver nanoparticles. Colloids Surf B Biointerfaces 111:680–687

Shao F, Yang A, Mei D et al (2018) Bio-synthesis of *Barleria gibsoni* leaf extract mediated zinc oxide nanoparticles and their formulation gel for wound therapy in nursing care of infants and children. J Photochem Photobiol B Biol 189:267–273

Singh H, Du J, Yi TH (2017) Green and rapid synthesis of silver nanoparticles using *Borago officinalis* leaf extract: anticancer and antibacterial activities. Artif Cells Nanomed Biotechnol 45:1310–1316

Singh J, Vishwakarma K, Ramawat N et al (2019a) Nanomaterials and microbes' interactions: a contemporary overview. 3 Biotech 9:1–14

Singh M, Srivastava M, Kumar A, Pandey K (2019b) Biosynthesis of nanoparticles and applications in agriculture. Role of plant growth promoting microorganisms in sustainable agriculture and nanotechnology, Elsevier Inc., pp 199–217

Sivaraj R, Rahman PKSM, Rajiv P et al (2014) Biosynthesis and characterization of *Acalypha indica* mediated copper oxide nanoparticles and evaluation of its antimicrobial and anticancer activity. Spectrochim Acta A 129:255–258

Sood R, Chopra DS (2018) Optimization of reaction conditions to fabricate *Ocimum sanctum* synthesized silver nanoparticles and its application to nano-gel systems for burn wounds. Mater Sci Eng C 92:575–589

Sriramulu M, Sumathi S (2018) Biosynthesis of palladium nanoparticles using *Saccharomyces cerevisiae* extract and its photocatalytic degradation behaviour. Adv Nat Sci Nanosci Nanotechnol 9:025018–025024

Srivastava S, Sinha R, Roy D (2004) Toxicological effects of malachite green. Aquat Toxicol 66:319–329

Suganya USU, Govindaraju K, Kumar GG et al (2016) Anti-proliferative effect of biogenic gold nanoparticles against b -7. Appl Surf Sci 371:415–424

Sulaiman GM, Hussien HT, Saleem MMNM (2015) Biosynthesis of silver nanoparticles synthesized by *Aspergillus flavus* and their antioxidant, antimicrobial and cytotoxicity properties. Bull Mater Sci 38:639–644

Suresh J, Pradheesh G, Alexramani V (2018) Green synthesis and characterization of zinc oxide nanoparticle using insulin plant (*Costus pictus* D. Don) and investigation of its antimicrobial as well as anticancer activities. Adv Nat Sci Nanosci Nanotechnol 9:1–8

Tankhiwale R, Bajpai SK (2009) Graft copolymerization onto cellulose-based filter paper and its further development as silver nanoparticles loaded antibacterial food-packaging material. Colloids Surf B Biointerfaces 69:164–168

Tanzil AH, Sultana ST, Saunders SR et al (2016) Biological synthesis of nanoparticles in biofilms. Enzyme Microb Technol 95:4–12

Teitzel GM, Parsek MR (2003) Heavy metal resistance of biofilm and planktonic *Pseudomonas aeruginosa*. Appl Environ Microbiol 69:2313–2320

Thivaharan V, Ramesh V, Raja S (2018) Green synthesis of silver nanoparticles for biomedical and environmental applications. Macabresque Hum Viol Hate Genocide, Mass Atrocity Enemy-Making, pp 387–439

Tripathi RM, Pudake RN, Shrivastav BR, Shrivastav A (2018) Antibacterial activity of poly (vinyl alcohol) – biogenic silver nanocomposite film for food packaging material. Adv Nat Sci Nanosci Nanotechnol 9:025020–025025

Tsuji JS, Maynard AD, Howard PC et al (2005) Research strategies for safety evaluation of nanomaterials, part IV: risk assessment of nanoparticles. Toxicol Sci 89:42–50

Vanaja M, Paulkumar K, Baburaja M, et al (2014) Degradation of methylene blue using biologically synthesized silver nanoparticles. Bioinorg Chem Appl 2014, Article ID 742346, 8 pages, 2014. https://doi.org/10.1155/2014/742346

Vanti GL, Nargund VB, Basavesha KN et al (2019) Synthesis of *Gossypium hirsutum*-derived silver nanoparticles and their antibacterial efficacy against plant pathogens. Appl Organomet Chem 33:1–9

Vijayan SR, Santhiyagu P, Singamuthu M, et al (2014) Synthesis and characterization of silver and gold nanoparticles using aqueous extract of seaweed, *Turbinaria conoides*, and their antimicrofouling activity. Sci World J 2014, Article ID 938272, 10 pages, 2014. https://doi.org/10.1155/2014/938272

Vladimirov V, Betchev C, Vassiliou A et al (2006) Dynamic mechanical and morphological studies of isotactic polypropylene/fumed silica nanocomposites with enhanced gas barrier properties. Compos Sci Technol 66:2935–2944

Wang Y, Qiao Y, Wang P et al (2018) Bio fabrication of silver nanoparticles as an effective wound healing agent in the wound care after anorectal surgery. J Photochem Photobiol B Biol 178:457–462

Wu H, Yin J-J, Wamer WG et al (2014) Reactive oxygen species-related activities of nano-iron metal and nano-iron oxides. J Food Drug Anal 22:86–94

Xia QH, Ma YJ, Wang JW (2016) Biosynthesis of silver nanoparticles using *Taxus yunnanensis* callus and their antibacterial activity and cytotoxicity in human cancer cells. Nanomaterials 6:160–175

Xu H, Xiao Y, Xu M et al (2019) Microbial synthesis of Pd-Pt alloy nanoparticles using *Shewanella oneidensis* MR-1 with enhanced catalytic activity for nitrophenol and azo dyes reduction. Nanotechnology 30:056507–056518

Yam KL (2010) The Wiley encyclopedia of packaging technology. Wiley, New York

Ying W, Tan J, Chen C et al (2019) Biofabrication of silver nanoparticles and its application for development of wound dressing system in nursing care for burn injuries in children. J Drug Delivery Sci Technol 54:101236–101243

Zangeneh MM, Ghaneialvar H, Akbaribazm M et al (2019) Novel synthesis of F*alcaria vulgaris* leaf extract conjugated copper nanoparticles with potent cytotoxicity, antioxidant, antifungal, antibacterial, and cutaneous wound healing activities under in vitro and in vivo condition. J Photochem Photobiol B Biol 197:111556–111569

Ziv-Polat O, Topaz M, Brosh T, Margel S (2010) Enhancement of incisional wound healing by thrombin conjugated iron oxide nanoparticles. Biomaterials 31:741–747

Index

B. A. Omran, *Nanobiotechnology: A Multidisciplinary Field of Science*, Nanotechnology in the Life Sciences, https://doi.org/10.1007/978-3-030-46071-6

The manufacturer's authorised representative in the EU is Springer Nature Customer Service Centre GmbH, Europaplatz 3, 69115 Heidelberg, Germany. If you have any concerns regarding our products, please contact ProductSafety@springernature.com

Printed and bound by CPI Group (UK) Ltd, Croydon, CR0 4YY

30/09/2026

02225504-0007